Our Kind of People

INSIDE AMERICA'S BLACK UPPER CLASS

ALSO BY LAWRENCE OTIS GRAHAM

Member of the Club

Proversity: Getting Past Face Value

The Best Companies for Minorities

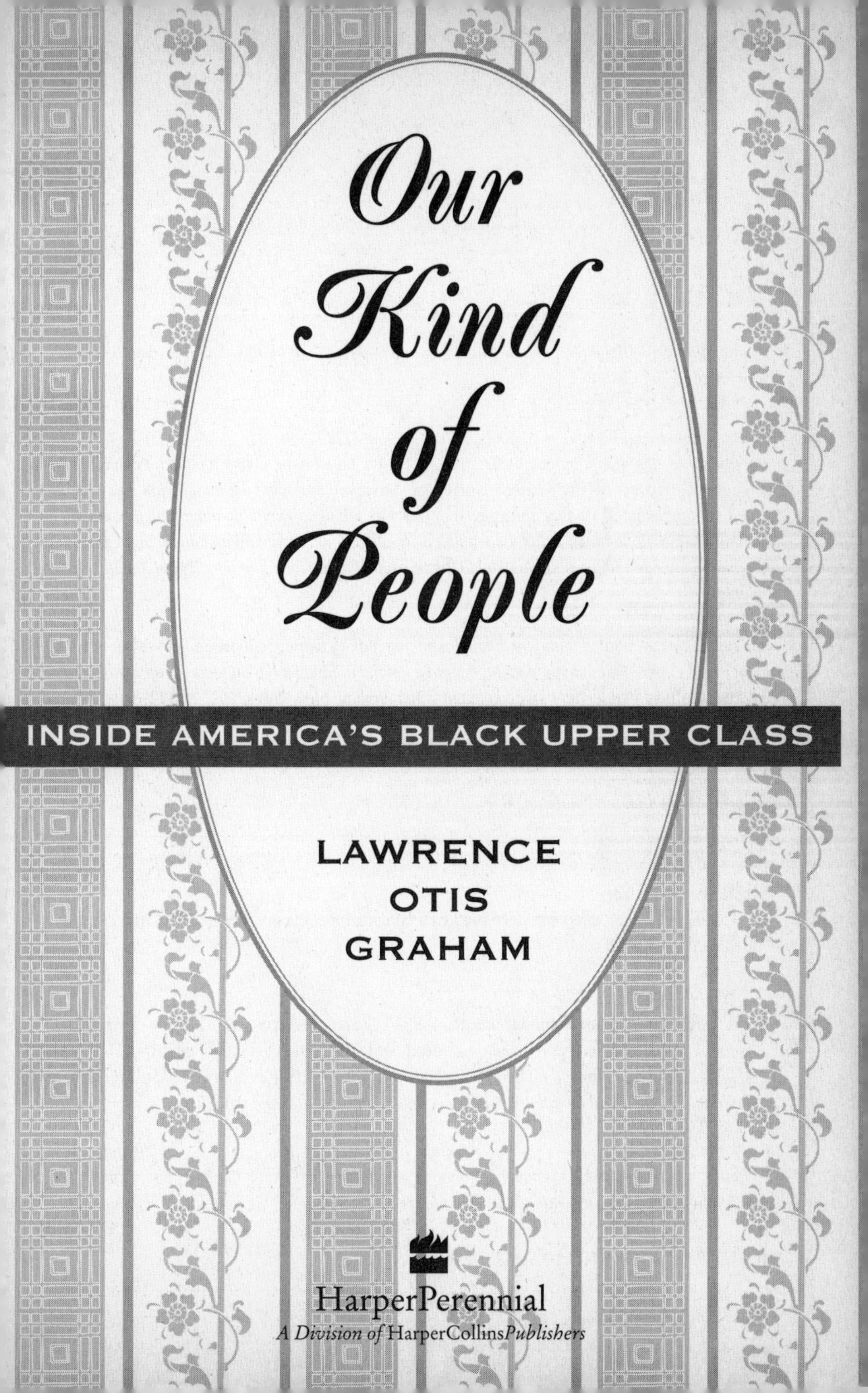

Our Kind of People

INSIDE AMERICA'S BLACK UPPER CLASS

LAWRENCE OTIS GRAHAM

HarperPerennial
A Division of HarperCollinsPublishers

A hardcover edition of this book was published in 1999 by HarperCollins Publishers.

HarperCollins books may be purchased for educational, business, or sales promotional use. For information please write: Special Markets Department, HarperCollins Publishers Inc., 10 East 53rd Street, New York, NY 10022.

First HarperPerennial edition published 2000.

Designed by Liane F. Fuji

The Library of Congress has catalogued the hardcover edition as follows:

Graham, Lawrence.
Our kind of people : inside America's Black upper class / Lawrence Otis Graham. —1st ed.
p. cm.
ISBN 0-06-018352-7
1. Afro-Americans—Social conditions—1975– 2. Afro-Americans—Social life and customs. 3. Upper class—United States—History—20th century. 4. Elite (Social sciences)—United States—History—20th century. 5. United States—Social conditions—1980– 6. United States—Race relations. I. Title.
E185.86.G644 1999
305.896'073—dc21 98-34046

ISBN 0-06-098438-4 (pbk.)

00 01 02 03 04 ❖/RRD 10 9 8 7 6 5 4

To my brilliant wife, Pamela Thomas-Graham.
Thank you for making my dreams come true.

CONTENTS

Photographs follow page 194.

INTRODUCTION

to the HarperPerennial Edition

Although I spent six years researching *Our Kind of People*, I could never have been prepared for the controversy that it elicited from various groups upon its initial publication. Although there is a constant cry for diversity in our media, our literature, our history books, and in our communities, it became obvious to me that there are certain narrow stereotypes—even within an integrated society—that people are simply unwilling to relinquish. The stereotype of the working-class black or impoverished black is one that whites, as well as blacks, have come to embrace and accept as an accurate and complete account of the black American experience. *Our Kind of People* upset that stereotype. And it upset many people—particularly blacks—who have been taught never to challenge a stereotype that we had been saddled with since slavery.

For many people, this book is a political or social hot potato in the sense that even though most blacks talk about the issues of elitism, racial passing, class structure, and skin color within the black community, they don't want to see it broadcast in a book. For a few black members of the media, the topic struck too close to their own past experiences of being excluded by snobbish members of the black elite. Some of them quietly told me that they were glad that I wrote the book because it was disproving the stereotypes, but that they could not publicly support the book because their white audiences would find the concept of rich, educated blacks too threatening and because their black audiences would find the subject too painful.

An extensive national tour uncovered many surprises during media interviews, personal appearances, and Internet dialogues, as well as at high-end cocktail parities.

A short time after *Our Kind of People* was first published, there was a series of discussion panels set up in different cities so that members of the black elite could come together and share their views on the book and address the controversy that had arisen around the issue of class within the black community. Each of these panel discussions was preceded by a somewhat formal cocktail party, where guests—many of them members of old-guard social clubs or prominent families—had the opportunity to mingle with old friends as well as representatives from the media. Although the largest event took place at the Harvard Club in New York City, the most memorable one occurred on a hot spring evening in Los Angeles.

"You'd better not show your face in Martha's Vineyard *this* summer," snapped an attractive Yale graduate as she remarked how she and some of her other black friends were responding to the controversy around *Our Kind of People*. She'd evidently seen me in Oak Bluffs in prior years and had heard that many people were disturbed either because their names hadn't appeared in the book or because they believed that wealthy blacks shouldn't be talking about their accomplishments.

I felt both groups were being unreasonable and I told the woman so. "Some black folks may be uncomfortable to learn that there are several generations of elite blacks who live in a separate world, but like white people, blacks also have to learn to accept the facts in our history," I remarked. "I don't think the black upper-class crowd should be ashamed of its success any more than the WASP elite, Italian elite, or Jewish elite."

"You opened a real can of worms," another woman remarked as she stood by listening. "Folks don't want to hear about rich blacks unless we're playing basketball, singing rap music, or doing comedy on TV."

As casual as that remark was, it was actually an accurate assessment of many black people's response to a book that I had spent six years researching. While a few whites expressed amazement that there had been black millionaires and black members of Congress as far back as the late 1800s (one white TV anchor told me on live television that if I had not displayed photos of the well-known entrepreneur Madam C. J. Walker and her family's 20,000-square-foot 1902 mansion in New York, he would never have believed that a black millionaire existed at the turn of the century), most were fascinated by how closely class structure and elitism among blacks mirrored that found in other racial and ethnic groups.

A large number of blacks, however, were not so comfortable with the 120-year history of the black upper class in America. Many with ties to these families, organizations, schools, fraternities, or summer resorts accepted this history with pride—so long as they did not have to admit their status in the company of non-elite blacks.

"I don't want the other black folks in Atlanta to think I'm looking down on them," explained a millionaire surgeon who attempted to avoid association with the old-guard clubs, schools, and institutions whenever they were mentioned in mixed company.

And this is why I have concluded that although every racial, ethnic, and religious group in the United States claims to want a piece of the American dream, there is no group that apologizes more for its success than black people. The cultural identity or integrity of a black millionaire rap star, basketball player, or TV performer will never be questioned. But an equally wealthy black professional with an upper-class background and a good education will earn the label of a "sellout" or a "Negro trying to be white."

The black Yale grad with the long mane of hair in Los Angeles was aware of that fact when she warned me about showing up on the beaches of Martha's Vineyard. In the cities that had preceded L.A., the people attending these gatherings had represented a rather insular circle. They were members of old-guard families and social clubs like the Links or Boulé; individuals who had attended the schools, camps, and cotillions that had been written about; and wealthy physicians or businesspeople who understood and embraced the world captured in the book.

But something about this gathering in L.A. seemed different—and intriguing. For the first time, there seemed to be people present who did not come from the world of the black elite. Yes, they were black, but they were not uniformly wealthy or old-guard. Since the book was gaining attention through its selection by the Book-of-the-Month Club, its appearance on various bestseller lists, numerous magazine excerpts, and many positive reviews and features in the *New York Times* and the

Washington Post, it was attracting an audience beyond the initial wealthy black and wealthy white readers. It was hitting the mainstream.

"Lawrence, I don't think these are our kind of people in this room," remarked a Los Angeles doctor friend of mine who had grown up with me in Jack and Jill and had summered in Martha's Vineyard.

"I'm hearing some pretty nasty stuff from these folks," remarked a friend of my mother's, who had brought some of her friends from the Links. She moved closer to me. "Some of us are leaving, so I don't know if you really want to be here."

That very week, the *Los Angeles Times* had run a front page feature story about the book and about the wealthy black people and organizations that had been profiled in it. Although it spoke glowingly about high-powered black Angelenos, the article and surrounding discussion clearly did not sit well with certain wealthy blacks who had not been included. And the middle-class blacks were enraged to discover that there was a circle of elite black people, institutions, and activities that excluded them. If talk about the right debutante cotillions and best black schools and summer camps hadn't already offended them, the accompanying photo of me standing on Beverly Hills' Wilshire Boulevard convinced them that the black elite was existing completely outside the mainstream black world. Of course this was not entirely true.

"Nobody wants to hear about rich black people," one angry attendee remarked.

"How come none of these supposed black folks are Baptists?" another asked.

"Why are all the people in the book light skinned?"

"What kind of black man is gonna go to Harvard anyway?"

I told him that the black intellectual W. E. B. Du Bois graduated from Harvard in the 1890s. He didn't want to hear it.

And neither did the members of the black elite who suddenly discovered that they were surrounded by middle-class and working-class blacks who were there to make the rest of us apologize for a class structure we had not created. A prominent author who sat with me on the panel suddenly announced that she could stay for only the first five minutes of our impending sixty-minute discussion. Rather than discuss her experience of being a black woman at Wellesley in the 1940s and her experience of growing up in a well-connected, light-complexioned black elite family, she neutralized the angry crowd by trotting out an unconvincing I'm-just-a-down-home-sister-like-the-rest-of-you introduction.

After a final statement that included such trite phrases as "giving back to the people . . . we're all the same . . . each one teach one . . . we're all in this together . . . don't forget where you came from," the blond-haired Wellesley graduate ran for the door with the security that her Links and AKA friends would keep her secret.

To my left was another member of the black elite who pulled a "one-eighty" at the last minute. Rather than hightail it to the door like our fellow panelist and a few of the other "club" people, she instead pandered to the audience with a revised personal résumé and an irrelevant speech on how awful it was to even be talking about the black elite when kids were being shot in the streets. Despite the fact that the hosts had invited the woman to the panel because of her academic credentials and her ties to one of the city's most aristocratic black millionaire families, she refused to speak to the topic of the evening.

By the end of the discussion, the room had divided itself into "non-elites" who hated black elites, black elites who remained silent about their backgrounds, and black elites who reinvented their backgrounds to placate the hostile attendees who felt that black people had no business owning summer homes or going to medical school.

I am sure I am partly responsible for the intensity of the response surrounding the book. Although I had not expected the reaction, the very first lines of the book upset many readers, including some of the most powerful black celebrities in the country. I should have explained the genesis of the opening lines, "Bryant Gumbel is, but Bill Cosby isn't. Lena Horne is, but Whitney Houston isn't. . . ." They were not my words, but remarks that were made to me as I traveled from one city to the next asking old-guard blacks to tell me which famous black celebrities from their city were members of old-guard families. In Philadelphia, Bill Cosby's hometown, I asked wealthy black socialites if the well-credentialed and wealthy Dr. Cosby was in their crowd. They told me, "No, he's not our kind of people, but his wife, Camille [because of her Spelman College background and her light complexion], is." In Detroit, I asked about Diana Ross. In Brooklyn, I asked about Lena Horne. In Chicago and New Orleans, I asked about Bryant Gumbel. Elite Detroiters told me that despite her current wealth, Ross was not accepted because of her less-privileged family background. The Horne and Gumbel families were "in" because of their families' relative wealth, academic ties, and connections. These conclusions were handed to me by others. They were not conclusions that I drew for myself.

Since the initial publication of the book, a certain amount of intellectual honesty has returned to the discussion of *Our Kind of People* and the black upper class. I have seen that more people are able to remove themselves and their personal experiences from the debate, while placing a greater focus on the facts that have been revealed through this social history. The attacks that were launched through anonymous letters and email have surely been outnumbered by the positive feedback and continued interest that has been elicited through reading groups, college courses, and panel presentations focusing on the black upper class.

As an increasing number of individuals have discussed this book and its subject matter, they have moved beyond the stereotypes that they had embraced around the issues of race and class. It has now become obvious that people are prepared to accept a broader definition of what it means to be black in America. It should be gratifying for all of us to discover that one can hold on to one's unique culture, history, and identity and yet still share in the American dream.

Lawrence Otis Graham

PREFACE

My decision to write a book about the black upper class was made several years ago, while I was in my first year at Harvard Law School. It was the result of an awkward but memorable encounter that I had during lunch with one of the country's most successful businessmen.

It was the mid-1980s, and the businessman was Reginald Lewis, the wealthiest black man in America. A Harvard Law alumnus who was then only forty-two years old, Lewis had recently purchased the $55 million McCall Pattern Company in a leveraged buyout and had begun appearing in national business articles. We had met several months earlier during a visit that he and his young daughter, Leslie, had made to the campus, and since that time, he had been offering me occasional academic and career advice.

None of us knew it at that moment, but within three years, Lewis was to become many times richer through his 1987 purchase of Beatrice Foods, the $2.5 billion international packaged goods company, in what was at that time the largest leveraged buyout in U.S. history. Within the next six years, he and his wife, Loida, would amass an enormous art collection, as well as homes in Paris and East Hampton and on New York's

Fifth Avenue. He would also give $3 million to Harvard Law School, $1 million to Howard University, $2 million to the NAACP, and hundreds of thousands to many other institutions and charities.

But on this particular afternoon, in 1984, we left his office at 99 Wall Street and headed for India House, a private club on Hanover Square in the financial district. We chatted about his role as chairman and owner of the 114-year-old sewing pattern company, his rising profile within the mostly white business world, and the Park Avenue law firm of Paul, Weiss, Rifkind, Wharton, and Garrison, a firm he wanted me to consider once I left Harvard.

Then he changed the subject.

"Now I need *your* advice on something," Lewis said, with his perpetually furrowed brow even more pronounced.

"Sure," I responded, even though I was quite certain that there was almost nothing about which I could counsel this man.

"You've grown up around upper-class whites, you've attended white schools, and you are comfortable around wealthy whites, yet you still seem to be balanced from a black perspective."

I shrugged.

"I'm saying that you seem to also have a black orientation that lets you mix among blacks and whites comfortably. I have two daughters and I want them to do that."

I suddenly knew where he was taking this conversation. Reginald Lewis was becoming wealthier and more powerful each year. Increasingly, with each new business deal, he was working, socializing, and living less among black people.

"I didn't grow up wealthy, but my daughters are growing up that way," he said while tapping his cigar into a porcelain ashtray, "and I don't want them to grow away from their black heritage."

I nodded with understanding.

"They don't interact with working-class or middle-class black kids, and I'm afraid they may get rejected by the white kids." He shook his head with mild frustration. "I'm thinking there needs to be someplace where they can meet other well-to-do black kids and not feel caught in between two worlds and rejected by both."

Lewis knew that my family did not have the incredible wealth or contacts he was amassing. But through our past conversations, he also knew that I had long-term relations with the well-to-do blacks I had grown up with and known all my life.

"I know there's a whole history and world of wealthy, professional black people," explained Lewis as we completed lunch and walked to his waiting limousine outside, "but I don't know how to crack it and introduce my kids to it."

Once I got over the initial shock that this multimillionaire was, on the one hand, hobnobbing with white millionaire investment bankers, yet was stuck outside of black elite circles, I told him about the elite black organizations and activities that he should immediately introduce to his children. I told him about the old-guard families dating back to the 1860s and about how they lived and where they lived today. I told him about the sixty-year-old Jack and Jill organization, a national invitation-only social group for black kids from well-to-do families. My brother and I had grown up in it and met our closest black friends there.

I also talked to Lewis about the Boulé and the Links, considered to be the most prestigious private social groups for, respectively, black men and black women. And though he was somewhat familiar with the topic, I talked in detail about the proper summer resorts, sleepaway camps, boarding schools, and black debutante cotillions.

"You oughtta turn all of this into a book," Lewis said as he jotted down some of my suggestions.

I laughed at the time because his remark seemed like a subtle mocking of my detailed knowledge of all the issues and groups he should consider. Then, as we rode north up Park Avenue in the black limousine, I thought that a book about the black elite and its history of thriving in the awkward position between two worlds—one black and one white—would be interesting indeed, but probably only to the wealthy blacks who already knew the stories and the rules. These black individuals already knew the distinctions between the old-guard blacks from Washington, Atlanta, Chicago, New York, or Los Angeles. They already knew about the obsession our group had with Episcopal churches, "good hair," light complexions, the AKAs, and Martha's Vineyard.

But as I thought more about the idea of this book and discussed it with people outside the insular circle who belonged, I discovered that there was a broader audience. I discovered that having grown up within the circle, I had been too close to see that the experiences really were unique. There was a story here that began as early as the 1870s with the nation's first black congressmen and the first black millionaires.

As I completed law school, I put the book project away for another time. I worked on other books and became a corporate attorney in New

York. Over the next couple of years, I lost touch with Lewis, even though I frequently read about him in the news, following his purchase of Beatrice Foods and the subsequent torching of his $5 million summer home on Long Island in 1991. During our occasional conversations, he informed me that his daughters had been enrolled in Jack and Jill as I had suggested, that Leslie had started at Manhattan's Brearley School, and finally that Leslie had been accepted into Harvard. He said that she and her younger sister had a strong black identity. It was good to hear and it reminded me again of the book project.

The real in-depth research on this book did not begin until the early days of 1993. Ironically, it was the same month that I was to receive the tragic news that this friend and mentor, Reginald Lewis, had died suddenly of a brain tumor. It was announced on the front page of the *New York Times* on the day of President Clinton's inauguration. As I began the first of more than 350 interviews, and the research that would take me throughout the country visiting libraries, manuscript collections, newspaper archives, and private homes, I had in my mind what Reginald Lewis had said to me several years earlier: There needed to be a chronicle of a community that was hidden from so many people.

Although it's a world I've known all of my life, and although it's an important part of our nation's history, it's a world that is filled with irony and conflict. This book was an opportunity to reveal a rarely discussed aspect of American history. It was an opportunity to capture the stories and lives of people like Lewis and many others, who have lived at the boundary of two worlds and been misunderstood by both.

Our
Kind
of
People
INSIDE AMERICA'S BLACK UPPER CLASS

The Origins of the Black Upper Class

Bryant Gumbel is, but Bill Cosby isn't.
Lena Horne is, but Whitney Houston isn't.
Andrew Young is, but Jesse Jackson isn't.
And neither is Maya Angelou, Alice Walker, Clarence Thomas, or Quincy Jones.
And even though both of them try extremely hard, neither Diana Ross nor Robin Givens will ever be.

All my life, for as long as I can remember, I grew up thinking that there existed only two types of black people: those who passed the "brown paper bag and ruler test" and those who didn't. Those who were members of the black elite. And those who weren't.

I recall summertime visits from my maternal great-grandmother, a well-educated, light-complexioned, straight-haired black southern woman who discouraged me and my brother from associating with darker-skinned children or from standing or playing for long periods in the July sunlight, which threatened to blacken our already too-dark skin.

"You boys stay out of that terrible sun," Great-grandmother Porter would say in a kindly, overprotective tone. "God knows you're dark enough already."

As she sat rocking, stiff-lipped and humorless, on the porch of our Oak Bluffs, Martha's Vineyard, summer home, she would gesture for us to move further and further into the shade while flipping disgustedly through the pages of *Ebony* magazine.

"Niggers, niggers, niggers," she'd say under her breath while staring at the oversized pages of text and photos of popular Negro politicians, entertainers, and sports figures who were busy making black news in 1968.

Great-grandmother Porter, the daughter of a minister and a homemaker, was extremely proud of her Memphis, Tennessee, middle-class roots. While still a child, she had worn silk taffeta dresses, had taken several years of piano lessons, and had managed to become fluent in French. Her only daughter had followed in her footsteps, wearing similarly elegant dresses, taking music lessons, and attending the private LeMoyne School a few years ahead of Roberta Church, the millionaire daughter of Robert Church, the richest black man in the South. She often reminded us that one of her sisters, Venie, then grown and married, had lived for years on Mississippi Boulevard next door to Maceo Walker, the most affluent and powerful black man in Memphis. Great-grandmother was proud of many things, such as being a Republican like the Churches and most other well-placed blacks in those early years. Like all blacks in racist southern towns in the early 1900s, she despised the insults, the substandard treatment, and the poor facilities that the Jim Crow laws had left for blacks. But like many blacks of her class, she was able to limit the interactions that she and her family had with such indignities. Rather than ride at the back of the bus and send her daughter to substandard segregated public schools, she and her husband bought a car and paid for private schooling. For my great-grandmother, life had been generous enough that she could create an environment that buffered her family against the bigotry she knew was just outside her door.

Even though it was 1968, a period of unrest for many blacks throughout the country, Great-grandmother—like the blue-veined crowd that she was proud to belong to—seemed, at times, to be totally divorced from the black anxiety and misery that we saw on the TV news and in the papers. In public and around us children, her remarks often suggested that she was satisfied with the way things were. She often said she didn't think much of the civil rights movement ("I don't see anything civil about a bunch of nappy-headed Negroes screaming and marching around in the streets"), even though I later learned that she and her church friends often gave money to the NAACP, the Urban League, and other groups that fought segregation. She said she didn't think much of Marvin Gaye or Aretha Franklin or their loud Baptist music ("When are we going to get beyond all this low-class, Baptist, spiritual-sounding rock and roll music?"), even though she would sometimes attend Baptist services. She was proud when a black man finally won an Academy Award, but was disappointed that Sidney Poitier seemed so dark and wet with perspiration when he was interviewed after receiving the honor.

An outsider might have looked at this woman and wondered whether she liked blacks at all. Her views seemed so unforgiving. The fact was that she was completely dedicated to the members of her race, but she had a greater understanding of and appreciation for those blacks who shared her appearance and socioeconomic background.

Disappointed and disillusioned by how little she saw of herself and her crowd in the pages of *Ebony* magazine, Great-grandmother looked up and once again focused her attention on me and my brother.

And then she thought about her hair.

Stepping back inside the house for her ever-present Fuller brush and comb, she was, no doubt, frustrated by the fact that her great-grandchildren were several shades darker than she, with kinky hair that was clearly that of a Negro person.

My brother and I noted her disappearance into the house and thus once again ran out of the shade and danced around the sand- and pebble-covered road, breathing in the sunshine and the fragrance of the dense pine trees that rose from the layers of sand and brush.

"Young men—young men," her voice called from the rear bedroom, "you aren't back in that sun, are you?"

"No, ma'am. We're in the shade, ma'am," my eight-year-old brother, Richard, called back with complete conviction as he stopped just out of my great-grandmother's range of vision, thrusting his bare brown chest and oval face into the ninety-six-degree July sun, boldly willing his skin to grow blacker and blacker in defiance of her query.

Even at age six, I knew the importance of class distinctions within my black world. As I moved quickly to the safety of the shade, beckoning my brother to protect his complexion from the blackening sun, I gave legitimacy to my great-grandmother's—and many of my people's—

fears. At age six, I already understood the importance of achieving a better shade of black.

Unlike my brother, I already knew that there was *us* and there was *them*. There were those children who belonged to Jack and Jill and summered in Sag Harbor; Highland Beach; or Oak Bluffs, Martha's Vineyard; and there were those who didn't. There were those mothers who graduated from Spelman or Fisk and joined AKA, the Deltas, the Links, and the Girl Friends, and there were those who didn't. There were those fathers who were dentists, lawyers, and physicians from Howard or Meharry and who were Alphas, Kappas, or Omegas and members of the Comus, the Boulé, or the Guardsmen, and there were those who weren't. There were those who could look back two or three generations and point to relatives who owned insurance companies, newspapers, funeral homes, local banks, trucking companies, restaurants, catering firms, or farmland, and there were those who couldn't. There were those families that made what some called "a handsome picture" of people with "good hair" (wavy or straight), with "nice complexions" (light brown to nearly white), with "sharp features" (thin nose, thin lips, sharp jaw) and curiously non-Negroid hazel, green, or blue eyes—and there were those that didn't. I had a precious few of the above, while many I knew and played with were able to check off *all* the right boxes. In fact, I knew some who not only had complexions ten shades lighter than that brown paper bag, and hair as straight as any ruler, but also had multiple generations of "good looks," wealth, and accomplishment. And, of course, I also knew some black kids who could claim nothing at all.

It was a color thing and a class thing. And for generations of black people, color and class have been inexorably tied together. Since I was born and raised around people with a focus on many of these characteristics, it should be no surprise that I was later to decide—at age twenty-six—to have my nose surgically altered just so that I could further buy into the aesthetic biases that many among the black elite hold so dear.

During my youth, it was often painful for me to acknowledge that I had one foot inside and one foot outside of this group. I never quite had enough of the elite credentials to impress the key leaders in the group, but my family and I checked off enough boxes to be embraced by a segment of this community. Sometimes I knew where I was lacking and sometimes I didn't. For example, I knew that my complexion was a shade lighter than the brown paper bag, but that my hair—while not coarse like our African ancestors'—had a Negroid kink that made it the antithesis of ruler-straight. I knew that we lived in the right neighborhood, summered at the right resorts, and employed the right level of household help, but that we had not attended the right private schools or summer camps. While my grandparents owned businesses and property that brought them and their children an unusually high standard of living, none of them had gone to the right schools. My mother was accepted by the old guard's most exclusive women's social club, but my father belonged to a fraternity that the elite group considered to be "distinctly middle-class." My brother and I grew up in the country's most exclusive black children's group, yet we were never invited to serve as escorts in the best debutante cotillions.

But whether it was mainly the skin color, the hair texture, the family background, the education, the money, or the sharpness of our features that set some of us apart and made some of us think we were superior to other blacks—and to most whites—we were certain that we would always be able to recognize our kind of people.

The characteristics of the black elite have roots that can be traced back almost four hundred years to when slavery began in this country. When the first Africans arrived on the shores of Jamestown, Virginia, in 1619, neither the white Dutch or Portuguese slave traders nor the white American plantation owners had any knowledge of, or interest in, the tribal and cultural differences between the Mandingo, Dahomean, Ashanti, Mbundu, Ewe, or Bantu blacks who were brought from different regions of Africa. They had no knowledge—or interest in letting the rest of white America know—that these blacks had come from established villages where they were already skilled in crafting iron, gold, leather, silver, and bronze into tools, artwork, and housewares, and where they were already weaving clothing; speaking different languages; growing tomatoes, onions, and fruits; raising livestock; practicing different religions; and establishing laws, banking mechanisms, medical treatments, and various other cultural traditions. The Ivory Coast, Guinea, the Gold Coast, and other areas of West Africa were simply profitable ports of call for the Dutch, Portuguese, Spanish, British—and later, American—slave traders and shippers who captured, dehumanized, raped, and sold more than fifteen million men, women, and children on blocks at boat docks up and down the East Coast of the United States. To the traders, a sale at an average price of $500 per man and $250 per woman or child (these South Carolina 1801 prices eventually rose as high as $1,300 or $1,400 by 1860) made the business of human bondage a profitable one that required no conscience and no need to draw cross-cultural connections.

Prior to 1442, when the Portuguese first arrived in West Africa to capture blacks and begin the four-hundred-year period of black slavery, Africans had their own class distinctions that were based on tribe, locale, and the individual tribal member's assigned role in his community. All those distinctions became moot in the New World.

When black slaves arrived on many southern plantations, they were ultimately divided into two general groups. There were the outside laborers who worked in the fields harvesting rice or tobacco, cutting sugarcane, picking cotton, or building roads and structures. Included among this outside laborer group were those slaves who worked at smelting iron, digging wells, and laying bricks. Many slaves built the very plantations and buildings around which they worked. Thomas Jefferson's Monticello and George Washington's Mount Vernon were both important homes built by slaves. And, for the most part, throughout the South, these outside laborers lived in windowless, unfloored shacks, or in structures close to their owners' livestock.

The second group of slaves were those who performed the more desirable jobs inside the master's house: cooking, cleaning, washing, and tending to the more personal needs of the owner's family around the home. While these laborers were also slaves, with no more or fewer rights than the outdoor workers, the distinctions between the slaves in the field and the slaves who served as butlers and "mammies" in the house were not at all subtle. The terms "house niggers" and "field niggers" grew into meaningful labels as generations of slaves in the master's house gained more favorable treatment and had access to better food, better work conditions, better clothing, and a level of intimacy with the owner's family that introduced the house slave to white ways, minimal education, and nonconsensual sexual relations.

As the caste system among the slaves was gradually instituted by slave owners and their families, the slaves themselves came to believe that one group was, indeed, superior to the other. The plantation owners began to place their lighter-skinned slaves in the house, thus creating an even greater chasm between the two groups—now based on physical characteristics, not just random assignment. Because these lighter-skinned blacks were perceived as receiving greater benefits and a more comfortable lifestyle, resentment among the darker-skinned field slaves only grew. "Although it was illegal to educate slaves," explains history professor Dr. Adele Logan Alexander of George Washington University, "it was far more likely that the house slave would learn to read, be introduced to upper-class white traditions, be permitted to play or interact with white family members than would a field slave. In fact, slave-owning families found they could run their homes more efficiently when their house slaves were more knowledgeable and educated."

Not surprisingly, both whites and "house niggers" came to consider the dark-skinned "field niggers" to be less civilized and intellectually inferior.

Since many white slave owners established clandestine and forced sexual relations with their female house slaves, the mulatto offspring (who were also assigned slave status) extended the size of the house slave staff. In fact, it was to the owners' benefit to mate with as many of these female slaves as possible, for each new child was a new slave. While none of these illegitimate offspring had any more rights than the unmixed African slaves, they became a part of a growing phenomenon of lighter-complexioned house slaves who separated themselves even further from the field blacks.

"It is evident that the fixation on skin color by both upper-class whites and blacks derives from the fact that light-skinned blacks were given a favored status by white slave owners from their very early interaction during the slavery period," explains Professor Alexander. In her book *Ambiguous Lives*, Alexander explores her own family's roots in Georgia as free, light-skinned blacks in the eighteenth and nineteenth centuries. The phenomenon was not limited to southern blacks. Although the South was the most notorious for its use of slaves, blacks were owned and used as slaves in the northern states well into the 1800s. For example, although Pennsylvania was the first state to establish a law abolishing slavery, with its 1780 Act for the Gradual Abolition of Slavery, the weak terms of the act allowed some blacks to remain enslaved until 1850, when the Federal Census ultimately found no slaves living in the state. Over time, then, the caste system took on a third dimension: free blacks versus enslaved blacks. For example, in Pennsylvania, the Federal Census of 1790 reveals that of the 10,274 blacks living in the state, more than one-third of them were slaves, with the remaining group labeled as "free blacks" or "free people of color."

There were many arguments used by white government officials, religious leaders, and highly esteemed landowners to justify the continued enslavement of blacks in North America. Just as they had done when attempting to enslave the Native Americans, the white population insisted that Africans should be enslaved because they were not Christians—and so long as they did not embrace such religious tenets, they needed to be ruled by civilized whites who did. Upon the conversion of some of the blacks to Christianity—and after a certain number of years of indentured servitude—a very small percentage of slaves were set free, thus creating a population of free blacks in North America. This practice of freeing slaves was sporadic and was ultimately halted in the South. In fact, it was as early as the 1660s, in Virginia, that blacks would legally be made slaves in perpetuity. Many white landowners and government officials elsewhere quickly came to agree with this decision as they realized how much the economies of their communities had benefited from the free labor.

While their numbers were inconsequential, free blacks in the South became an elite group with a status somewhere between their enslaved brethren and white citizens. For example, in the state of Georgia in 1800, there were 1,019 free blacks, and by 1860 that number had grown only to 3,500. In the city of Charleston, South Carolina, in 1860, out of a population of 40,522 people, close to 14,000 were black slaves, and only 3,200 were free blacks. Free blacks in the South were generally required to carry papers proving that they were not slaves and were required to register annually in their counties, listing their white guardians. However, they were permitted to work for money and to own property, thus creating the first opportunity for blacks to establish their families with some moderate wealth. The nucleus of the black elite was formed around these families.

One of my first encounters with an early pioneering family in the black upper class was meeting members of the aristocratic Syphax family from Virginia. I had grown up hearing my father tell me about their family history, as one of his father's business associates had known several family members. Talking to Evelyn Reid Syphax at a Links meeting that my mother had brought me to, I learned one way in which some black families—including her own—gained wealth and a place among the upper class. "My family had owned fifteen acres of the land where the Arlington National Cemetery now sits," Syphax explained as she recounted the history of her family, which can be traced back to Maria Custis, the mulatto child of First Lady Martha Washington's grandson, George Washington Parke Custis, who owned the mansion that sits on the cemetery today. "Custis fathered Maria with Ariana Carter, one of his female house slaves," explained Syphax, a well-to-do, retired real estate broker who lives in Virginia, "and when Maria asked her father, who was also still her owner, for permission to marry Charles Syphax, a black slave who worked for her father, he released both from slavery, gave her a wedding in the mansion, and offered her and her new husband fifteen acres of the Arlington estate. That mansion and the surrounding property—minus Maria Syphax's fifteen acres—was later given by Custis to his white daughter, Mary Custis, who eventually married Confederate soldier Robert E. Lee—thus making the house a famous building in the southern state."

On my first trip to Washington with my family in the late 1960s, I remember my parents remarking that the Syphax story was evidence that a black family's genealogy could be as relevant to our country's history as any white family's that I might read about in school. The fact that the Syphaxes were still residing in the community today made their story even more real to me at that age. Since that time, the Syphax family has continued to gain wealth through other real estate holdings and through businesses that involved commercial and residential development and sales.

Whether they began as free people of color or as enslaved house servants, those blacks who came to make up the black aristocracy were typically those who were able to gain an education and various professional skills. Access to a college education was clearly the earliest and surest method for earning respect among progressive whites who were willing to teach blacks various trades and offer them limited work.

A college education was made possible for blacks around the time of the Civil War by the generosity of religious groups like the American Missionary Association, an association of liberal whites who had initially opened elementary schools for blacks in the South before the war. As the abolitionist group raised funds in the North and attracted northern white and black teachers to the South in the 1860s, they established secondary schools and colleges like Fisk University in Nashville, Hampton Institute in Virginia, Tougaloo College in Mississippi, and many other universities that exist today. By 1865, the American Missionary Association had recruited more than three hundred teachers to its black southern schools.

Another group that made it possible for blacks to receive formal education at this time was the Freedmen's Aid Society, which was established by the Methodist Episcopal Church in 1866.

By the 1870s, not even a decade beyond slavery, handfuls of blacks were breaking into their own middle class by attending the earliest established black colleges like Howard University, Fisk University, Atlanta University, and Morehouse College—all founded in the 1860s—and later Spelman College, founded in the 1880s for black women. Not coincidentally, these are among the six or seven schools that the old-guard black elite still consider to be most appropriate for their children today. Specifically not included are dozens of other black colleges—schools like Alabama State and Tougaloo College—because they were state-sponsored or populated by poorer students from the Deep South. There are also examples of blacks attending white northeastern universities like Harvard and Amherst, and prep schools like Phillips Exeter, as early as the late 1800s. Another white school that was also known to admit blacks in this early period was Oberlin College in Ohio, which African Americans attended in small numbers as early as the mid-1860s.

Although Oberlin is still spoken of fondly by well-to-do blacks who went to college in the 1950s and prior to that time, its popularity among the black elite has been eclipsed by the top black colleges and by the more prestigious and better-endowed white East Coast colleges like Harvard, Brown, and Wellesley. Although Oberlin was notoriously abolitionist during the 1800s and remains extremely liberal to this day, there is a preference among upper-class blacks to attend schools that have more of a worldwide reputation.

Another factor in the elevation of a handful of black families was the possibility of attaining political office. This happened initially in the South during the reconstruction that lasted from 1865 until 1877. In southern states where large black populations had suddenly gained the status of free men and women, black men began to run for seats in the House of Representatives and the Senate. The Reconstruction Act of 1867 emboldened them to attend political conventions and join political clubs.

In 1870, Hiram Revels of Mississippi became the first black elected to the U.S. Senate. Senator Blanche K. Bruce, a black U.S. senator elected in 1874, also from Mississippi, had been born a slave in Farmville, Virginia, in 1841. While rebuffed by some in Washington's white society, he and his wife, Josephine Willson Bruce, a prominent dentist's daughter, were early leaders in Washington's black society during the late nineteenth century.

Other blacks who were elected to the U.S. Congress at this time included Robert Smalls and Joseph Rainey of South Carolina, Jefferson Long of Georgia, Benjamin Turner of Alabama, Josiah Wells of Florida, and John Lynch of Mississippi. Between 1870 and the late 1890s, nearly two dozen blacks served in the U.S. Congress. And while elected office did not promise wealth or acceptance within the white social structure, it did bring lasting prestige to certain family names.

Prominent families socialized with one another, built businesses with one another, and intermarried, establishing well-to-do and well-respected dynasties. These included names like Terrell, Pinchback, and Grimké in Washington; Herndon and Rucker in Atlanta; Minton and Purvis in Philadelphia; Bishop, White, and Delany in New York; Wheeler and Williams in Chicago; and many others.

Although very large post–Civil War black populations were found in the rural sections of southern states like South Carolina, Virginia, Georgia, Mississippi, and Louisiana, these were not necessarily where the black upper-class families built their foundations for work and education. Instead, they built their futures in more progressive cities, close to black universities and removed from rural poverty. The cities where they established and built businesses and institutions in the South were Atlanta, Washington, Nashville, Charleston, and Memphis, where my father's family began a variety of businesses.

And both preceding and following the path of the post–World War I and II migrations of blacks to the North, the black elite also established themselves in Chicago, Detroit, New York, and Philadelphia. There were also pockets of the black upper class—particularly physicians—in Tuskegee, Alabama, that grew up around the nation's only black Veterans Administration hospital and Booker T. Washington's Tuskegee Institute; in Nashville, centered on the campuses of Fisk University, Tennessee State, and Meharry Medical College; and later, to a lesser extent, in Los Angeles.

Although some prominent families established their roots in positions with important church congregations or universities, the more common paths to success in the black elite have been through careers in medicine, dentistry, and law. In addition, examples of early and continuing black wealth can be found in individual entrepreneurship that served either the black or the white communities. For example, in most cities with large black populations, black funeral home owners formed a core part of the elite, along with the founders of local black banks, insurance companies, and newspapers. Wealthy and influential Atlanta families like the Herndons and the Scotts, for example, can point to Atlanta Life Insurance and the *Atlanta Daily World*, respectively, as the source of their wealth and influence.

Radcliffe graduate and Washington, D.C., TV news producer A'Lelia Perry Bundles is the great-great-granddaughter of Madam C. J. Walker, the nation's first self-made woman millionaire. "Many people are surprised to learn that a black woman who was developing hair care products and cosmetics in the late 1800s would actually be the first self-made woman millionaire in this country," says Bundles, as she looks through photos of Walker standing with Booker T. Washington and other prominent black businesspeople. In addition to building homes in Indianapolis and Harlem and a twenty-thousand-square-foot stone mansion that still stands near the Hudson River in suburban Westchester County in New York, Walker used her millions to help the NAACP's antilynching campaign.

Members of the black elite have historically given generously to charities like the Urban League, the United Negro College Fund, and the NAACP, a ninety-year-old organization that has its roots among this group. The irony is that while today most of these successful people will still write checks for the NAACP, they do not consider it part of their social circle in the way they would have done before the 1960s. "We'll see what Myrlie Evers can do with it, because she has a lot of class, but that NAACP is a bit too grassroots for people like us," explains a Chicago matron as she sits in the dining room of a large home in the city's Hyde Park neighborhood. The woman holds a graduate degree and belongs to several boards and elite black organizations. "The NAACP did wonderful things for us in the South—and up here too—and I'm happy to give to them because they help all of us, but you just don't find a whole lot of professional blacks socializing among the NAACP. We've got our own groups."

As the woman pulls out her family photo albums, she points to herself, her parents, her husband, and their friends who hold memberships in the Alphas, AKAs, Deltas, Links, Drifters, Girl Friends, Boulé, and Guardsmen. There are photos of gatherings at the Ink Well in Martha's Vineyard; relatives playing tennis at a summer home in Idlewild, Michigan; photos from an important Detroit wedding reception that included nearly one thousand guests; and, of course, snapshots from a family member's black-tie debutante cotillion. "I have a graduate degree and my friends have degrees from Howard, Spelman, Morehouse, and

Meharry. Why would I be socializing with some caseworker or mailman who goes to NAACP events? I'd have as much in common with them as a rich white person has with his gardener."

And it is because of these real or imagined differences that the black elite in every major city and suburb has its own churches as well.

"The black upper class has most often been associated with the Episcopal Church," says Rev. Harold T. Lewis, the author of *Yet with a Steady Beat: The African American Struggle for Recognition in the Episcopal Church* and rector of Calvary Episcopal Church in Pittsburgh. Despite earlier affiliations with the Baptist and Methodist denominations and the larger numbers of blacks who currently make up those congregations, the black elite have often selected the more formal high Episcopal Church or Congregational Church.

The Episcopal faith was attractive because of its formality, and both faiths were appealing because they were known for having well-educated clergy and a small number of members. Well-to-do black Americans with roots in the West Indies had natural historic ties to the Episcopal Church, which had served a major role in Jamaica and other former British colonies for several generations. The Congregational Church's popularity among the black elite grew from the fact that it was the denomination that had given the greatest support to the American Missionary Association's efforts in establishing secondary schools and colleges for southern blacks in the late 1800s.

And for some of the most cynical and status-conscious members of the black elite, the two denominations were particularly appealing simply because most blacks were not of that faith.

In every city where there are members of the black elite, there is an Episcopal or a Congregational Church that dominates the upper-class black religious scene: In Chicago, it is St. Edmund's or Good Shepherd; in Detroit, St. Matthew's; in Philadelphia, St. Thomas; in Memphis, Second Congregational; in Charleston, St. Mark's; in Washington, St. Luke's; in Atlanta, First Congregational; and in New York, St. Philip's. Some say that the black upper class disdains the open display of emotions that are often shared in Baptist and AME churches, while others say that Episcopal and Congregational denominations have better-educated church leaders.

For whatever the reason, the choice does keep the elite separated. And just as there have been special churches for the black upper class, so are there special social groups that separate men, women, and children of different classes.

As a child, I grew up in Jack and Jill, the ultimate membership-by-invitation-only social group for black children and their families. Founded in 1938 by seven well-to-do black women who were establishing a play group for their kids, the elite national organization now has 220 chapters throughout the United States and Germany. "The purpose of the group," explains Dr. Nellie Gordon Roulhac, a Jack and Jill mother who served as the organization's national president from 1954 to 1958, "is to introduce black children and their families to other children who are interested in social service activities, educational programs, and other projects that improve the lives of people in the black community."

As a child going to Jack and Jill weekend excursions, fund-raisers, black-tie dances, or backyard tennis parties, I was introduced to the fact that there was a national network of black children who grew up smart, ambitious, well-to-do, and proud of their black heritage. Whether or not the group was intended to serve as such, it also introduced young boys and girls as future dates and spouses. "I grew up in Jack and Jill during the 1940s," explains New York resident William Pickens, whose mother, Emilie, was a national president of the group, "and I eventually married a Jack and Jill alum in the 1960s. It was a tight circle of people who all had a lot in common. Our fathers were lawyers or doctors and our mothers had gone to college together. This was where we were expected to find our friends and dates." I knew exactly what Pickens meant, because while I was a student at Princeton, I dated women at Harvard, Brown, and Mount Holyoke—all girls I knew through Jack and Jill. Armed with Jack and Jill's annual yearbook *Up the Hill*, I had a coast-to-coast directory of boys, girls, and parents who were just like me.

Other popular groups among the black elite are separated on the basis of gender. For the women, there is the nine-thousand-member Links, which was founded in 1946 and is populated by professionals, socialites, educators, and well-to-do matrons. Several years ago, my mother was inducted as a member, along with Dr. Betty Shabazz, the widow of Malcolm X, and a host of other accomplished women. The chapters—which allow no more than fifty-five members—are highly selective and include such women as former Washington mayor Sharon Pratt Kelly, Children's Defense Fund head Marian Wright Edelman, former Spelman College president Johnnetta Cole, Secretary of Energy Hazel O'Leary, and many other prominent names. Another highly selective and socially elite group is the Girl Friends, founded in 1933. The black-tie debutante cotillions the group sponsors in different cities for charity and for the introduction of young girls to society are among the most sought-after tickets on the black elite social calendar. Other groups include the Drifters, the Northeasterners, and the Smart Set. And for college women, there are two particular sororities that are favored above all others. They are the AKAs and the Deltas—primarily because they were founded over eighty years ago at Howard, and also because they took only the right girls from the right families.

For men, there are also exclusive, hard-to-crack social groups. At the top of the list is Sigma Pi Phi—known simply as the Boulé, which admits professional men once they have distinguished themselves in society. It was founded in 1906, and its early membership included W. E. B. Du Bois; chapters have since been established in most major cities, initiating such men as Washington attorney Vernon Jordan, Atlanta mayor Andrew Young, Urban League president Hugh Price, Virginia governor L. Douglas Wilder, and American Express president Kenneth Chenault, as well as most black college presidents and the most prominent black physicians, attorneys, dentists, and corporate leaders in any major city. When I joined the group, the once-secret organization became an important bond for me and a wide circle of friends.

Other elite groups for men include the Guardsmen, the Fellas, the Comus, and One Hundred Black Men. And as with the sororities, there are select college fraternities that are held in highest esteem: the Alphas, the Kappas, and the Omegas.

As I was growing up amidst this social world, there were certain things of which I became acutely aware. One was the importance of tradition. Another was pride in family background and accomplishment. While some outsiders—both black and white—dismissed our traditions as bourgeois or as being solely "white aspirations," members of the black elite still feel that there is nothing odd or inappropriate about the fact that they have created traditions that include a specific summer camp for black elite kids (Camp Atwater in Massachusetts), a specific boarding school for students of old-guard families (Palmer Memorial Institute in the South, and Northfield Mount Hermon for an integrated experience in the North), specific social clubs, specific summer resorts, specific family taboos ("passing" for white, putting down blacks in front of whites), specific colleges (among the black schools as well as the white ones), specific fraternities, specific churches, specific attitudes about whites (emulate them, but don't marry them), specific neighborhoods or streets within certain cities, specific suburbs, and even specific funeral homes and cemeteries in which to be memorialized.

When I recall the rules of the group, I think of a remark I once heard a friend of my parents make one afternoon when we were on my cousin's yacht on the Long Island Sound. I had asked her about sororities when Mrs. Jenkins (not her real name) began explaining the distinctions between the group of people who joined either the Deltas or the AKAs, and the very different group of college girls who might join the other sororities. "Now, I'm an AKA, and the only other natural sorority for me would be the Deltas. The other sororities were just not a consideration. It's not that I'm looking down on them, but the fit wouldn't have been right for me or them. They were a tad bit darker than us. And they really didn't come from our background. We were daughters of doctors, teachers, dentists, pharmacists, and such. They were doing good if their fathers were even Baptist preachers." Mrs. Jenkins stared off into the water for a moment. "We just came from different worlds and we weren't like-minded people. And why spend time with people who aren't similar? Of course they were all quite polite and well-behaved. After all, we were at Spelman."

"Polite and well-behaved." Although spoken with an air of condescension, that is probably the most disrespectful remark that a woman like Mrs. Jenkins—Spelman graduate, wife of a lawyer, mother of a debutante—ever would have said publicly about those blacks who were not of her class. People like her have equal respect for all blacks, regardless of their socioeconomic background and professional aspirations, but as she confessed, there is always a preference for being around like-minded people.

During my first year at Harvard Law School, I had an experience that I've always considered to be an epiphany in my journey toward greater racial awareness. I had made the requisite first visit to Wellesley College by hiring a private car to drive me the thirty minutes from Cambridge to Alumnae Hall at the women's school for a Saturday night party. It was sponsored by a black student organization called the Sisters of Ethos, one of the few small societies (along with ZA, TZE, and Shakespeare) on the sprawling suburban campus.

When I arrived at the circular drive and marble entry of the building sometime around 10 P.M., it immediately occurred to me that the scene looked like every Jack and Jill party of my childhood. We were all still students, yet the parking lot was jammed with expensive cars. There was a completely dark ballroom and a brightly lit foyer, jam-packed with well-dressed, light-skinned young people in expensive yet conservatively tailored clothes. With long, streaked, straight—or straightened—hair flying behind them, the Sisters of Ethos were running in and around the tall French doors, inspecting college IDs as they approved or turned away male partygoers who either passed or failed the ubiquitous "brown paper bag and ruler test."

I knew many of these girls and their families, and I also knew the game, but it was still a bit disconcerting to see them practice it so well against male classmates of mine who were pounding at the doors trying to overturn their quick rejections.

As the night got later, and as the sisters got hungry for more visitors, the rules were loosened and the darker-skinned guys (along with some darker-skinned non-Wellesley women) were admitted. By 11:30, the party was in full swing and we all turned our attention to the inner ballroom where the party was taking place. As I walked into the crowded, nearly pitch-black room, I could feel the beat of some new top forty crossover semirock, semirap song pounding away. As I circled the room, I saw reminders of my childhood. The "dark outer circle" was very much apparent. I remembered parties like this—where you'd customarily find the darker, less affluent, less popular, and less attractive guys standing against the wall, dateless and unpartnered. This was where one found the geri-curled guys and the dark-skinned women with "bad hair" and bad weaves.

As I walked farther into the room, I found a young woman who was a daughter of one of my mother's grad school classmates. She was an outgoing student whom I had last seen at a cotillion in Washington. As my reunited dance partner and I made our way to the center of the room, somewhere beneath a large unlit crystal chandelier, I found the creamy center that existed at every one of these blue-vein parties.

Dancing around us, taking up the central core of the ballroom, were the long-haired, yellow-skinned leaders of the young elite. Some with nose jobs, some with thinned lips, some with naturally green eyes and naturally straight hair. Tania from Atlanta, Crystal from St. Louis, Julie from Washington, Brooke from New Rochelle, Cheryl Lynn from Los Angeles. Each a daughter of some light-skinned doctor, banker, or judge who belonged to some chapter of some organization and who had gone to some school before marrying some woman who belonged to some sorority, and who lived some kind of nice life that had "our kind of people" written all over it.

As many privileged black kids come to discover once they leave the safe confines of their parents' home, I then realized that several hundred miles away from my own home and the people I knew—here on a small campus in suburban Massachusetts—were my kind of people all around me, all intact.

As I stood there dancing to the crossover semirock, semirap song, I found myself feeling a little bit smug and a little bit scared. Although I should have been used to all of this after so many years of seeing it, it frightened me—perhaps because it seemed so predictable, or maybe because I felt there was so much to keep up with and there were so many others being shut out.

One can find both pride and guilt among the black elite. A pride in black accomplishment that is inexorably tied to a lingering resentment about our past as poor, enslaved blacks and our past and current treatment by whites. On one level, there are those of us who understand our obligation to work toward equality for all and to use our success in order to assist those blacks who are less advantaged. But on another level, there are those of us who buy into the theories of superiority, and who feel embarrassed by our less accomplished black brethren. These self-conscious individuals are resentful of any quality or characteristic that associates them with that which seems ordinary. We've got some of the best-educated, most accomplished, and most talented people in the black community—but at the same time, we have some of the most hidebound and smug. And adding even further to the mix are those of us who feel we need to apologize to the rest of the black world for our success and for being who we are. For me, the black upper class has always been a study of contrasts.

Jack and Jill: Where Elite Black Kids Are Separated from the Rest

"Well, what I'm really hoping is that she'll become a litigator or maybe some kind of judge," uttered a woman, with total confidence, as we looked out over Central Park from a Fifth Avenue apartment in the East Seventies. Dressed in a conservative linen suit, the forty-three-year-old mother adjusted her Chanel scarf in the reflection of the large living-room window. "We scrapped the idea of banking because the industry still won't be ready for a black woman, and we're steering her away from medicine because she's only getting B's in the sciences."

"What about something more creative, like advertising or a career in the entertainment business?" asked one of the mothers.

Although I didn't know the woman in the scarf well, I instinctively knew that she would find this a ridiculous question.

"Entertainment? We have enough black girls doing that. I expect more from Laura than advertising or entertainment," she added disdainfully.

Looking across the room at twelve-year-old Laura, I shrugged my shoulders in sympathy. This was exactly why we had all been enrolled in Jack and Jill as kids. So that we would quickly accept the fact that each of us was supposed to do great things.

"Don't get me wrong," the woman with the scarf added as she glanced over at the other mothers, who were sipping ginseng tea and eating cookies with their adolescent sons and daughters. "I'm not turning my nose up at advertising or entertainment, but Laura needs to be a professional. I'm a professional, her father is a professional, and three of her four grandparents were professionals. Everybody she knows is at the top. She's not going to start her life off by setting us back two generations. She'll like being a litigator."

It was a Saturday afternoon and we'd all just arrived from a Jack and Jill career day held in a series of classrooms at Collegiate, a private school for boys on Manhattan's Upper West Side. The day's events had begun with a presentation by Jack and Jill alumni who discussed their different professions with current Jack and Jill children and parents. The children—all of them from private day schools like Horace Mann, Spence, Riverdale, and Dalton—then stood up and told about their high school activities, followed by a description of their career ambitions.

Afterward, the students lined up to collect business cards from the group of alumni, of which I was one—two lawyers, two physicians, a state assemblyman, a *Fortune* 500 executive, an architect, and an investment banker. "So, did you really know what you wanted to be when you were fourteen?" asked a ninth-grader who was attending Horace Mann. For half a second, I considered avoiding the question in order to prevent the increased anxiety that my affirmative response would, no doubt, elicit. But, then I remembered where I was. Of course I'd known what I wanted to be at fourteen. In Jack and Jill, it was expected. Not only had I known I wanted to be a corporate lawyer, I knew where I wanted to attend law school. The same had been true for all of us, including Carla, the Jack and Jill girl I dated back then. She wanted to be a physician. And now, after four years at Harvard and four more at Yale Medical School, she'd done what we'd all expected.

"Okay, now I want all the alumni and the children to mix," announced one of the mothers, who was acting as moderator for the career day event. "I want each of you children to pick one alumnus—three kids to an alumnus—and spend five minutes with that person to learn as much as you can about them. Find out what they do, where they went to school, how they became successful, the steps they took, the mistakes they made, and what they like most about their careers. Take notes if you want, but ask *lots* of questions!" said the woman, who happened to be a clinical psychologist and the mother of a rather gregarious fourteen-year-old boy.

"When I ring this bell in five minutes," she continued in a voice not unlike an announcer at a prizefight, "I want you—all of you—to switch to the next alumnus and ask the same kinds of questions. Over the next thirty-five minutes, you will have gotten to meet all seven of them up close. So don't miss this opportunity to ask what you want. This will benefit you immensely."

As I looked around at the twenty teenagers, I breathed in about thirty-five minutes' worth of air and smiled faintly at the state assemblyman standing next to me.

The bell sounded. "Okay, children, go!"

A fourteen-year-old boy in a green cotton turtleneck raised his spiral notebook toward me as he and two fifteen-year-old girls approached to shake my hand.

"Did you know you wanted to be a lawyer before you started college?" asked yet another boy, who approached me from behind. "Or did you make up your mind once you got there?"

"Well, you really don't have to make all those decisions when you start. You can make up your mind after you get there."

"That's not what my father said," responded one of the girls, who eyed me suspiciously. "He said that if I wanted to be an obstetrician, I should start setting up my summer internships at hospitals now, instead of waiting until everybody else gets in line at college."

I nodded. "That may be true, but you may get to college and discover you don't even want to be an obstetrician."

"And what?" the other girl asked. "Be a podiatrist?"

"Or a *dentist*?" the boy asked with a snide laugh.

It took me about six seconds to recognize the sarcasm in these kids' remarks. Their sophistication and elitism were mixed so adroitly that it was like listening to conversations between me and my friends at age thirty-five.

The second girl cleared her throat. "What would really be helpful is if you have some friends who could give us advice on one of those six-year or seven-year premed programs where you get your bachelor's degree and your medical degree from the same school without taking the MCATs and having to apply to medical school too."

Over the next hour, I shared advice on how to write a cover letter to a future employer, how to rehearse for a college interview, and how to deal with a bigoted guidance counselor. One or two of our participants seemed exhausted by the details and the suggestions, but most of the students were just as I had been when growing up in Jack and Jill—determined to show our parents that we would leave no question unasked, no teacher unchallenged. Determined to live up to the ideal Jack and Jill child, we had all been convinced that we were as good as, if not better than, the smartest white kids at school.

Later that afternoon, the whole group moved on to one of the Jack and Jill members' apartments on Fifth Avenue, where mothers and kids were having tea and dessert snacks while partaking in a discussion of black role models and getting a tour of this member's family art collection. A cynic would dismiss this as a pretentious, nouveau riche display. But these parents and kids were sincere, and the purpose of the visit was genuine and necessary. Just as my mother and father had done in the 1960s and 1970s, these parents were cramming as many experiences and as much information as they could find into their young black children. Nobody questioned whether it was too early for the kids to digest it. The attitude was, "Show it to them now so they'll be ready for it later." Everyone smiled and pretended to be easygoing, but just underneath the surface was ambition raging to break through.

As I stood there with those kids and parents, it occurred to me that *I* was barely ready for what was on display in that apartment: this was the first time that I'd seen a Renoir, a Beardon, and a Picasso in someone's living room. It was the first time I'd been served by a butler who poured tea with white gloves. The whole day was a consummate Jack and Jill experience. It was one of the reasons that Laura's parents had wanted her to be a part of Jack and Jill, and it was one of the reasons that my parents had made the same decision for me and my brother over thirty years ago.

Founded on January 24, 1938, Jack and Jill of America has long been one of the defining organizations for families of the black professional class. It has 218 chapters throughout the United States and Germany, and its membership includes more than 30,000 parents and children. A nonprofit service organization, it focuses on bringing together children aged two to nineteen and introducing them to various educational, social, and cultural experiences. In addition to sponsoring public service projects in their communities, the various Jack and Jill chapters raise money for local nursing homes, shelters, hospitals, and educational institutions.

"One of the major reasons why parents want their kids to grow up in Jack and Jill is the social and educational benefits," says Shirley Barber

James, the group's fifteenth national president. "Even if you send your children to the best private schools and colleges, it doesn't mean that they will get to meet black role models who inspire them and make them feel that they can succeed in a white world. That's what Jack and Jill can do," adds James, a Savannah parent whose son, Robert, grew up in Jack and Jill and later graduated from Savannah Country Day School, Howard University, and Harvard Law School.

Because Jack and Jill is very selective and admits members by invitation only, it provides a great opportunity for professional parents to introduce their kids to children of similar families. Whether one is in Boston, Atlanta, Houston, or Beverly Hills, the children of a community's most prominent black families will be found in the local Jack and Jill chapter. For generations, it has served as a network for parents who want play groups for their children, as well as a network for young adults who want companionship, dating relationships, and ultimately marriage partners.

William Pickens, who was first enrolled in Jack and Jill as a two-year-old in 1938, says he wasn't surprised that he eventually married a Jack and Jiller. "My parents wanted me and my brother in the organization so we'd be introduced to other black kids who came from similar backgrounds," explains Pickens, while recalling the friendships he has maintained from the organization. Pickens, whose father, William, was a lawyer and whose mother, Emilie, was national president of the group in the early 1950s, came from the kind of family lineage for which Jack and Jill became famous. His father was Langston Hughes's roommate at Lincoln University, and his grandfather, a class-of-1904 graduate of Yale, was the first black dean of Morgan State University and a friend of W. E. B. Du Bois. His mother was a class-of-1923 graduate of the University of Washington and a descendant of one of the oldest black families in Pennsylvania.

Now president of the Paul Robeson Foundation in New York, Pickens recalls that many of his friendships, as well as his marriage, grew out of ties to Jack and Jill. "My wife, Audrey Brannen Pickens, had been teen president of her chapter in Queens. I had been social chairman in my New York chapter. We had a lot in common. Like many kids in the group, we grew up spending our summers in Sag Harbor and going to the same cotillions. And the Jack and Jill tradition continues. All three of our kids were members while growing up, and now we also have grandchildren in the group." Since Pickens's sister-in-law, Barbara

Brannen Newton, is the group's national secretary, he still feels close ties to the organization.

Few can deny the early educational and cultural benefits that one gains from growing up in this organization. Most of the kids who grew up with me in my chapter of Jack and Jill lived an almost completely white existence during the week. Unlike most of our parents, who had grown up in segregated towns and school systems with black neighbors and teachers, we lived in all-white neighborhoods and attended classes that had never seen a black face beyond the custodian's closet. We played with white kids who claimed us as their "favorite" (read "only") black friend. In fact, so nonintegrated was our Monday-to-Friday schoolday experience that we rarely heard racial issues discussed outside of the house.

My white friends knew that I was different, but except for the occasional slip on their part, there was an almost unspoken gentleman's agreement between me and my white friends that nobody would remark on my blackness except to ask me for a translation of the occasional black slang words they heard on *The Flip Wilson Show* or to answer some dumb question like whether we tanned in the sun or put starch in our Afros. In a sense, we were all on our best "guarded" behavior—politely answering naive kids with nonthreatening responses that carefully disguised our annoyance. We were black kids living in a white world that had strict rules on conformity.

But when Saturday came, there was a collective sigh of relief as we could remove the masks and settle into our black upper-class reality. We had our Jack and Jill gatherings: carpools of black mothers and fathers pulling up to museums or riding stables in Ford and Buick station wagons, or groups of kids getting off a plane in order to visit our congressmen in Washington. No matter how different all of our weekend activities were, we were around other black kids, where we black children could be ourselves and tell each other exactly what we thought. We could talk about racist incidents just the way we talked about baseball and coin collecting. We could talk about how we were sick of all the *What's Happening!!* ghetto dwellers on TV while also talking about some new remote-control airplane we were building at home.

Shortly before she died, former national president Nellie Thornton told me, "Jack and Jill is especially important for children who are growing up in suburban communities who have a foot in two different worlds." Thornton, an elementary school principal who had been one of my mother's closest friends, was concerned about keeping herself, her husband, and her two daughters connected to the black community and black history. Living in a large white stucco mansion in a Westchester county suburb, she wanted her daughters to know blacks like themselves. A graduate of Fisk, Thornton divided her time between being president of her Links chapter, membership with the AKAs, and bringing her kids to monthly Jack and Jill events.

Black suburbanites have leaned on Jack and Jill for generations because the bucolic streets outside their home failed to reflect the racial diversity of the cities. Elsie Ashley remembers being inducted into a suburban Connecticut chapter at the same time as Rachel Robinson, the widow of baseball player Jackie Robinson. "It was 1958, and there were just no other opportunities for my children to meet black role models in suburban Stamford, Greenwich, or Norwalk," says Ashley as she sits in her spacious kitchen. Even though she and her husband were also eventually to buy a summer home in Martha's Vineyard, like many other Jack and Jill parents, she and her kids got some occasional reality checks from her white suburban neighbors. "Even though the world saw our community as a progressive part of New England," explains Ashley, a native of Connecticut's Fairfield County, "my children and I still remember the day the neighbors burned a cross on the Robinsons' lawn."

Ashley's daughter, Yvonne, doesn't allow such incidents to cloud her memory of being black in such settings. As she recalls the Jack and Jill picnics they used to enjoy on the Robinsons' five-acre estate, she points out, "Four of my closest girlfriends are girls that I met through that organization. That makes it all worth it."

Even though the Jack and Jill kids were, in general, a bit wealthier and far more cynical and sophisticated than my well-to-do white friends in my neighborhood, I felt a lot more secure around the Jack and Jill kids. I felt I knew where I stood with them: even the ones whom I didn't like and who didn't like me. There was no hidden racial agenda, as there was among some of my white classmates at school. In Jack and Jill, we all said what we thought. Not as at school or in the neighborhood, where whites were curious about racial issues but talked around them by asking instead about some black sitcom character or making a reference to some new dance move they saw on *Soul Train*. As white classmates avoided the real issues, much of what we thought went unspoken.

For example, if a white kid at school didn't invite me to a birthday party or bar mitzvah, I didn't know if it was because he didn't like me or if it was because he and his parents didn't want a black in their house or at their club. If someone didn't want to trade sandwiches in the lunchroom, I wasn't sure it was because he didn't like tuna fish or because he thought black food would poison him. If someone didn't want to bob for apples after me at a school PTA bazaar, I wasn't sure if he was tired of the game, or if it was a fear of mixing black and white germs. These are all real incidents that happened to me, and although they occurred thirty years ago, they still hurt as if they had happened yesterday.

At school and in the neighborhood, race was always part of the unspoken subtext. Maybe it was paranoia, but it was clearly something the other Jack and Jill kids saw Monday to Friday too. On the weekend, we could all escape. In our elementary and junior-high years, it was not something we could articulate, but as we got older we became more aware that our experiences—no matter what city or town we lived in—were not that different. And being able to talk about them was a kind of salvation.

Savannah's Shirley Barber James says that Jack and Jill kids are introduced to a variety of important issues and experiences through the group's programs. "Many of our chapters have taken the children to Africa and other places where they can learn about our people's history," says James, who raised three children in the organization. "My husband and I sent our kids to the best private schools in our city, but we realized that their socialization would not be fully realized in an academic setting. We needed the church, the family, and Jack and Jill." Because James's children attended the academically competitive Savannah Country Day School, she was comfortable with their academic experiences, but she wanted to strike a balance with leadership and cultural programs that included greater racial diversity. "And I also wanted them to attend cotillions and parties with other black children," the Savannah native adds.

While the organization operates with local chapters or branches, Jack and Jill is actually controlled by a national office, with each local chapter chartered under the group's constitution and bylaws. Each chapter, in turn, is broken down into special age groups consisting of toddlers aged two to four; young children aged five to eight; preteens aged nine to eleven; junior teens, who are twelve to fourteen; and senior teens, who are in high school. Age-appropriate activities are provided for each of the different groups. The parents and the officers (usually the mothers, although fathers are allowed to participate) chaperone their child's groups to museums, theater performances, sporting events, and overnight trips to historic places in the United States and abroad. There is an activity at least once each month and there is often a focus on learning about black history. Most of what I know today about the early accomplishments of blacks I learned by growing up in Jack and Jill.

Depending on the size of each chapter's separate age groups, monthly organizing meetings are held at members' homes or at local clubs or hotels. During the year, there are usually two or three parties just for the kids (at least two will be semiformal or formal). And at least twice a year, there will be a semiformal or formal party given at a private club or hotel just for the parents and their invited guests. Many of the Jack and Jill chapters also sponsor debutante cotillions for those young girls who have reached their senior year in high school and who will simultaneously graduate from their chapters. Regardless of their age, the members get a heavy dose of tradition and organizational history. When I was growing up, I was always aware of the group's history. At local conferences, there were always photo displays of past officers. At national conventions, there were always speeches by officers. Knowing who started the group and who was still around was something we were compelled to know and remember.

When Jack and Jill's official founder, Marion Turner Stubbs Thomas, first developed the idea of the organization with Louise Dench in Marion's Philadelphia home with several other mothers in her city, she had the same concerns for her children as did other well-to-do black women who saw their kids growing up during a period when blacks had access to few positive black images and to few activities that brought them into contact with other middle- and upper-class black families. The daughter of Dr. John P. Turner, a surgeon who was the first black on Philadelphia's board of education, Marion was born in Philadelphia and graduated from the University of Pennsylvania in 1930. A child of affluent and worldly parents, she attended the Sorbonne, then returned to the United States and married a thoracic surgeon, Frederick Douglas Stubbs, in 1934. After Stubbs died, Marion married Detroit physician Alf Thomas, and the two became a celebrated couple in that city's black elite. They raised three daughters in Jack and Jill—and the daughters went on to raise their own children in the organization as well.

Like Thomas, the other women in Jack and Jill's organizing chapter of 1938 came from well-to-do families. Among them was Helen Dickens, a physician who later became a professor and associate dean at the

University of Pennsylvania School of Medicine; Sadie T. M. Alexander, an attorney who had also earned a Ph.D. from the University of Pennsylvania in 1921 and whose husband, Raymond Pace Alexander, was a Philadelphia judge; and Dorothy Wright, who was married to Emanuel Crogman Wright, son of the founder of the Citizens and Southern Bank and Trust Company. Dorothy Wright, who was also pivotal in the later creation of the elite black women's group, the Links, was the first national president of Jack and Jill.

Today, most parents are attracted to the organization for the same reasons that inspired these founding mothers. "Of course, some of the parents see it as a way to establish themselves among the black elite in their communities," says a former Oakland Jack and Jill mother as she tells me about the culprits in her own chapter, "but most simply recognize the need to introduce their children to role models, since many of them live in or utilize schools in predominately white settings and want their kids to meet other blacks who understand their experience." Anita Lyons Bond of St. Louis agrees. "Our three kids were in a predominately white private school—St. Louis Country Day—and they needed the socialization skills that this group teaches." Bond, a former chapter president, and her husband, a prominent surgeon, both believe the group teaches confidence.

"Jack and Jill was extremely important to my sisters and me because it constantly reinforced the role of education and excellence in the black family," says Ilyasah Shabazz, daughter of slain civil rights leader Malcolm X and Dr. Betty Shabazz. "When you are attending competitive schools that are predominately white, and when you are raised in a world that only profiles white success stories," says Shabazz, who grew up in the same Jack and Jill chapter as me while she attended the exclusive, all-girls Masters School in suburban Dobbs Ferry, New York, "you need to be exposed to black success stories."

Many non–Jack and Jillers would probably be surprised to learn of the number of "celebrity" families who participated in Jack and Jill. But for a lot of the kids of famous blacks, a place in Jack and Jill was a safe harbor. Even though some of us whispered about the celebrity of the Shabazz family, for the most part they were just another group of kids in our play group. It was a situation that offered much more normalcy and much less scrutiny than they would ever find in white schools and neighborhoods.

Now working as a public relations director in the New York area,

Shabazz recalls her Jack and Jill years fondly and continues to give back to black children through her activities in the Links and its various programs that support young people. "It was a challenge growing up in the spotlight," says the attractive and popular Shabazz, "but thank goodness my mother realized that Jack and Jill would expose us to educational activities and to role models and other black families who wanted the best for their kids."

The impeccable résumés of the organization's past presidents demonstrated to local chapter members and others that professional accomplishments were crucial to Jack and Jill's success. I never heard this stated explicitly, but I was always aware that the women who had been running the organization were completely and fanatically attached to what the elite valued most. The fact that these women had the best social connections was another indicator that the organization wanted to represent the most elite families in each city. Not coincidentally, almost all of the past national presidents had been members of the exclusive Links group as well as the Deltas or AKAs—long considered the choice college sororities of the black elite. As Nellie Gordon Roulhac's history book on Jack and Jill, *The First Fifty Years*, reveals, many of the women held doctorates in medicine or teaching and most were members of prominent families.

For example, Dorothy Bell Wright, the group's first president, was an accountant with the IRS whose great-grandfather, Private John Henson Swails, joined the Union Army in 1864. Her husband's family had founded one of the most respected black banks in the country. Dorothy's daughter, Gwynne Wright, remembers growing up in a home where famous names and faces were always passing through. "Because of my parents' activities in black organizations and businesses," says the Philadelphia native, who also belongs to the Links, "many influential people like Mary McLeod Bethune visited or stayed with us. Jack and Jill attracted an incredible network of leaders and families."

Emilie Pickens, the second president of the group, was descended from John Montier, a Philadelphia resident who appeared in Benjamin Franklin's 1790 census and built a historic home that has stood since the 1770s. A class-of-1923 graduate of the University of Washington, she was married to a successful attorney. Her son, Bill, still visits with Jack and Jill kids whom he met in the late 1930s and 1940s. "Most of my oldest friends are the children of the Jack and Jill mothers that my mom worked with," says Pickens, a resident of New York City.

Other early national presidents of Jack and Jill attracted intellectual prestige: a quality that brought depth to the group. The next three presidents, Alberta Banner Turner, Nellie Gordon Roulhac, and Ruth Brown Howard, all held doctorates: Turner had been the first black to earn a Ph.D. in psychology from Ohio State before serving as professor of psychology at several universities, including Wilberforce and Lincoln Universities. Roulhac, who sat on the board of directors of the United Cerebral Palsy Association, had earned master's and doctorate degrees from Columbia University and the University of Sarasota. She was the child of an old Philadelphia family, and her father and grandfather also had doctorates. Her successor, Ruth Howard, was a graduate of Wellesley and Columbia and received her Ph.D. from Catholic University before writing several college textbooks and joining the faculty of San Francisco State University.

The leadership's progression from socially elite to intellectual elite and then to professional elite was evident by the late 1960s. Ninth and tenth presidents Mirian Chivers Shropshire and Pearl Boschulte, both graduates of Howard Medical School, were physicians who practiced, respectively, internal medicine and ophthalmology.

Others like Lillian Adams Park, Margaret Simms, Eleanor DeLoache Brown, Ramona Arnold, Nellie Thornton, Shirley Barber James, and Eva Wanton were all educators with master's degrees or Ph.D.'s. Not surprisingly, most of them were also members of other black groups such as the Links, as well as predominately white groups like the Junior League.

"The credentials of the people that have been running this organization can be unique and quite intimidating," says Jack and Jill alumnus Henry Kennedy, a judge in Washington, D.C.'s superior court. "The families that belong to the organization and the mothers that hold the offices reflect a unique mix of black and white social credentials and political activism. They represent the very best that black America has to offer, and this is a great standard to set for kids who are looking for role models," adds Kennedy, who grew up in the Washington chapter before leaving for Princeton and Harvard Law School. As one of three siblings who went to Princeton, he knows that being a member of the group made a difference in their lives.

Like Kennedy, I would agree that the credentials of Jack and Jill members in particular, and the accomplishments of black people in general, were dished out at all times to us as kids, no matter what events we participated in.

Membership

So how does one get in? By the time I was in the fifth or sixth grade, I was pretty well aware that Jack and Jill was something that distinguished me from the other kids—even the black kids—at school. Because many of the black kids, in addition to the white students and the white teachers, had never even heard of the group, it was like belonging to a secret club. It was something that I could rarely talk about at school because only about two other kids in my class belonged to my chapter of the organization. One was the son of a physician and the other was the son of an IBM executive. The other black kids at school—some who lived in public housing or modest homes—were outside the Jack and Jill circle, and their absence made me quickly draw the obvious connection between wealth and membership. The conversations I overheard between my parents and their friends confirmed my suspicions. But these are conversations that few will own up to today.

"There are no applications or request forms," says a mother who has two children enrolled in the Los Angeles chapter. "Membership is by invitation. You get in by being asked to join. But it's not about money."

In more recent years, I've discovered that many parents who are on the outside of Jack and Jill find the admissions process highly frustrating. For well-to-do professionals, who are accustomed to paying for any house, car, or summer camp that they want, it is particularly disturbing to be told that admission is not for sale. Membership in Jack and Jill is coveted by these parents because they know that the organization will provide their children and themselves with a network of activities and friendships that will last for a lifetime. But it is a most exclusive club.

The way one gets into Jack and Jill is by knowing someone who already belongs. And that is how we got in: my parents' attorney and his wife, a pharmacist, proposed us in the early 1960s. But it's even more complicated because after one is sponsored by a current member one may not even be interviewed or considered if there is no open space for a new family to join. In fact, the only individuals guaranteed membership are those whose parents were Jack and Jill members as children. To this day, my mother gets great pleasure in recommending names of young parents who would be great candidates for membership. Even though she is now more than a generation older than these young mothers and fathers, I have seen her display great zeal as she extols the virtues of Jack and Jill membership to these couples—most of them physicians: "You'll never find a better safety net for your children if you're going to raise them in non-integrated neighborhoods and schools."

But wanting in and getting in are two different things. As I have been told repeatedly, there is no minimum family income threshold required of those being considered for membership. But because members are often required to entertain fellow chapter families in their homes, in between larger events that take place at country clubs, hotels, and restaurants, it is no surprise that the most successful applicants are the ones who have the space, the money, and the time to entertain and host parties and fund-raisers.

My own experience in the group belies the notion that the wealth of members isn't noticed. Even if it wasn't explicitly focused on by parents, I definitely remember its significance to me and my peers. I still remember the day I came home feeling bereft after a backyard Jack and Jill picnic at a friend's house. I was approximately twelve years old at the time, and the friend had been bragging that his family had two live-in workers—a maid and a caretaker—in addition to a part-time worker who was on hand at their private dock for the family yacht moored at the base of their Tudor estate on New York's Long Island Sound.

"How are we going to compete with that?" I asked my father when he suggested that we host next year's picnic in our own comparatively small backyard.

What I now recall was lacking—at least in my own Jack and Jill experience—was any real sense of the anger and dissatisfaction that the rest of black America was expressing in the late 1960s and early 1970s. Martin Luther King had been shot, cities had been burned, Nixon and Agnew were in the White House, and yet we were learning how to ride horses, make leather belts, or commandeer a small yacht.

In retrospect, it all seemed as if we were operating in a world that was separate from what we saw on TV news or in local newspapers. Of course parents would make vague remarks about "giving back to the community," "appreciating the struggle," or "advancing the black cause," but to suburban kids like me and my brother, the "community," the "struggle," and the "cause" were just terms we nodded at before we turned to the Jack and Jill kid next to us and politely said, "It's my turn to drive the boat."

Politeness reigned supreme in the preteen divisions of Jack and Jill. Since most of the mothers stood on ceremony—regardless of the activity planned for the day—children were expected to be polite no matter how dull a project might be. The "black anger" and "black slang" that we heard in the halls from other kids at school had no place at Jack and Jill events. I remember one afternoon in the contemporary home of a Jack and Jill family who lived in ritzy Chappaqua, New York. The mother was an artist and she was teaching us something I had no interest in—something like the history of origami or Japanese flower-arranging.

As we sat around the living room of the mostly glass house, the mothers sat quietly with the children until an unruly set of twin brothers started arguing. Rather than be embarrassed in front of the group, the mother took the two boys out a sliding glass door, and I don't remember them coming back inside that afternoon. As I came to learn, politeness was the paramount virtue.

Many a Jack and Jill activity dazzled me as an adolescent. Although it has left me feeling proud to have known so many successful black families, at the time the effect was to distract me from the many privileges my parents had bestowed upon me and leave me feeling that I was somehow disadvantaged. Of course, I could have made these same comparisons with some of the white kids I grew up with, but their wealth seemed less relevant to me. For one thing, the yardstick for white success was far different: they talked about country clubs, third-generation family businesses—accoutrements that were impossible for a black person to gather given the de jure discrimination that had ended only twenty years earlier.

Attracting high-profile families to the group was probably not an end in itself, but I always got the sense that it was an extension of the group's overzealous focus on its reputation and stature in the hierarchy of black organizations.

"They say income and professional status don't matter," remarked a woman who told me that her kids were members of a Dallas chapter, "but I distinctly remember us turning down someone whose husband wasn't a professional. Everybody was polite about it—because there are no rules on it—but we all voted 'no' because we all kind of knew that the family didn't have the kind of money to support the activities, dues, and social demands. Depending on the chapter, you can find some really tough members out there. I know of people whose standards have gotten higher as black people have gotten richer. More and more, they are looking for people with summer houses, country club memberships, large yachts, or ties to corporations who can donate to the group. They want the works!"

Each year, the organization publishes an annual yearbook called *Up the Hill,* which features photos and reports from the local chapters as they detail their service, cultural, and social activities of the prior year. The pink-and-blue book is as thick as a big-city phone directory and serves as a chronicle of growing up in the black elite, coast-to-coast. Many think of it as a black children's version of the *Social Register*.

"Frankly, I think those *Up the Hill* books help to stir up competition between the different cities," says a second-generation Jack and Jiller who is a student at Spelman College. "It was always a big deal each year they came out. We'd see my mother rush home from the monthly meetings where they handed them out. She and my father would then, out loud, read the three or four pages that had been dedicated to our chapter—specifically noting the accomplishments of the various kids who were growing up with me and my brother. Then Mom and Dad would flip to the front of the book and go through, page-by-page, city-by-city, trying to see how their chapter and their children measured up to the other chapters and children around the country."

The Spelman student, who says she liked most of her thirteen years in the organization, adds, "I'm sure they intended it to be an innocent little yearbook, but it brought out the worst in my parents. They'd sit there, at the kitchen table, with a box of paper clips, marking pages in the book of kids from other cities who were winning awards we hadn't won or going to camps they wanted us to attend, or being presented at balls that they wanted me to debut at. And we didn't even know half of the people." The nineteen-year-old laughs as she sits in her dorm room staring up at the Jack and Jill commemorative plaque that she had received upon graduation. "I guess you can call it healthy competition, but sometimes I wanted to burn that book."

By the time I was in the ninth grade, I too began to comb through the data listed in *Up the Hill*. Finding attractive girls to meet and reading about interesting summer experiences or unusual internships to apply for were just a few of the benefits to be gained from the book.

Notwithstanding the annual *Up the Hill*, the most comprehensive chronicle of the group was written by Philadelphia grande dame Dr. Nellie Gordon Roulhac, in her book *The First Fifty Years*. Roulhac, who served as the group's fifth national president from 1954 to 1958, says, "Jack and Jill has a rich history of contributing to generations of families, as well as to important social and civic causes." While relocating the family twice with her husband as he moved to different positions, Roulhac raised her children in Jack and Jill chapters in Tennessee, in Pennsylvania, and then in Georgia, where she helped found a chapter in 1950. "My husband and I found ourselves in Albany, Georgia, a southern segregated town, in 1950 and suddenly realized how different it was from Philadelphia," says the University of Pennsylvania graduate, who recognizes that the Pennsylvania city's black elite had always had the major black society groups like Jack and Jill, the Links, and the Boulé. "We liked Georgia, but we knew our kids needed organized social programming in that segregated environment—so we pulled together a group of parents. I'm proud to see our grandchildren in it too. They have an incredible network of role models and mentors at their fingertips."

I could always identify with the suburban kids I met in Jack and Jill. They were always rather bland and "safe," with little interest in testing the authenticity of my blackness. The members who gave me a difficult time were the ones who had grown up in all-black neighborhoods in large urban areas. Initially, I was surprised that their parents saw a need for the group, since their kids were already surrounded by neighbors who valued their black identity.

Berecia Canton Boyce's experience demonstrated the need. "After my husband and I left a mostly white, affluent section of the Upper West Side and moved to Harlem," explains Boyce, a graduate of New York University and Columbia's Graduate School of Arts and Sciences, "we found a great diversity of black people—working-class, middle-class, and affluent. But instead of dealing with white kids who didn't understand him, my youngest son found himself around black kids who teased him by calling him 'the rich kid' or 'the professor.' Moving to Harlem was the best thing we ever did, but I also still needed to surround my children with as many high-achieving kids as possible."

Boyce is president of the New York chapter, fifty-nine years old and famous for the annual dance it sponsors at the city's Copacabana Club each year for about one thousand Jack and Jill teenagers as well as for its contributions to Hale House and the Studio Museum of Harlem. The chapter roster has included graduates of all the best private schools and selective public schools (and lists the children of Mayor David Dinkins and Manhattan borough president Percy Sutton as alumni). Boyce has three children: one, a graduate of the Bronx School of Science; a second who graduated from Brooklyn Polyprep; and the youngest, Brent, a student at the exclusive Collegiate School, who has been there since first grade and is setting his sights on attending a top college. "Our kids have done everything from summering in France and the Caribbean to skiing in Switzerland," explains this proud mother as she considers the benefits of belonging to a group where such activities are the norm.

Although Jack and Jill's earliest chapters began in urban areas similar to Philadelphia and New York City, the greater trend today is toward developing chapters within suburban communities where black parents are hard-pressed to find black friendships for their children, who are reaching a crucial social development stage in their lives.

"I grew up in an all-white neighborhood, I attended a virtually all-white suburban elementary school, and it became obvious to my parents that I knew nothing about black people or other black kids by the time I hit the third grade," explained a Jack and Jill alumnus who recently graduated from Yale and admits that he had so internalized white attitudes that he was afraid of blacks at his elementary school. "I'm sure my parents were also looking ahead to that time when kids start separating on the basis of race, and they knew I'd be left out by the white kids. If it wasn't for Jack and Jill, I wouldn't have known any black kids until I got to college."

"My oldest daughter didn't get asked on a single date until she was a senior in high school," explains a mother whose daughter was the only black in her suburban Virginia high school's class twenty years ago. "And when she did go, it was with a white boy. It just killed me, but how could I tell her to stay home?"

Interracial dating is still an uncomfortable issue for even the most open-minded black families. Many black families living in predominately white settings like the one I grew up in point to the fast-growing chapters in suburban communities as a means of addressing the conflicts of interracial dating. Or—more frankly—as a means of *avoiding* interracial dating.

"I got so tired of hearing my friends tell me about their older black sons—boys from good families—who were bringing home these white girlfriends from school because they supposedly didn't know any well-bred black girls," explains a mother who had just returned from a teen summit conference in Oklahoma City. "I got busy and put my son in Jack and Jill. I know it sounds desperate, but it was the best move I could make in order to let him know that there are black girls who are smart, attractive, well-to-do, and well-bred. In Jack and Jill, we've got debutantes, intellectuals, athletes, politicos, artists, everything you could imagine."

Whether one makes this argument because of racial pride or just outright bigotry, it has always been an uncomfortable subject for Jack and Jill members. Of the girls I dated while growing up, all of them were black, and all but two were in Jack and Jill. I now question the wisdom of such a narrow experience. Given many of the parents' initial aversion to interracial dating, the group's early response to children of interracial parents is unsurprising. Some chapters have taken in mixed-race families who want to raise their kids around black children. It remains an unusual phenomenon that is only rarely attempted in most chapters—it is usually done only when the *mother* is black. I never knew interracial kids in Jack and Jill when I was a child, and while I'd like to think they would have been welcomed, I'm not sure. Today, the attitudes are more progressive.

"Sometimes they make me feel uncomfortable," said a black tenth-grader who has gone to teen conferences and met fellow Jack and Jillers who are biracial. "This group's focus is a problack one, and I feel like I've got to walk on eggs when the interracial kids are around because although I'm not saying anything negative about whites, I always feel like I can't say anything that sounds too positive about blacks."

Since it is not a political organization, the group has never taken a public stand on controversial issues such as interracial relationships in the way that groups like the NAACP or the Association of Black Social Workers will. Many members do say that if the group is going to see any changes in the near future, it will be in deciding how far it will go to address the needs and concerns of the interracial family.

Taking the Bad with the Good

"I was always the last girl to be asked to dance, or to be invited to parties," says a late-1980s Jack and Jill graduate from northern Virginia. "I can't tell you how many Jack and Jill gatherings I went to where somebody would say something like, 'You're pretty attractive for somebody so dark. You have nice white teeth.' It was excruciating, and it happened not just at the Jack and Jill events, but it would happen when I saw these same people at other places—like at the Howard homecoming, the Tuxedo Ball, even at the AKA cotillion where I came out. The Jack and Jill crowd was so into the long, pretty hair–light skin–nice features scene that it really devastated me."

As disappointing as it was, this young woman's story rang completely true to me. Even though the purpose of the organization was to create a support system for the black child who might otherwise feel like an outsider in a predominately white world, it nevertheless reflected some of the best and worst characteristics of the privileged class. From the time I entered the group at age six or seven until I graduated at seventeen, I saw moments that were truly inspiring, as well as some that were the most heartbreaking of my childhood. There were good kids and there were mean kids, thoughtful parents and jealous parents.

Even though we kids were constantly told how we were all "in this together," there were often cliques among the kids and the parents. Sometimes they divided along geographical lines, with certain kids sticking with kids from their own town, or hanging out with the kids of parents who went to the same black colleges or boarding schools. But more often they divided along economic lines, with the doctor families sticking together with the affluent lawyer families, and the more middle-class clique consisting of kids whose fathers were educators, entrepreneurs, or successful civil service people. Because many of the events and meetings took place in people's homes, the kids were acutely aware of each other's homes and cars.

"Your house is so small, I walked in the front door and fell out the back."

"Your kitchen is so tiny, I lit a cigarette and your mother asked, 'Who turned up the heat?'"

"Your garage is so small, you need a shoehorn to get your car inside."

The four-bedroom home I grew up in didn't compare with the large Victorians or stately colonials of some of my fellow Jack and Jill friends. And I can still remember the hurtful remarks that some of them made to me when we were no more than eleven or twelve years old. Even though the rest of the world would have seen our house as roomy and comfortable, I suddenly saw it as embarrassingly small during these Jack and Jill get-togethers. Fortunately, the house sat inside a neighborhood that matched or surpassed most of my other Jack and Jillers'. But in our world, you were supposed to have the whole package in order: a good neighborhood or two fancy cars didn't get you off the hook. As new kids were inducted into the organization, they too were subjected to house jokes ("Your house is so small . . .") or neighborhood jokes ("Your street is so tacky . . .") until another aspect of their lifestyles or families was found to put them to shame. Virtually everything was up for discussion and comparison: your allowance, your father's job, your spring vacation, your stereo, your mother's car, your housekeeper, your complexion, and even the size of your family's summer home.

In its early years, Jack and Jill—like many groups that catered to the black establishment in the first half of the twentieth century—attracted a negative reaction from many blacks who lacked the resources, the pedigree, or the physical appearance to be considered for membership. History shows that some chapters, particularly the ones in the larger southern cities, were clearly guilty of placing a great emphasis on these characteristics, but others were unfairly attacked for doing the same thing when what really was happening was that they were just nominating people who were in their social circle, their church, their bridge club. And not surprisingly, these darker, less-pedigreed people had long before been shut out of those institutions.

"Maybe there *was* something elitist about creating a group that catered to the concerns of well-educated and affluent people," says Portia Scott, who grew up in the Atlanta Jack and Jill and whose family owns the *Atlanta Daily World* newspaper group, "but why is it okay for well-educated whites to be ambitious—and then not okay for blacks? We are not a monolith. Just as many whites and other ethnic groups want one generation to improve on the last generation, blacks want to do the same." A member of one of Atlanta's oldest and most influential black families, Scott, who is general manager of the sixty-five-year-old company, feels Jack and Jillers should not apologize for their ambitious agenda and aspirations to be successful. I understood her point after attending years of Jack and Jill events where the kids were practically berated for settling for the mediocrity that we might have encountered from the non–Jack and Jill black kids we saw at school.

"You don't contribute to black achievement by knowing how to dance or play basketball," one of our chaperones used to tell us when I was vice president of our teen group. Her remarks were filled with well-meaning yet elitist overtones. "You are the ones who are supposed to be setting the example for the rest of those kids at school. Just because you look like them doesn't mean you have to act like them."

"Sure," says Boyce, as she and three other parents stared out at the Atlantic Ocean in the backyard of a Jack and Jill mother's home on East

Hampton's Lily Pond Lane. "There are still people that want to stigmatize us as being into elitism. They also say we're into straight hair. But here I'm president of the second oldest chapter in the country—a group with 108 families—and I'm not light-skinned, straight-haired, or green-eyed. So that old stereotype is long gone. What they should really be judging is the fact that this organization's goal is to put children first."

Ilyasah Shabazz agrees. "A Jack and Jill child has nothing to apologize for. Today's black child should want to be introduced to everything that the smartest white child has access to. To be ambitious, well-rounded, and bright are things we should celebrate."

When Ilyasah and other Jack and Jill alumni of my generation look back, our hindsight is twenty-twenty because we now realize the futility of trying to please the detractors—both black and white—who might question our blackness when we display characteristics that are uncommon for blacks.

But a part of me envies the Jack and Jill kids who have come through since the 1980s. I grew up when the black middle and upper classes were small and therefore nearly invisible. We felt outnumbered in school, in the country, in statistics, and on TV. We seemed equally unreal to whites and other blacks. We were often mocked by the other black kids because our lives looked like the *Brady Bunch* kids during a period when the only black families on TV were *Sanford and Son* junk dealers, *Good Times* project dwellers, and *Flip Wilson* jive talkers.

I envy the Jack and Jill kids of the 1980s and 1990s because they are more readily accepted as valid representatives of black America by whites and other blacks, thanks to *The Cosby Show, Oprah,* Bryant Gumbel, and Colin Powell. Because of these TV images, it is now at least believable that black families can be well-educated, intact, and articulate. My Jack and Jill existence took place during a transition period that left me feeling somehow betrayed by a world that said we couldn't possibly be real. We were fifteen years ahead of our time: living like the Cosby kids before mainstream America had been formally introduced to the notion.

Another Jack and Jill member who is able to contrast the old and new Jack and Jill is Judge Henry M. Kennedy Jr., who grew up in the Washington, D.C., chapter in the late 1950s and the 1960s. Growing up during the final years of segregation, he saw the group's effect strictly within the boundaries of the black community. Today, with two children who live in a more integrated world, he sees Jack and Jill's role and stature in a black community that is more diverse and more spread out.

A member of the superior court of the District of Columbia, Kennedy says, "Our kids are in the same chapter that my parents and I first joined in 1956, and we have them there for the same reasons I was put there." Kennedy acknowledges, however, that even though he and his wife feel the group is necessary, "we meet more and more parents in black elite circles who don't feel Jack and Jill is so necessary any more."

A graduate of Princeton and Harvard Law School, Kennedy acknowledges, "When I was growing up, I was in a mostly black environment and Jack and Jill was coveted to a much greater extent than it is today. Years ago, it was a necessary sign that showed that a person was a member of the black elite. It put a small group in touch with others who were like them. At that time, almost all of the parents were doctors. An affluent family would never have passed up the opportunity to join. Today, there is more diversity in that we do have many members who are affiliated with universities or governmental agencies. And furthermore, there are now some Washington doctors who, for whatever reason, simply decide not to join the organization."

Ronald Walter, station manager of WREG-TV, the CBS affiliate in Memphis, remembers growing up in the Memphis Jack and Jill and eventually serving as teen president of the Memphis chapter. "My three children are now in the same chapter where I grew up," comments Walter as he sits in the living room of his sprawling brick home on a tree-lined section of Parkway, a boulevard that is famous for the affluent Jack and Jill families who have populated it for three generations.

With the perspective that comes from seeing two generations of members, Walter believes Jack and Jill is becoming more progressive. He remarks on the way in which it differs from other elite black organizations. "Jack and Jill is much more fluid and progressive than other groups today because there are constantly new officers and members joining each year." As Walter points out, the parents who serve as officers of their children's chapters are allowed to hold positions only while their kids are still members. Once the parents' youngest child graduates from high school, the parents have to move on to the Associates division of Jack and Jill, which is an extremely active but nonvoting group of members.

"That's an excellent point," comments a mother in the Columbus chapter. "One of the reasons why Jack and Jill remains progressive and continues to stay on top of important legislation like children's rights and education is that there is no stagnation. You will not find officers who have stayed in positions for years and years like those you will find in other groups." I remember several mothers in my parents' generation who reacted to this rule with great consternation. They were old-world people whose agenda was distinctly social and elitist. Their agenda was focused on tennis parties, fashion shows, reading clubs, and horseback riding.

A look at the agenda at a recent national conference demonstrates how the group has evolved and kept pace with the changing needs and concerns of its members. In addition to workshops on "Building Your Child's Home Library" and "Parenting in a Media Culture" and "Investing Your Family's Wealth," there were seminars entitled "Single Parenting" and "Addressing National Health Concerns."

At a recent Jack and Jill convention, national president Shirley Barber James spoke of the importance of staying on top of the changing political climate. Her son, a 1995 graduate of Harvard Law School, had been extremely active in his Jack and Jill chapter in Savannah. "Like my two daughters, Robert developed a strong identity through Jack and Jill." As a student at the very exclusive and mostly white Savannah Country Day School, James's son was able to mobilize his black and white classmates to insist on establishing a Martin Luther King Jr. holiday in the face of the school's reluctance to observe the event. In my day, a Jack and Jill child would have been perceived as a troublemaker to have pursued such an agenda. Not by the 1980s.

As a part of its attempt to become more socially and politically relevant, in 1968 Jack and Jill created a foundation that continues to give scholarships and offer grants to national and international programs to address such issues as illiteracy, job training for Vietnam Veterans, and early childhood education.

I would imagine that most former members would agree that the Jack and Jill of today is less status-conscious and less color-obsessed than the Jack and Jill of my childhood. Of course it still sponsors the cotillions, black-tie dinners, and student trips that were de rigueur in the 1950s and 1960s, but the organization has also placed a priority on training young people to be unafraid to function outside their social class. As the group moves toward its fourth generation of family members, it is developing a greater number of programs that will embolden black youth to participate in and change the surrounding less-advantaged black community. Jacqueline Parker Scott, a Jack and Jill mother from southern California, founded the organization's national teen leadership development program, a project offering workshops on public speaking, leadership qualities, and personal management skills. The group is also establishing teen summits, where the teen presidents from each of the nation's chapters will talk about such formerly Jack-and-Jill-taboo topics as affirmative action, abortion, interracial marriage, and the politics of South Africa. These are all subjects that would have drawn uncomfortable reactions one or two generations ago.

Because the organization's original founders and many of its subsequent leaders were light-skinned professional blacks with money and position, some detractors still sound quite credible when they accuse certain chapters today of choosing their membership only from light-complexioned, well-educated, well-to-do families. That accusation is often fueled by individuals who have been unable to gain admission or by those who simply ignore the fact that the group has become far more diverse over the years. In fact, these detractors have failed to notice that some old-guard blacks are distressed that most of their elitist traditions have been eradicated from the group today.

"At one point," a Jack and Jill mother from Atlanta admits while standing in a Cleveland hotel lobby when I attended the group's recent annual convention, "I used to be a little hesitant to tell certain black friends that I was in Jack and Jill because in some parts of the country it has this elitist reputation of only taking parents who are professionals or socialites. You know, the doctor-lawyer, high-church, high-yellow, Episcopalian crowd."

I nodded at the woman.

She continued. "But then I realized that what I was really apologizing for was this group's focus on shaping successful kids. I was embarrassed that I believed in ambition. And frankly, every other group—Jews, Asians, and other ethnic persuasions—values families and individuals that accomplish a lot. Why shouldn't we? This is supposed to be an elite group."

"I agree with you," I added, almost sure that we were on the same wavelength.

"And I think we can simultaneously infuse Jack and Jill with a more diverse group of black people at the same time," the woman added. "If they are hardworking and ambitious, who cares how wealthy they are, what their parents do, what they look like, or how big their house is?"

Even though I was sometimes victimized by the socioeconomic narrow-mindedness of the kids I grew up with in Jack and Jill, a part of me recoiled at the suggestion that the group could or should be more diverse or open-minded. Having survived the tough economic and social standards of the group, I wasn't sure I wanted one of its members suggesting that it was changing or even *needed* to change. I had always thought that its strength and its value were based on establishing an ideal black family model. How could it change, become more diverse, and still be a place where we could find the elite families? How could it remain an important credential for me, for other alumni who had belonged, and for our children, who would one day need to be introduced to kids who were just like them?

As she walked back into the ballroom where a group of teenagers—half of them with complexions darker than my own—were lining up for a presentation of gospel songs, she pointed out with pride in her voice, "Now there's the new Jack and Jill. I never heard gospel music at a Jack and Jill event when I was growing up."

I paused and thought for a moment. It was obvious that we had a lot to learn from this younger generation of Jack and Jill members.

The Black Child Experience: The Right Cotillions, Camps, and Private Schools

Within the black elite, parents aren't the only ones expected to have impeccable résumés: Kids are too. The most cynical members of the old guard would argue that a child's social and academic credentials are intended not only to prepare the child for a life filled with competition and high standards but also to fill out a family's already stellar résumé or shore up a less-than-perfect one.

Since my mother had not been a debutante and my father was not a physician or an attorney, others would indicate to me, even when I was only eleven or twelve years of age, that there were serious holes in my nuclear family's résumé. Each of my parents and many of their siblings—as well as my grandparents—had *some* good connections to respected groups and schools. But I think they didn't push us toward the black cotillion-and-boarding-school crowd because the gaps in our social résumé might be even more closely scrutinized—and make us feel even more like outsiders—in settings where the most elitism occurred. Nevertheless, the members of the old-guard elite quickly made it clear that it was up to my brother and me to affiliate with the right organizations and schools if we were ever to be accepted by the upper ranks of this group.

Getting their kids into Jack and Jill was not the only strategy that the black elite used to ensure that the next generation was starting off on the right foot—it just generally happened to be the first step. Since most kids were placed in the organization by their fourth or fifth birthdays, membership preceded their entrance into the designated debutante cotillions, private schools, clubs, and black summer camps whose sole purpose was to cater to the affluent black child. Like Jack and Jill, these groups have become a social screen, an educational strategy, a network, and a cultural reinforcement mechanism.

"When my parents told me I was coming out, I knew I might as well give in right then and there. It had always been an important event in their crowd," says Phyllis Murphy Stevenson, who was one of the most celebrated black debutantes in New York in 1952. "I don't think I really understood the significance at the time, but being a debutante gave me the additional confidence that most teenagers of any color lack when first put in public situations," explains Stevenson, who was presented at a Girl Friends black-tie ball during the same year that her mother was elected national president of the organization.

"Sure, some people might have thought of it as elitist, but my mother and father just saw it as an additional part of my exposure as a child," says Stevenson, whose parents also insisted that she be exposed to annual summer trips to the elite black Sag Harbor resort community. "My parents started going there in the 1930s."

"I don't think any child actually *asks* to be a debutante or an escort at a cotillion," says Alberta Campbell Colbert, a Washingtonian whose daughter, Doris, was presented in 1962 by the Bachelor-Benedicts with twenty other girls. "Quite simply, she did it only to please her father."

Historically, because of segregation or because northern Junior Leagues and other white groups—prior to the 1970s—shut them out with their unwritten rules of bias, the black elite were limited to presenting their daughters at cotillions that were sponsored by their own clubs and organizations.

But of course the best families wanted their daughters to be presented only by certain groups within the black community; the local church group or YWCA would never have been satisfactory. The favored national groups have always been the AKAs or the Deltas, the Links, the Girl Friends, the Smart Set, Jack and Jill, or certain local or regional clubs like the Royal Snakes in Chicago, the Co-Ettes in Detroit, or Washington's Bachelor-Benedicts.

"When I was presented by the Smart Set in Houston in the early eighties, I didn't really find the event that intimidating," says Kimberly Webb, an attorney who graduated from Wellesley and Duke Law School and grew up in one of those stellar families where the parents held memberships in the Boulé, the Links, and the Girl Friends. But even if a debutante is as unfazed as Webb was, there is another component to the debutante scene: the escort.

Whether a debutante's parents are hunting for the right young escort, or a young man's parents are jockeying to steer their son to the right debutante, the whole scene can be both maddening and exciting for the sons who get thrown into the mix. Many of them are used like pawns to curry favor between families and friends. Because I came of age in the late 1970s—a time when the New York area groups had abandoned debutante balls—I missed playing the role. My parents received numerous requests for me to be sent to Memphis to serve as an escort for the neighbors of my relatives, but it would have required no fewer than three plane trips to attend the escort training sessions that preceded the actual ball.

"I got stuck acting as an escort for three different balls simply because my parents had to pay back favors they'd gotten when my two older sisters had come out two years earlier," says a Morehouse graduate from Cleveland. "It's not that I didn't like the three girls, but they were all so afraid that I would embarrass them—so afraid I wouldn't know how to do all the bowing, parading, and dancing. It just wasn't any fun after all the lessons and practicing. Just a lot of profiling."

"It wasn't such a bad experience," says one of my friends, John Evans, now a New York dentist, who recalls serving as an escort at a Girl Friends ball in the mid-1970s. "There was a lot of rehearsing on the weekends leading up to the cotillion. We had this humorless old guy who drilled us on how to hold a lady's hand, how to bow, and how to lead in a waltz."

Finding a good escort for a debutante is de rigueur for parents who want to look good in front of the guests they've invited to the cotillion—a black-tie event where tickets sell for around $150 per person. The parents of debutantes rely on the sons of their college roommates, the boys who belong to their Jack and Jill chapter, even cousins and nephews. "Just so long as he has one or two good credentials you can brag about, and a different last name," says a mother who had her nephew escort her daughter to her debut in Dallas, "you can pull it off." Still mostly a southern tradition, the black debutante balls were not very different from the balls I have heard my white college friends describe: lots of tulle, black ties, patent leather shoes, and female angst.

Some boys are dream escorts because they are good-looking, credentialed, and completely unfazed by the whole routine of dance lessons, formal dress, and social fanfare. Such was the case for Henry Kennedy, who was a favored Washington escort back in 1965—before he went off to Princeton and Harvard Law School and became a judge in the superior court of Washington, D.C. "Most of us didn't admit it, but it was a badge of honor to be an escort at certain cotillions—particularly at the Girl Friends Cotillion," acknowledges Kennedy, who escorted Virginia Brown, the daughter of the archbishop of Liberia. "She was a student at the National Cathedral School and we met at a Jack and Jill skating party."

Escorts, like the debutantes, are required to go through weeks of dance classes learning to waltz, bow, and present. In many cities, the black elite go to a particular dance studio that also teaches etiquette for formal parties. In fact, a new component has been added to the boys' regimen now that some organizations have established "beautillions" where young men are presented to society at extravagant black-tie affairs.

"When I was an escort, the debutante scene consisted of a year-long chain of parties," says Judge Kennedy, who remembers that he and the other teenage escorts were required to wear not just black tie, but tails, to the Girl Friends Cotillion, which was held in the main ballroom of the Washington Hilton.

"For the young women to get accepted into a prestigious cotillion, they usually had to be sponsored by a respected organization or social group," adds real estate developer E. T. Williams, who grew up in New York and was an escort in both New York and Washington for the different Girl Friends cotillions. Long before he became a board member at New York's Museum of Modern Art and joined the Comus Social Club and the Boulé, he remembers forming a social club that kept a careful watch over and sponsored some of the popular debutantes like Esquire cartoonist E. Simms Campbell's daughter Liz, who later married photographer Gordon Parks.

Other popular debutantes during the heyday of cotillions included New York's Carol Coleman, who attended boarding school at Northfield

Mount Hermon; Chicago's Karen Gibson, whose grandfather was Truman Gibson, the chairman of Supreme Liberty Life Insurance Company; Norma Jean Darden, whose parents were Montclair, New Jersey, socialites; Chicago's Diane Dickerson, daughter of attorney and businessman Earl B. Dickerson; and Detroit's Gail Burton, the daughter of hospital founder Dr. DeWitt Burton.

Mary-Agnes Miller Davis of Detroit still remembers Gail Burton's debut from the 1950s. "The whole event was so elegant, *Ebony* magazine put Gail on the cover," says Davis, who has been an important figure in the Detroit cotillion world since the 1940s, when she founded the Co-Ette Club. "Gail Burton's debut included a formal at the Book Cadillac Hotel as well as an elegant tea party at their home on Arden Park where there were five hundred guests."

Davis acknowledges that cotillions should be seen as more than just lavish parties. This is why she established the Co-Ette Club in 1941. "I wanted to establish an actual club for high-school-aged girls where we would emphasize community volunteerism and scholarship," says Davis, who started the exclusive group while she was still a student at Wayne State University. Now operating in other cities as well, the Co-Ette chapters have up to thirty-five high-school-age members, who are required to maintain a B-plus GPA and to volunteer a certain number of hours on community projects. "We now have more than two thousand former Co-Ette Club members," says Davis, "and many of them are judges, physicians, attorneys, and politicians."

Barbara Anderson Edwards's parents placed such importance on a debut that they sent her from her hometown of Greenville, South Carolina, to Raleigh, North Carolina, for the AKA cotillion. "My father was president of the Alphas and my mother was in the Links as well as president of the AKAs in Greenville," explains the New York attorney, "but there was no local AKA cotillion, and they weren't going to let that fact get in their way. In retrospect, I am glad they did it for me."

Marianne Savare Walter of Memphis is glad that her Links chapter has brought back the debutante cotillion for the young ladies in her city. "There are a lot of parents—particularly in Jack and Jill—who want the debutante experience for their kids, so we brought the cotillion back seven years ago," says Walter, whose 1987 society wedding to TV executive Ron Walter drew over one thousand members of the city's elite. As the second black to join the Memphis Junior League, Marianne has insight on both the black and the white elite cotillion crowd. "When we had our first Links cotillion at the Peabody Hotel, we had only ten young ladies. Now, we actually have a waiting list of fifty," she says as she stands in the sprawling living room of her redbrick colonial in Memphis, "so it is obvious that people are hungry for the experience in this town."

Judge Kennedy evaluates the Washington scene and says, "Although the popularity of those events has faded somewhat in D.C., a lot of kids develop confidence and build self-esteem by participating in such activities. I'd be surprised if my daughter would want to be a debutante when she gets to high school. I would be happy to support her because I saw the benefits, but I realize that era has passed."

E. T. Williams agrees that in spite of the confidence that cotillions offer, he sees that today's new generation of elite young people are less interested in the tradition. He was not surprised that his two daughters, both graduates of Manhattan's Spence School and Harvard, turned down the opportunity. "My daughters had no desire to be debutantes even though my wife, Auldlyn, was presented with big fanfare in 1957 as the queen of the Me-De-So Cotillion in Baltimore," he says. Auldlyn was the daughter of one of Baltimore's most famous black surgeons, Dr. I. Bradshaw Higgins, and was an alumnus of the Westover School, Bennington College, and Fisk, so it was expected that she would be presented. Today, such prestigious credentials might, instead, lead one to other activities.

In recent years, certain elite white organizations have opened up a space for a black debutante who often has credentials that equal or outshine those of white participants. Such was the case when Harvard graduate Candace Bond became the first black to be presented by St. Louis's exclusive, all-white Veiled Prophet debutante ball; when Yale graduate Elizabeth Alexander became queen of the mostly white Azalea Festival in Washington; and when Auldlyn Higgins became the first black in the Baltimore Junior League. Candace and Auldlyn were the daughters of prominent surgeons, and Elizabeth was the daughter of Clifford Alexander, who not only was a graduate of Harvard and Yale Law School but was also serving as secretary of the United States Army.

While there are some well-to-do blacks who would strive to have their daughters debut in a prestigious white cotillion in order to enhance their own professional ties to the larger white community, these opportunities are rare and typically involve the daughters of downtown attorneys and bankers rather than the typical black entrepreneurs, physicians, and dentists who depend less on ties to white institutions.

In the rare circumstance where a black debutante debuts in a white cotillion, it often takes place because the white sponsoring organization has specifically sought out one black candidate to fill out a "black space" in the cotillion. When St. Louis debutante Candace Bond became the first black to be presented at the Veiled Prophet Ball in the early 1980s, some blacks said it was done so that the previously all-white event could avoid accusations of bigotry. Others acknowledged that the Bond family's prominence and Candace's academic success—she later went to Wellesley—were the credentials that precipitated her selection for the event.

Among many old elite families, there simply is no value in introducing a daughter to society through a nonblack club or group. Explains a black mother in New Orleans, "It makes no sense to be introduced by them if you're not going to be a meaningful part of the white community."

In recent years, the coming-out scene has changed dramatically in ways that go beyond race. First of all, because fewer daughters are interested and because more mothers are working at busy jobs and have little free time, fewer groups—particularly those in the Northeast—are sponsoring debutante balls. Some say it's sexist and paternalistic, while others say that young women should be focusing on academic achievement rather than social ambition. Another change that seems to be having an impact on the black elite's outlook on the black society cotillion is that more and more traditionally white balls are welcoming young black women from prominent families into their cotillion presentations. While most black elite parents would never choose a white cotillion in favor of one sponsored by one of the cherished black groups, scheduling conflicts are making some young women decide to forgo the tradition altogether.

Perhaps the best example of how old-guard parents from around the country have unified to preserve, yet update, the black-tie cotillion concept for their children is seen in the Tuxedo Ball that was started in the 1980s in Washington. Dr. Carlotta "Buff" Miles, a Washington psychiatrist, has worked to establish committees of parents in each major city so that a variety of young people and parents from elite families will attend the elite Washington-based ball.

"The Tuxedo Ball might be compared to a cotillion, but it's actually a formal dance for young people and their parents," says Fredrika Hill Stubbs, who has attended the ball with her husband and sons. "Each year the ball issues a directory with the names of the participants along with their home and college addresses so that the kids can keep in touch with each other," explains Hill, who sees the importance for young black adults to network with each other before they move so far apart.

"Many of the participants are between eighteen and twenty-five," says Paul Thornell, who has served on the organizing committee and attended the ball since graduating from Sidwell Friends School in Washington. His grandmother, Frances Vashon Atkinson, was one of the nine charter members of the Links in 1946. "Dr. Miles wanted an event that brought together young people from many cities along the East Coast, and we get as many as six hundred people each year." The invitation-only ball has become such a strong support system that the group has begun offering workshops to participants on practical subjects such as improving college study habits and preparing for a first job.

Cotillions continue to be important events in the lives of young adults in the black elite even if the formality of the occasion has lessened over time. My wife and I expect that our children would participate in these events with even less enthusiasm than the prior generation of young people, but we feel that such experiences offer a view of the past that is worth repeating.

Summer camps are an important complement to the school-year activities of Jack and Jill. Founded in 1921 with forty young campers by the late Dr. William DeBerry, Camp Atwater has historically been the only "right" camp for children of the black elite. Listed on the National Register of Historic Places, the seventy-acre camp with its twenty-four buildings and a three-acre island is located on Lake Lashaway in North Brookfield, Massachusetts, just an hour west of Boston.

Held in high esteem for more than three generations, the camp has a dazzling list of people on its national advisory board, including Vernon Jordan, Senator Ted Kennedy, Philip Morris executive George Knox, former Walt Disney president Dennis Hightower, National Council of Negro Women founder Dorothy Height, Abyssinian Baptist Church minister Reverend Calvin Butts, and former assistant secretary of state Clifton Wharton.

"Kids came from all over the country to go to that camp when I was growing up," says Gladys Scott Redhead, who attended Atwater in the 1940s and then sent her sons while they were growing up in Scarsdale, New York, in the 1960s and 1970s. "My husband and I knew lots of childhood friends who went to Atwater. Some of them I saw again when I went to Howard, and some I run into today at conventions and social get-togethers."

The daughter of a successful dentist, Redhead was typical of the privileged children who returned each year to the camp from all over the United States, Africa, and the Caribbean. One of her fellow campers, Eileen Williams Johnson, is also still a fan of Atwater. "We met new girlfriends from all over the country at that camp," says Johnson, whose father was a New York physician when she was attending. "And the staff ran it like true professionals." A Los Angeles physician who sent his daughter there in the early 1960s after she had first gone to a mostly white camp in Connecticut says, "My wife and I made the biggest mistake when we passed up Atwater the first year our daughter, Michelle, went away. We figured that a white camp would introduce her to the real world. The only problem is that she had no protection at all. She called us crying every day about some new racial incident. Either they were making fun of her hair or laughing at the idea that she could tan in the sun, or telling jungle jokes, watermelon jokes, Martin Luther King jokes. And the last straw was the midsummer dance where all the boys were white and no one would ask her to dance. White counselors at white camps don't know how to deal with this. And we were crazy to think that they could," says the father. "A camp is supposed to help build your kids' esteem and give them freedom. Michelle didn't find that experience until she got to Atwater."

Redhead and Johnson, who are, respectively, active in the Girl Friends and the Links, had no illusions about the racial benefits of Atwater. "It was nice to be around kids who understood who you were and who could make you feel like you belonged," says Johnson.

Just as it did when Redhead and Johnson were visitors, Camp Atwater still maintains the same schedule of running a boys' season from June 30 to July 27 and a girls' season from July 28 to August 24, so that the two groups are never at the camp at the same time. Each season, two hundred campers ranging in age from six to fifteen years old participate in such activities as soccer, swimming, tennis, studying African history, horseback riding, golf, and building leadership skills. Redhead and Williams remember Atwater's famous seventy-five-year-old "Moho Story," which is a nighttime tale about a mythical monster that haunts the woods around the campers' cabins.

"Many of our campers are accustomed to spending part of their summers in Martha's Vineyard or Sag Harbor," says Atwater's executive director Wanda Johnson, "so a lot of them get here and renew friendships that they had already formed during previous summers in those resort areas. Nobody feels like an outsider."

Although the Vineyard and Sag Harbor crowd are still present, the camp now shies away from the old label of being designed for well-to-do black children and, in fact, works very hard to ensure that one-third of its campers are not from middle- or upper-income families. Some members of the old guard say the camp lost its upper-class cachet after the late 1950s, when it got "too egalitarian." Since the camp is run by the Urban League of Springfield, there is a commitment to including less-advantaged children in the camp's programs and activities. For some members of the elite, this is a negative factor because they like the idea of insulating their kids from inner-city children who might confirm certain negative stereotypes, or worse, mock their children, but for many other parents the diversity helps them introduce their affluent children to the "real world."

Today, many of my friends are raising young children and simultaneously hoping that there will be some way to re-create what Camp Atwater used to be. They agree that it is wonderful to provide less-fortunate children with a camp experience, but they are still left with the desire to have what affluent white parents have: camps for their children that are filled with children who are just like them.

The right private school is another mechanism that has aided the black elite family's child-rearing goals. The very earliest members of the elite usually attended private schools for one of two reasons: either they lived in southern cities with segregated school districts that didn't offer a decent—or in many cases *any*—school for blacks, or their parents wanted to introduce them to other children from black elite families and create further distinctions between their kids and the local blacks who were not considered to be from prominent lineage. For the most ambitious black parents, a private boarding school was the solution.

The idea of the black boarding school grew out of the segregated South's refusal, in many communities, to build public high schools for black students. When many non-upper-class black children advanced beyond the eighth grade in the late 1800s or the early 1900s, they would usually do so by traveling great distances on a daily basis to the closest city that offered blacks a high school education. Another popular alternative was to send the children out of town for the entire school year so that they could live with a relative who resided in the North, where high schools were integrated, or with a relative who resided in a large, more cosmopolitan southern town with a wider selection of black secondary schools.

While there were a handful of affluent blacks who sent their children to such prestigious northeastern white boarding schools as Phillips Exeter, Northfield Mount Hermon, and Phillips Academy as early as the late 1800s, most upper-class blacks living in the South—and a few in the North as well—sought out top black boarding schools for their children. With plans of sending their children to top colleges like Howard or Spelman, and with an ultimate goal of preparing them for graduate degrees and promising careers, these families wanted schools that not only taught what the local white high schools were teaching but also offered lessons in refinement. Many wanted both the academic challenges of chemistry or Latin and instruction in such stereotypically "upper-class subjects" as opera, dance, foreign literature, and the like. And they wanted an environment that would produce ladies and gentlemen. Many members of the old black elite presumed that such results could never be produced in a private day school where kids were present only a few hours each day. They wanted their child's entire experience to be managed and cultivated. The boarding school that did this better than any other was Palmer Memorial Institute in Sedalia, North Carolina.

"As kids, we were all a little intimidated by our headmistress, Charlotte Hawkins Brown, because she expected so much from us," says Manhattan resident Herman Robinson, who entered the Palmer Memorial Institute boarding school in 1930. "Not only did she want excellence from us in the classroom but she wanted us to be setting a good example in the dormitory and in the dining hall. She modeled the school after Andover, Mount Hermon, and her other favorite prep schools, and she was determined to outdo them."

Originally from Cambridge, Massachusetts, Brown was a black woman who named her school after a white friend and benefactor, Alice Freeman Palmer, who had been president of Wellesley College. Brown's school began at a time when the American Missionary Association was creating schools to teach black children throughout the South, and Brown operated it on a forty-acre campus with a philosophy that differed dramatically from the other black boarding schools. It was more expensive, more selective, and decidedly elitist in its approach to training students.

Herman Robinson remembers his first impressions of the school. "Although it was an imposing campus with a long driveway that led up to redbrick school buildings and dormitories, I don't think that many of us initially knew how important the school was. But within a few days you really got the idea," says Robinson, who began boarding at Palmer after leaving a private school in Vermont that his family felt he had outgrown intellectually. "Mrs. Brown lived in a canary-yellow, wood-framed house on campus with her three nieces. One of them, named Marie, later ended up marrying the singer Nat King Cole and sent her own kids to boarding schools. From that house, Mrs. Brown would make periodic visits to our classrooms, the dining hall, and dormitories—always commenting on what it took to be a successful lady or gentleman."

Robinson remembers it all as if it were last week. In addition to giving lessons on dining etiquette that were reinforced by the teacher who sat in the dining hall at the students' assigned tables ("Elbows off the table," "Don't create a bridge with your utensils from the plate to the table top," "Don't cut all your meat at once," "Don't blow on your hot food," "Never push your plate away when done eating"), there was dancing etiquette for boys ("Always keep an open handkerchief in your right hand when dancing with a lady so your palm doesn't touch her back"), as well as specific rules on how to properly open and close a door when assisting a female student.

"After I graduated, Mrs. Brown published her rules of etiquette in a book called *The Correct Thing,*" says Robinson, who remembers that the school's endowment, reputation, and popularity among affluent blacks grew particularly quickly in the late 1930s and throughout the 1940s and 1950s. During this period, Brown became extremely aggressive in her approach to raising the children of black society. She implemented mandatory piano and French lessons and the formal tea service, and she designed several different uniforms for the boys and the girls. Weekly formal dinners required that boys be in black tie and girls wear long white dresses and gloves.

At the time Robinson was attending Palmer, it had approximately one hundred students in the ninth, tenth, eleventh, and twelfth grades—with girls outnumbering boys three to one. "Although Mrs. Brown was a real stickler for good manners, she did emphasize the importance of learning about blacks like Crispus Attucks and Sojourner Truth. We didn't call it 'black history' back then, because Mrs. Brown never liked the term 'black,' but the teachers always taught us about blacks who had contributed to society," recalls Robinson, who later graduated from City College in New York.

Over the years, many well-to-do blacks attended the school, including A'Lelia Robinson, daughter of Madam C. J. Walker, the country's first self-made woman millionaire; and relatives of actor Paul Robeson as well as singer Carol Brice and her two siblings, who were there in the early 1930s. "I remember that several of the kids who grew up in wealthy Atlanta families were sent up to Palmer Memorial," recalls Ella Yates, who grew up in Atlanta and graduated from Spelman in the late 1940s.

Although Brown stepped down as head of the school in 1952, the school stayed in operation until 1971, when it was still attracting children of the black elite. "We all called her 'The Big Wheel' behind her back because there was nothing that could stop that woman," says Robinson as he recalls how hard she worked to win over supporters from the North or among wealthy white families in nearby Raleigh or Greensboro. "I have great respect for her because she wanted the best for us, and that was a hard thing to get as blacks in the South during those times."

Dressed in a blue skirt, white blouse, and white stockings and shoes like the rest of her female classmates, my godmother, Mirian Calhoun Hinds, spent seven years at the private Mather Academy, which served as a breeding ground for smart children of families who wanted their kids to attend top schools. I grew up hearing stories of Mirian's experiences, and because the school remained open until the late 1980s, lived under the fear that she might convince my parents that I or my brother should go there. By the time I reached boarding school age, though, most members of the northern black elite had access to top public and private day schools in our own communities. And we were living in much more integrated school experiences than our parents. "I went away to Mather Academy when I was in the sixth grade because my hometown of Orangeburg, South Carolina, didn't have a high school that blacks were allowed to attend," says Hinds, who now holds a Ph.D. from New York University, "and my family decided that my brother and I should attend a boarding school that set high standards for blacks."

Run by a white headmistress, Lulu Belle Brown, Mather Academy had been founded by the United Methodist Church and established in Camden, South Carolina, about thirty-two miles outside Columbia. "My class had forty-seven students—and we came mostly from southern states, but there were also kids from New York and New Jersey," says

Hinds, who recalls how protected she and her brother were from the real world on the fully fenced and gated campus. "Mrs. Brown made that campus feel like a community. We were escorted to church every week, we grew many of the vegetables that we ate, we studied together in our dormitories, and we helped each other choose our majors," says Hinds, who remembers that the teaching staff was racially integrated and very serious about the students' futures. "We weren't as ritzy or as socially conscious as Palmer—with all the white gloves and tea service—but everyone took us seriously and worked hard to motivate us," she adds, recalling that students were expected to attend college and graduate school. Like herself, Hinds's brother, Thomas Calhoun, continued beyond college and is now a surgeon in Alabama.

My generation and the generation of children that followed in my godmother's family—particularly if they lived in the South or in large northern cities—were much more likely to serve as the transition group that integrated white boarding schools in the northeastern part of the country. When my godmother's niece Dolly was sent from Tuskegee, Alabama, up to Northfield Mount Hermon and my cousin Bill was sent from Memphis to Phillips Exeter in the early 1970s, my brother, myself, and our other young northern cousins pitied them for having to serve as guinea pigs in what we felt was a twenty-four-hour, seven-day-a-week white environment with no "black buffer zone": no black parents, no black church, no Jack and Jill, and no visible means by which they would be protected from verbal or physical bigotry.

"I went to Northfield Mount Hermon in the 1970s primarily because my dad and uncle had come up from Louisiana to go there in the 1940s," explains Eric Chatman, one of my brother's closest friends, who grew up in Chicago and attended the private Latin School of Chicago before leaving to board at the exclusive prep school in Massachusetts.

Before the two schools merged, Northfield and Mount Hermon had been among the first white boarding schools to accept blacks with any regularity. In fact, one family—the Dibbles of Tuskegee and Chicago—has sent more than two dozen family members spread across three generations to the exclusive school. "I was there at the same time as a couple of the Dibble children," says Chatman, who grew up with members of the family in Chicago, "and we all knew that the school had a different history for blacks than the other boarding schools. We expected it to be a friendlier place for blacks because it had always been very international and very liberal."

Some black families have so completely embraced boarding school education that they don't even consider public or private day school education beyond junior high. Sam Watkins, a graduate of the Groton School and Harvard, practices law in New York and grew up in a family where five children attended white northeastern boarding schools. "I grew up in Queens, where the public schools were fairly good, but my father felt we could do better," explains Watkins, who acknowledges that his father was concerned about his kids losing a positive black self-image. "My father was very conscious of the racial differences that would be exaggerated at these schools, and he worked very hard to create a support system by visiting every month and talking to us about our black identity," says Watkins, whose oldest sister, Theresa, was the first black woman to attend the elite Milton Academy in Massachusetts in 1965.

The Watkins family experience is what most blacks are seeking for their children when they choose the private school route. Each one of the children became a private school success story, outdoing many of his or her white classmates. For example, oldest brother LeRoy, who is now an attorney, entered Groton in 1965, then went on to Yale. Theresa, who is now a pediatrician in Washington, graduated from Milton Academy, Harvard, and Howard Medical School. She is now on Milton's board of trustees. Brother Joe, who has run for Congress, graduated from Middlesex School and the University of Pennsylvania. He, too, is on his prep school's board of trustees. Younger sister Dorothy graduated from Milton Academy before going to Wellesley.

"Of course it was an awkward transition for us to move from a grammar school with other black kids to schools and dorms with virtually no other blacks," says Watkins, who remembers being the only black student in his eighth-grade class at Groton, "but parents have to think of the educational benefits of getting a top education. I can't tell you how much the boarding school network helps in college and in the workplace."

Few blacks debate the academic and career benefits that come out of a white boarding school experience, but even the most integrated blacks acknowledge the gap that remains between their children and the overwhelmingly conservative white classmates who leave them out of social groups and fail to share their interests, politics, hobbies, or music. In her book *Black Ice*, black journalist Lorene Cary wrote about the rocky time she experienced as a student trying to fit into the ultra-WASP St. Paul's School. She points out that hair, clothes, music, and family background were all reasons for the barriers between herself and her white classmates. "I did not, however, tell the girls what I was thinking," says Cary about her white female classmates. "We did not talk about how differently we saw the world. Indeed my black and their white heritage was not a starting point for our relationship, but rather was the outer boundary."

"Of course she didn't fit in," snaps a Boston physician, who sent his son to Hotchkiss in Connecticut and dismisses Cary's strikingly poignant story. "She was a scholarship student from Philadelphia. If your black child grows up in a privileged home, the transition is not so hard. My son had no problem fitting in or maintaining his identity because he grew up with everything those white kids had. Of course, he has yet to date a black girl—but this interracial stuff could have happened in the local public school."

For the majority of the black elite, however, the stories fall somewhere in between, with a lot more tending toward Lorene Cary's experiences. More than a few say that the affluent black child has no problem adjusting to settings with affluent whites but that many of the white kids who attend these schools come from conservative families who are actually trying to "buy their way" out of integration. "One white mother expressed outrage when her daughter was assigned to room with my daughter at her boarding school," says a black parent who received a call from the parent within a day of their daughters moving in together. "'Now don't get me wrong,' the woman said to me after about five minutes of friendly conversation, 'but my husband and I are paying seventeen thousand dollars a year for our daughter—and for her to end up with an African American roommate was a little shocking.' She expected me to appreciate her disappointment. She assumed that my daughter must be some inner-city scholarship child. So the net result is that you may find that the school works hard to encourage their students to mix, but many of these boarding school kids grew up in homes that are far more insulated and bigoted than that of the kid who goes to the run-of-the-mill private day school."

Some black parents go to great lengths to prepare their children for an experience that they acknowledge may be fraught with negatives. For example, Groton alumnus Sam Watkins recalls how his father sent him to the expensive, all-white Camp Timanous in Maine for two summers to prepare him for the experience of being at a white boarding school. "My dad didn't want me to be shocked or intimidated by those kids, so he created every possible safety net."

But no matter what the outcome, it is wrong to suggest that most of the black kids who go to these schools—and the parents who send them—are running away from their black identity. In fact, many of the families who choose these schools are people who are extremely active in the black social and civic worlds and maintain a strong sense of their black culture. Dr. William Curry, a prominent vascular surgeon at New York Hospital, sent his sons to a northeastern boarding school, but the boys were also spending their summers in the black Sag Harbor resort community. Similarly, Curry and his wife Katherine spend much of their own time as members of black civic groups like the Reveille Club and the Links.

When it came to elite public schools for black society, no school could out-perform Washington, D.C.'s Dunbar High School, which, prior to 1915, had been known as "the M Street School." During its heyday from 1900 to the 1950s, the school sent a large number of students to Ivy League and Seven Sisters colleges. Among its graduates were attorneys William Henry Hastie and Charles Hamilton Houston, who went on to Amherst; Judge Robert Terrell, who graduated from Harvard; and historians Rayford Logan and Carter G. Woodson.

"My husband, Frank, was in the same class as Senator Edward Brooke," says Alberta Colbert, who recalls the significance of Dunbar High School when she was attending Howard University. "There was a very large contingent of Dunbar graduates in many of Howard's different departments." Because of Dunbar's prestige and the inability to get job offers from white universities in the North, many of Dunbar's teachers were black scholars who had received advanced degrees at northeastern universities and used Dunbar as a training ground or waiting spot before they went on to college teaching positions at universities like Howard, Fisk, Atlanta, Morehouse, or Spelman.

Other well-known public schools popular among the old black elite during this same period were Booker T. Washington High in Memphis, which was headed by Dr. Blair T. Hunt in the 1930s and 1940s; DuSable High School in Chicago; Girls High School in Brooklyn; Central High School in Philadelphia; and Booker T. Washington High School in Atlanta. Following the increased integration of school districts and the breakup of prestigious black high schools in the middle to late 1950s, the black elite left many of these schools, concluding that private day schools and boarding schools were the best solution for their children.

Whether they were using certain public or private schools, cotillions, day camps, or other special activities, the black elite have always been focused on providing their children with the kind of experiences and environments that prepare them for lives that will always be slightly different from those of whites or other blacks that they encounter.

Today, many black professional parents are discovering that their children can usually find acceptance among the black elite even if they don't check off all the boxes.

As the daughter of a physician who belonged to the Boulé and a mother who belonged to the Girl Friends, Paquita Harris Attaway says it was not out of the ordinary for her to have spent summers in Martha's Vineyard, attended Camp Atwater, belonged to Jack and Jill, and debuted at the AKA cotillion in the 1950s. "In my generation," explains the Washington, D.C., resident, "it was both common and expected that we would be tied to these institutions and events."

As I consider Attaway's experience and her ongoing ties to prominent black organizations, I wonder if the kids of today's generation are going to feel equally rooted to the other members of black elite society if the black boarding schools, camps, and cotillions continue to die out or become irrelevant. Even though I did not personally participate in many of these institutions, I had family and friends who did—so I was left with the pride that one receives in knowing that people of your race have some respected traditions that are repeated generation after generation. By having that knowledge and by having some close and some distant ties to those traditions, it kept me optimistic that one day I could play a greater role in that community.

Howard, Spelman, and Morehouse: Three Colleges That Count

"There is probably no group of alumni more loyal than Howard alumni," says Washington resident Bebe Drew Price, who grew up on the Howard campus when her father, Dr. Charles Richard Drew, was a professor there in the 1940s.

As a college sophomore, I found the same to be true. I visited the campus for the first time during Howard's biggest weekend, the annual homecoming. It was the biggest multigenerational gathering of black professional families that I'd ever seen. And I was met by folded arms.

"So I understand you go to one of those white colleges." This father had welcomed me into his daughter's home earlier that afternoon, but now stared at me blankly from his seat at the head of the dining room table.

I waited for someone to save me, but no one did. "Yes, Princeton," I answered.

"Well, I guess you've been told this is a Howard family. Three generations—all Howard College, law school, and medical school. Everybody went to Howard." The man paused and looked around at the nine other people sitting around the table. "Yes, all of us Howard. And proud to be black."

I quickly looked over at the friend who had invited me for the weekend of festivities, but she quickly dropped her head. She wasn't going to weigh in after her father offered the strange remark. I was also left without words. As we sat in the silence of that beige dining room in an upper Sixteenth Street house, I was simultaneously impressed and offended by the loyalty this man exhibited toward the school. For me, his remark underscored an attitude that pervades many black elite families: If you don't have ties to Howard, you're a little bit inferior. And if you didn't go to a black college, you're off the radar screen altogether.

Today, one of the complexities that are beginning to face more and more upper-class black children is choosing between black and white colleges after they have spent the last twelve years of their lives getting accustomed to being one of a handful of blacks in a predominately white suburban public or private school. Some parents are surprised or offended that their kids aren't following in their footsteps, but others have resigned themselves to the idea that their children's integrated childhood experiences will have a lasting effect on the child's future decision to abandon traditions and institutions that had been accepted and taken for granted only one generation earlier.

My brother and I were part of a wave of students who had been raised during the 1970s in white neighborhoods and predominately white public and private schools, somehow splitting our values between what whites felt was important and what our family and other blacks felt was important. Some of those values—college selection, for example—were overlapping among our black and white friends. Given that fact, it was expected that we would have selected white colleges. Our white friends were doing so, and the greater percentage of our black friends—except for some third- or fourth-generation Howard legacies—were also heading to white schools in the Northeast. My best friend from across the street had gone away to Groton for boarding school and was heading for Yale, like his four older siblings. So little interest did we have in southern schools like Howard or Morehouse that neither my brother nor I even sent away for a black college catalog. In fact, not only did we have older cousins from the South who had come north to attend white boarding schools and colleges like Brown, Harvard, Tufts, Wellesley, and Columbia but we were also surrounded by older Jack and Jill friends who were headed for white universities.

Today I recognize that although I had many relatives who had gone to good black colleges a generation before me, I had chosen a school on the basis of what was respected most by my white classmates and white neighbors. As the only black student in my honors classes and in my neighborhood, I was convinced that my college selection was going to reflect not only on me and my family but also on the entire black race, as it was viewed by white onlookers. I had endured sufficient skepticism from white neighbors that I was compelled to select a school that held the greatest universal appeal. Not surprisingly, this was to be a northeastern Ivy League university. Some would say that such a mentality—one that continually sought white approval—is yet another inevitable response to growing up in a mostly white school environment, and I am not sure they would be wrong. But I would also argue that my black friends from Jack and Jill—the black kids I most identified with—were making the same choice and were therefore a part of my decision-making process.

The decision to attend an Ivy League college was, therefore, in my eyes, a race-neutral decision. It was just a part of my quest for universal approval—and that included approval from people of all colors. But the decision to attend Princeton, in particular, was probably completely based on race. Brown University in Providence had been my first-choice college. My tennis partner from high school (he was white) and my best female friend, whom I knew through the NAACP (she was black), were planning to attend Brown. Just before high-school graduation, after I had settled on attending the Ivy League school, a white neighbor asked me where I would be going to college in the fall. I remember proudly answering "Brown."

The woman returned a blank stare.

The moment was devastating for me because it suddenly made me realize that the universal approval that I always craved was not about to be bestowed.

"Brown?" the woman asked. "I'm not sure I've heard of it. Is that one of those black schools?"

Now, I am ashamed to admit how crestfallen I felt. But even with seventeen years of valuing my black culture and history, I suddenly felt offended that this person would have thought that I, who had spent his adolescence proving to white kids that he could compete in their white-dominated arena, was now relegated to the less respected arena of "one of those black schools." This is why I chose Princeton that afternoon. Today when people ask me if I missed out on anything by not attending a black university, I tell them "yes" for one reason: Although a vast number of career opportunities have opened for me as the graduate of an Ivy

League school with a worldwide reputation, I missed out on a black college friendship bonding experience that has been impossible to replicate in any activity or organization that I have joined since.

"There are many black families today—regardless of their wealth—who consider black colleges because they recognize the overall benefits to their children," says Dr. Joan Payne McPhatter, a Howard professor and alumnus who has seen four generations of her family attend the 132-year-old university. "This is particularly true for blacks in the South."

Although two buildings at Tennessee State University are named after her father, it is Howard that fills the family résumé. McPhatter's son went there before going to Georgetown Law School. She went there before becoming a professor in the communications department. Her mother, who held a doctorate, was Howard's dean of women in the 1940s. And her grandfather went there before entering a career in accounting.

Just as the Roosevelts and the Kennedys had Harvard and the Buckleys and the Basses had Yale, old families among the black elite have selected certain colleges for their children and their descendants. While northern blacks and some free southern blacks certainly chose to come north during the 1800s to attend white colleges and universities like Amherst, Harvard, or Oberlin College in Ohio, most members of the black elite attending college during the immediate post–Civil War period preferred to establish their family roots at the black southern universities founded by religious organizations.

Today, there are a total of 117 historically black colleges and universities, but only a few of them play a role in the upper-class black résumé. The three most prestigious in this group are Howard University in Washington and Spelman College and Morehouse College in Atlanta.

"I never say it to people, but it's really the black man's Harvard," says an Atlanta lawyer who was the third generation of his family to attend Howard University. "I laugh when whites walk into my office and see the framed picture of Founders Hall. 'Oh, is that Harvard?' they ask," says the attorney as he points to the photo of the imposing redbrick building that serves as a centerpiece of the eighty-nine-acre campus that sits on top of a hill in northwest Washington.

Founded in 1867, Howard is the most popular school among America's black elite, with many of its students third- and fourth-generation alumni. Because of the reputation of the undergraduate, law, medical, dental, and other professional departments, Howard lays claim to more prominent black alumni than any other college or university. In fact, nearly one-third of all black physicians and dentists and about one-fifth of all black lawyers in the United States are graduates of Howard.

Included among the alumni are such people as Supreme Court justice Thurgood Marshall, Pulitzer Prize–winning writer and Princeton professor Toni Morrison, United Nations ambassador and Atlanta mayor Andrew Young, Virginia governor Douglas Wilder, New York mayor David Dinkins, U.S. senator from Massachusetts Edward Brooke, National Urban League president Vernon Jordan, TV actress Phylicia Rashad, Yale Medical School psychiatrist Dr. James P. Comer, and Nobel Peace Prize–winning United Nations undersecretary Ralph Bunche. Although it is the largest predominately African American university, it has always had whites and other nonblack students among its enrollment.

"Howard is not only the center of Washington's black society; it has long been at the center of this *country's* black society," says Alberta Campbell Colbert, who received a pharmacy degree from Howard in the 1950s and became a part of the elite Howard circle that represented the most accomplished blacks in medicine, business, and government. "My husband, Frank, had graduated from Howard and studied under Dr. Charles Richard Drew, who invented blood plasma storage, as well as so many other famous black intellectuals. The campus and LeDroit Park were teeming with famous faces when we were in school," she adds. Colbert's circle of Howard friends included some of the affluent physicians and businesspeople who belonged to Washington's Bachelor-Benedicts club with her husband. One of her oldest friends is Howard graduate Cynthia Mitchell, whose husband, Doyle Mitchell, also a Howard graduate, was chairman of Industrial Bank of Washington, one of the country's largest and oldest black-owned banks.

"The Howard name has long been an important credential in every part of the country," says Howard graduate Bennie Pratt Wiley, who heads the Partnership, an association of corporate executives in Boston. Wiley, whose sister, former Washington mayor Sharon Pratt Kelly, also went to Howard, still interacts with her circle of Howard friends, whether it is through her activities in the Links or summering each year in Martha's Vineyard. "When I was growing up here," adds Judge Henry Kennedy of the superior court of Washington, "Howard was an indicator of reaching the top. The people that went there were the people that really counted in black society and American society."

Not only has the school produced a dazzling array of graduates, but Howard has always attracted the top scholars in both its undergraduate and its graduate departments. Within that group are such individuals as Dr. W. Montague Cobb, who has degrees from Amherst and Howard and later became chairman of the Howard Medical School anatomy department; legal scholars William Henry Hastie, Spottswood Robinson, and Charles Hamilton Houston; historians Carter G. Woodson and John Hope Franklin; marine biologist Ernest Everett Just; chemist Dr. Percy Julian, who developed the synthetic cortisone used to treat arthritis; and religious scholars like Benjamin E. Mays, who later headed Morehouse and earned forty-three honorary degrees—to name just a few.

"Most of us Howard graduates knew all the major black Ph.D.'s in the country because they would always end up teaching there," says Colbert, who also sent her daughter, Doris, to the school. "But one of the reasons why we had the best professors was that these top professors weren't hired or taken seriously by white universities. Of course the times are different now."

Named after Oliver Howard, a commissioner of the Freedmen's Bureau, the school was originally established to educate recently freed slaves. But instead of identifying itself as a school that pushed an aggressive civil rights agenda, the school spent most of its early years apolitically, maintaining a conservative approach to leadership. While never stated, its goal was to become a black version of Harvard or Yale: The university wanted only the most accomplished intellectuals on its faculty and sought out students from the most ambitious and well-to-do families. Its goal was to duplicate the curriculum of northern white schools and produce black students who could compete in the same arenas.

The school was not without its detractors. In fact, some were disappointed that the university was so fixated on avoiding the emphasis of problack messages in the curriculum. So cautious was the black university about avoiding a strong black image that it continued to appoint white men to preside over the school for the first sixty years of its existence—passing up its own black scholars and other blacks who had applied for the position. In fact, the rumblings got so loud in 1912 when three black deans—George Cook, Kelly Miller, and Lewis Moore—got passed over in favor of a white minister from Maryland that the school was forced to increase the deans' powers in order to appease them and other blacks who thought it was time for this black institution for black leadership to actually employ a black leader.

Today, it is surely regarded as an embarrassing fact that it was not until 1926 that Howard selected its first black president. It was Mordecai W. Johnson, a light-skinned black who could easily pass for white, who received the honor of serving as the university's first black president. With degrees from Morehouse, the University of Chicago, Rochester Theological Seminary, and Harvard, and a doctorate from Howard, he was clearly qualified for the appointment. One of the best-known presidents in the school's history, he is credited with increasing Howard's national prestige by attracting contributions and top scholars to the campus. Both because of his physical appearance and because of his intellectual accomplishments, he was effective at winning access to, and support from, wealthy white donors who had previously expressed no interest in supporting black institutions.

James M. Nabrit, a graduate of Morehouse College and Northwestern Law School, became president of Howard in 1960, after teaching at its law school for twenty-four years. He retired from the presidency in 1969, when the position was given to thirty-seven-year-old James Cheek, who had a Ph.D. from Drew University. It has been said that it was during this very late period that Howard finally accepted its role as an important voice in the civil rights movement.

"The light-skinned, old-guard elite had run the school for so many years," says a Los Angeles attorney who attended the college and law school in the 1960s. "It was about time that we got a young president and a student population who were willing to sacrifice their ambition to emulate Harvard so that they could become moral leaders in the midst of the discrimination and riots that were going on around us." The attorney says he remembers one particular Howard alumni homecoming event that underscored how much Howard had finally changed. "For years, we had been this very accomplished yet smug school, filled with smart, light-skinned, good-looking kids from comfortable homes. All of a sudden, in 1969, we had this young president who looked like the rest of black America and who hadn't come from a family with Ivy League credentials—it was great!"

Howard continued to change even after the civil rights period, with its students becoming more and more activist and less concerned with creating an image of genteel, satisfied members of the black upper class. After it was announced in 1988 that Republican National Committee head Lee Atwater would be joining the school's board of trustees, students staged marches and campus sit-ins, expressing their repugnance that a conservative race-baiting politico who had worked to undermine programs such as affirmative action might be setting the agenda for their university. This incident caused many angry conservatives to focus on the fact that Howard had been receiving millions of dollars each year—40 percent of its operating budget, in fact—from the federal government because the school was chartered by the U.S. Congress. Not long after, the House Budget Committee was to end the government's annual subsidy to the school.

Many of the school's old guard felt that the militant voices on the campus sometimes went too far in the direction of offending whites or potential supporters. One example occurred in the early 1990s, when a campus group hosted controversial anti-Semitic speeches given by a Nation of Islam member, then turned away a Jewish Yale professor who had originally been invited to discuss the issue of black-Jewish relations.

The school's most vocal critics focused on its fourteenth president, Franklin Jenifer, who assumed the presidency in 1990 and lasted less than five years, during which the university suffered a severe downturn in finances and enrollment and in morale among faculty and students. Many alumni were upset with the administration led by the former biology professor and blamed him for a series of embarrassing incidents that have lingered in the popular press. People unfamiliar with the school are unlikely to know that under new stewardship the school's direction has shifted significantly.

Today Howard has approximately eleven thousand students, with a quarter of that group enrolled in the graduate and professional schools. Reemphasizing its interest in taking the most ambitious and accomplished students, current president H. Patrick Swygert recently announced that Howard's class of 2,000 included more National Achievement Scholars—seventy-nine—than any other school in the country—including the University of Virginia and Harvard, which came in second with sixty-nine of the scholars.

One of the most shocking facts about Howard is that despite the success of its many prominent alumni, a painfully small percentage of them—only 5 percent—contribute financially to the school.

"I've never understood why so few of us give back to our top black colleges," says a Howard alumnus who graduated in the mid-1940s. "When you look at top schools like Howard or Spelman, there are two possible explanations for the lack of contributions. First, many of us think that because they are the wealthiest of the black colleges, they must not need our money. And second, there is a culture within the black community—even the wealthiest of the group—that contributes a greater portion to church, fraternity, and civic groups like the NAACP and the Urban League: groups that are going to advance the black political agenda and aid impoverished blacks."

It is true that even among the well-educated black elite, the black church is perceived as a much more influential institution than the black university when it comes to effecting change in the black community. It is also true that the strongest ties that blacks maintain to their college past is through their fraternities and sororities. Unlike whites who saw their fraternities as purely social groups that lost relevance after graduation, black fraternities and sororities play a much more important role later in life and serve as a vehicle for black alumni to contribute money and time to civic projects, scholarships, and other programs to aid disadvantaged blacks in the United States and abroad.

Dr. Chester Redhead, one of Howard's longtime board members and one of its most generous alumni contributors, says, "I contribute to that school because I owe so much of my success to it. I went to college and dental school there. I met my wife, Gladys, there, right in front of Frazier Hall. My sons went there, and many of the doctors I work with got their training there. I wish more of our alumni realized that it is our money that will keep the school vital and relevant. We cannot rely on the government or anyone else to keep it going."

When talking to people like Chester Redhead, it becomes clear that the Howard circle is more tight-knit than any of the other black schools. Everybody knows everybody, and they know each other's parents, as well as their children, through their memberships in the elite organizations that attract the Howard crowd. Redhead serves on the board of trustees with actress Phylicia Rashad, as well as with close friend Elizabeth Graham Early, who is a member of Gladys Redhead's Girl Friends chapter. And Early's husband, Robert, is in the same Boulé chapter as Chester. They both own homes in suburban New York, and they throw parties for each other in the popular Sag Harbor resort community.

The daughter of a New York dentist, Gladys remembers considering Fisk because her sister had gone there, and Vassar because of a friend who was there, "but I knew that Howard was the gold standard." A retired schoolteacher who later received a master's degree at Columbia, Gladys remembers the affluence of her fellow classmates, as well as the superior scholarship and research that were present on campus. "The school is an oasis for children of black families who value intellect," Redhead adds, as she points out that the school's Moorland-Springarn Research Center has the largest collection of books, manuscripts, and artifacts on Africans, African Americans, and Caribbean people in the world.

Because of such intellectual "assets" and because of the name recognition it enjoys, Howard has no problem remaining relevant to today's young black elite. Even though it loses some top student applicants to the most competitive northeastern universities, the college and its many graduate schools remain the leading choice for old-guard families.

"If her father was a doctor, and her mother was well-educated, and she was from an attractive, socially ambitious family, then it's likely that she would have been sent to Spelman. She was like a half-step above the women at Bennett or Fisk," says a Spelman graduate who was attempting to describe the kind of black woman that went to the school prior to the 1960s. "After women's lib and the integration of the large white schools in the late 1960s, Spelman women were not as easy to categorize."

Founded in Atlanta in 1881 as Spelman Seminary, it was the first college for black women in the United States. Honoring its most important benefactor, the Spelman name came from the maiden surname of Mrs. John D. Rockefeller. Set on a traditional, bucolic campus of redbrick buildings on thirty-two acres in southwest Atlanta, the school is just steps from Morehouse College, where many Spelman women ultimately find their equally socially elite male counterparts.

Atlanta natives are very proud of pointing out that the school was first conceived at a meeting that took place in the basement of that city's historic Friendship Baptist Church.

From its founding, Spelman had traditionally been headed by white women. Many of the older alumni from that early period grudgingly admit that powerful white Atlanta politicians, residents, and dignitaries gave the school more money and more respect because it was run by a white person. The school named its first black president, Dr. Albert E. Manley, in 1954, and he was ultimately one of the longest-serving presidents at the college. He tripled the school's enrollment and raised millions for the school through the Rockefeller family during his twenty-two-year tenure. The brother of Michael Manley, prime minister of Jamaica, he held a doctorate from Stanford University and had previously served as dean of arts and sciences at North Carolina Central University.

One of the reasons for Spelman's prestige among many well-to-do black families was the school's strict rules governing academic and social life. Fourth-generation Atlantan Ella Yates remembers attending Spelman in the late 1940s, when Florence Matilda Read was president. "Miss Read ran the college like a small boarding school, and that's what put parents at ease," says Yates, who recalls some of the rules that governed students' daily activities. "Not only did we have to sign in and out of our dormitories after getting permission from our assigned 'house mothers' but we had to get permission to leave the campus at all times. We were not allowed to wear slacks unless an extracurricular activity required it, and we could not wear shorts unless we were in gym class." Yates also remembers that every student was required to attend services at Sisters Chapel every day, regardless of the student's religious beliefs. She adds, "And don't even talk about boys. To avoid awkward situations with the opposite sex, we were told to always travel in twos or more when we went shopping. Miss Read knew us and our parents by name, and she never hesitated to fire off a letter to them if she spotted inappropriate behavior. If a student got married while in school, she would have to withdraw, because Spelman felt that serious dating and marriage would distract us from our studies. That's why they also did not allow sororities at the time."

Yates continues, "Miss Read not only implemented a lights-out rule at 10 P.M. but she also monitored our hair, makeup, and jewelry."

Other graduates during the 1940s and early 1950s remember that male visitors were not allowed after dinner, and when they came earlier they had to remain downstairs in the common room. By senior year, the rules were even more strict, says Yates. "When I was in my last year, Miss Read told us that seniors would not be allowed to visit their parents more often than once every six weeks. She felt that this was the best way to teach us independence. She didn't want to see a class full of spoiled young women who would not know how to live their lives once they graduated. It was very clever."

Although Spelman was in the South, many well-to-do families sent their daughters from New York, California, Detroit, Chicago, and other distant locations because of the school's strict rules and because of whom the school attracted.

"My parents liked the fact that I would be surrounded by other girls from good families and by famous dignitaries who regularly visited the campus," adds Yates. "Marian Anderson was constantly there, and so were politicians, journalists, and African leaders. And because there was always at least one Rockefeller on the board, we'd always see them on campus." To ensure that the students benefited from these visits, the college required the visitors to meet with the students in both formal and informal sessions while they stayed on campus in the president's home.

It was during the tenure of President Manley that several local black institutions like the Waluhaje Ballroom and the Auburn Casino's Rainbow Room opened up in the area near Spelman. But these were places where one rarely found Spelman girls.

Among Spelman's more recent presidents are Donald Stewart, who headed the school starting in 1976. An active member of the Boulé, Stewart, along with his wife, Isabel, who heads the national organization Girls Incorporated and belongs to the Links, is a regular guest at Atlanta and New York fund-raisers. It wasn't until Stewart stepped down that the black women's college actually appointed a black woman to the presidency. That woman was Johnnetta B. Cole, who raised a record $113 million during her last capital campaign. Each year during her presidency, I saw the popular anthropology scholar and her husband, Arthur Robinson, as they were regular guests (and a major attraction) at the annual Labor Day tennis party held a few doors down from my parents' home, at the Tudor mansion of Spelman board chairman Bob Holland. Holland was the first black partner at the global consulting firm McKinsey and Company before later becoming CEO of Ben and Jerry's Ice Cream. Both of Holland's daughters attended Spelman during Cole's presidency. It was also during the Cole presidency that Bill and Camille Cosby gave a record twenty million dollars to the school. Following her term, Cole stepped down to teach anthropology at nearby Emory University.

The newest president, Dr. Audrey Manley, was appointed in 1997, after leaving her job as acting U.S. surgeon general in President Clinton's administration. A 1955 graduate of the school, she is the widow of former Spelman president Albert Manley.

A major center for Atlanta's black society, the school continues to receive support from some of the city's most prominent families, who are either active alumni or powerful trustees. Maynard Jackson, whose mother and five aunts graduated from Spelman, is an important supporter, as is board member Marge Yancey, who was one of the first blacks on the Atlanta school board and is the wife of renowned surgeon Dr. Asa Yancey.

While the old Spelman stereotype was that of a wealthy civic-minded wife of a doctor or attorney, its alumni include many women who have made major professional contributions of their own, including Children's Defense Fund founder Marian Wright Edelman, Pulitzer Prize–winning author Alice Walker, TV entertainers Esther Rolle and Rolanda Watts, and U.S. ambassador to Kenya Aurelia Brazeal.

Today, because of its faculty, its wealth, and its ties to the large Atlanta University system, Spelman is still able to compete successfully for the same students who are being accepted by Wellesley, Smith, Bryn Mawr, the other Seven Sisters colleges, and Ivy League universities. At Spelman, unlike most women's colleges, more than 80 percent of its faculty hold doctorates, and more than a third of the students major in mathematics, engineering, or the sciences.

When I encounter Spelman women of my own generation or of my mother's generation, I recognize a sisterhood and a camaraderie that I have never seen among black women who attended predominately white women's colleges or who have joined sororities at other historically black schools. Because it very much identifies itself to the world and to its students as the ultimate college for black women, these students and alumni are fiercely loyal to, and respectful of, this Atlanta institution.

Founded in 1867, Morehouse is the only historically black four-year college for men in the United States. Located in Atlanta near Spelman College and the Atlanta University consortium, it was originally established in Augusta, Georgia, by a group that included a former slave, a Baptist minister, and other religious leaders who were focused on educating former slaves. The funding for the school came from the white religious members of the American Baptist Mission Society, which operated in order to bring education to recently freed slaves. The Society followed in the footsteps of the more established American Missionary Association, a group formed in 1846 to encourage white southerners first to abolish slavery and then to invest in the education of freed slaves. Both the Society and the Association recruited black and white teachers from the North in order to found black southern schools like Morehouse and nearby Atlanta University.

Although it has long been a prestigious institution, Morehouse was not a school that traditionally sought white support and approval to the same extent as Howard and Spelman. In fact, by appointing John Hope to the presidency in 1906, the school preceded most of the other black colleges in selecting a black leader.

Some of Morehouse's best-known alumni are Dr. Martin Luther King Jr., Atlanta mayor Maynard Jackson, movie director and actor Spike Lee, former U.S. secretary of health and human services Dr. Louis Sullivan, *Ebony* magazine editor Lerone Bennett, and Georgia politician Julian Bond. Its most famous president was the religion scholar Dr. Benjamin Mays.

Ties between Morehouse and Spelman are so tight that not only do Morehouse men often date Spelman women but a look through the marriages column in *The Alumnus,* Morehouse's quarterly alumni magazine, reveals that an overwhelming percentage of Morehouse men marry Spelman alumni. Spelman alumna Ella Gaines Yates remembers marrying Morehouse alumnus Clayton Yates in the late 1940s. "Who else would we marry when they were right across the campus?" Mrs. Yates says with a laugh. "We were basically told that they were the only men good enough for us."

In recent years, Morehouse faced a scandal that rocked Atlanta when its president, Leroy Keith Jr., was accused of using the school's funds for a $700,000 home and for a compensation package worth $460,000. Keith resigned shortly after an audit was released. Replacing him in 1995 was Dr. Walter Massey, a class of 1958 Morehouse graduate who quickly launched a $60 million fund-raising campaign.

Like Massey, Morehouse alumni are diehard supporters of the school who can rattle off statistics that alumni of other colleges would never know. They announce that 74 percent of its faculty hold doctorates and that it is the only black university with a graduate who has won a Rhodes scholarship.

"There are many Morehouse families who would never consider sending their sons to another school," says Morehouse graduate Keith Chaplin after bumping into former Morehouse president Hugh Gloster at a Links brunch at the Atlanta Ritz-Carlton. The Atlanta-based CNN executive frequently runs into fellow Morehouse alumni in his tight society circles. "The Morehouse credential is real currency among the black elite," he adds, "because the school has admitted some of the most accomplished men in our community."

"I was absolutely furious," says a woman who graduated from Fisk in the mid-1960s. "My mother called me up in October of my sophomore year and said, 'Your girlfriends are getting engaged to future surgeons and you haven't even been over to Meharry to date one of those boys. We're not spending all this money for nothing.'"

For generations, women who went to Fisk for college have been insulted by friends and relatives who suggested that they were not using their time wisely if they had not yet left their campus to go across the street to find a male medical student to date at Meharry.

"It's an unfair characterization," says my cousin Robert Morton, a surgeon who graduated from Meharry Medical School and met his wife, Anna, while she was a Fisk student, "but there were a lot of Fisk-Meharry couples that came about simply because the campuses were so close." Founded in 1866 in Nashville, Tennessee, Fisk University is a coed college that has attracted some of the most prominent black intellectuals, including W. E. B. Du Bois, James Weldon Johnson, Harlem Renaissance figure Arna Bontemps, historian John Hope Franklin, sociologist E. Franklin Frazier, and poet Nikki Giovanni. Other well-known graduates include Congressman John Lewis, choreographer Judith Jamison, Solicitor General Wade McCree, New York judge Constance Baker Motley, and U.S. secretary of energy Hazel O'Leary.

Although Fisk is not as high-profile as Howard, Morehouse, or Spelman, many consider the school to be more intellectually driven because of its history: It was the first black school to have a Phi Beta Kappa chapter, the first black school to be awarded university status, and the first black college to gain full accreditation by the Southern Association of Colleges and Schools. In its early years, its reputation was also boosted because Booker T. Washington served on its board, married a Fisk alumnus, and sent his children there. Furthermore, it houses the private papers of many prominent individuals such as Langston Hughes and Marcus Garvey.

Located on a forty-acre campus, Fisk is among the smallest of the elite black schools—with just nine hundred students in its undergraduate and graduate programs. In 1946, the school appointed Charles Spurgeon Johnson, a social scientist who had previously established a race relations institute at Swarthmore, as its first black president.

"One of the reasons why I went to Meharry," says Morton, "is that it was one of the few places that was committed to training black doctors. When I looked at New York University Medical School in 1939, I heard that they had a quota which kept them from admitting more than two blacks and eight Jews per class. That was surely no way to begin my career."

Meharry Medical College has graduated more black physicians and dentists than any other school in the United States except Howard, and nearly 40 percent of the black faculty members in American medical schools are Meharry alumni. Founded in 1876, it was one of the last black institutions—waiting almost eighty years—to put a black at the helm. But even though its first few presidents were white, the bulk of the professors have been black for at least three generations. "We had a few white students while I was there in the early forties," says Morton, "but all of the professors were black. Of course that made for a stark contrast when I graduated and came to see the New York hospitals."

Like other Meharry graduates, Morton discovered that it was virtually impossible to find a job at a big-city hospital. "In fact," says Morton, who was to eventually land a residency at New York's Montefiore Hospital, "at that time in New York, there was only one institution that would allow a black to perform surgery. That was the Edgecombe Sanitarium—an ancient building without elevators. Black surgeons weren't even allowed to operate at Harlem Hospital."

Originally established by members of the Methodist Episcopal Church, Meharry now offers the M.D., D.D.S., and Ph.D. to an enrollment of 880 students. Its board of trustees includes Dr. Beny Primm, a New York anesthesiologist who heads an addiction research organization; and Hector Hyacinthe, a high-profile New York businessman who was also an appointee of President George Bush. Further evidence of the tightness of this elite group is the fact that the daughters of Morton, Primm, and Hyacinthe are all childhood friends who grew up in Jack and Jill together in New York.

In addition to Howard, Morehouse, Spelman, Fisk, and Meharry, other black schools that have attracted a number of members of the black elite are Tuskegee University, Bennett College, Hampton University, Lincoln University, Clark-Atlanta University (which was created by the merger of Clark College and the Atlanta University graduate schools), and Xavier University—for those who come from Louisiana or are staunch Catholics.

Although the Hampton credential has become important to the black elite during the last three or four generations, for a long time it was a stepchild to other colleges because it was founded as a trade institute intended to train blacks in various industrial and agricultural professions. In the early 1920s, it elevated its status by focusing more on the sciences and liberal arts that were offered at the more prestigious black colleges.

What is ironic is that even though the school's original teachings were quite different from other respected black colleges, its organizers had the same reluctance when it came to placing black educators in control.

"My father was only the second black to serve on Hampton's board," says Dr. James Norris with amazement in his voice. "In the 1930s, when my father joined their board, they had never even had a black president."

Norris is a Hampton graduate who now practices plastic surgery on the Upper East Side of Manhattan, and his loyalty to his school is as strong as that of his late father, Dr. Morgan Norris, but he is still mystified that a leading black university—founded in 1868—could have resisted appointing a black president or black board members for so many generations.

Playing a paternalistic role for the blacks that he was instructing, Hampton's white founder, Samuel Armstrong, was most determined to attract important white trustees who could bring prestige to the private college in Virginia. Among its early trustees were such U.S. presidents as James Garfield and William Taft, who served as chairman from 1914 to 1930. The school, which now has 5,700 students on its 200-acre campus, has graduated many prominent alumni, including Spencer Christian, the TV personality; John Sengstacke, publisher of the *Chicago Defender* and other major black newspapers; George Lewis, vice president and treasurer of Philip Morris; and Booker T. Washington, who later used the school as his model for establishing Tuskegee Institute. Because of their roots as industrial—rather than liberal arts—institutions, both Hampton and its successor, Tuskegee, are sometimes looked at as second-tier when compared with their Atlanta and Washington counterparts.

Founded by educator and activist Booker T. Washington in 1881 as Tuskegee Institute, the 4,500-acre Alabama campus is the only college in the country to be designated a national historic site and a unit of the National Park Service. Modeled after Hampton University, from which Washington had graduated in 1875, Tuskegee was often criticized and snubbed by the black elite in its early years because, like Hampton, it encouraged blacks to accept their second-class status in the postslavery period and assume roles as tradesmen in the industrial and agricultural professions rather than pursue careers in medicine, law, or liberal arts areas.

Because of Washington's calculated obsequiousness to powerful whites and a famous 1895 speech that he made in Atlanta at a major convention, Washington was able to attract more funds for Tuskegee from affluent and influential whites than any of the other black colleges. This alone allowed the school to become a force and an important institution among the black elite—even though many disagreed with his philosophy. In spite of Frederick Douglass's demands for equality and W. E. B. Du Bois's insistence that blacks ignore Washington's pacifist "go along to get along" approach, Washington was able to successfully promote his black self-help message to wealthy white donors such as Andrew Carnegie. The more he discouraged his students and other blacks from entering politics and government or from insisting on social equality, the more popular Washington and his school became among white contributors, who helped turn his campus into a showplace, attracting such researchers as scientist George Washington Carver.

But even though Washington's message was one that accepted segregation, his students all embraced his message about excellence. "My father graduated from Tuskegee in 1915—the last year that Dr. Washington signed the diplomas," says Patsy Campbell Petway as she pulls out yellowed pages from the thirty-fourth annual commencement, when her father, Emmett Cadwalader Campbell, graduated. "Like many of my father's classmates," explains Petway, who is an active member of the Nashville Girl Friends and Links, "he was an engineer and saw the positive messages of self-reliance and choosing an education that gave you a practical skill."

Today, Tuskegee's mammoth campus includes buildings catering to a curriculum that places a far greater emphasis on liberal arts, but it continues to offer degrees in architecture, construction science, engineering, and other areas that its founder felt were most important.

"They called us Bennett Belles because it was an all-women's college that produced intelligent southern ladies," says Orial Banks Redd, whose family has lived in the affluent white New York suburb of Rye since 1912. She remembers arriving at the Greensboro, North Carolina, campus and feeling as if she had never left home because of the school's provincial rules. Members of the black elite have always looked upon Bennett as the women's college that was the most likely to enforce rules of decorum. Founded in 1873, the liberal arts college was still sticking to its rules of etiquette through the 1940s and 1950s when Redd, now a newspaper publisher and Democratic political activist, was a student.

"Most of my classmates were the daughters of southern doctors, lawyers, and influential ministers," explains Redd, whose brother attended the less etiquette-conscious Hampton Institute in Virginia. "We had to wear gloves and a hat and carry a purse whenever we were going to town. We didn't have sororities and we were not allowed to wear earrings that dangled—they watched for that. And we had to be chaperoned by an adult or faculty member whenever we went shopping."

David Dallas Jones was president of the society-conscious school while Redd was a student. Like many of the presidents of the elite black colleges at that time, Jones had a complexion that was light enough to "pass" for white if he had chosen. He unified two important black dynasties when his son, David, married Anna Johnson, the daughter of Howard University president Mordecai Johnson.

"Bennett was a wonderful school," says Beatrice Moore Smith-Talley, who graduated in the early 1940s, "but the campus and the students were operating in a protective, almost predictable environment." Like many of her ambitious classmates, Smith-Talley was the daughter of a Spelman graduate who had high expectations for her future. "Like my mother, I went on to Columbia University for graduate school. And like her, I was expected to compete with the best."

When I look at educated members of the black elite today, I make constant comparisons about whose decisions were best: those who attended historically black colleges and graduate schools or people like me, who attended white schools like Princeton and Harvard Law School, or still others who compromised by attending a top black college and then a top white graduate school. I know of happy, well-adjusted adults from each of these groups. The group I worry about most is the offspring of parents who have no ties to black educational institutions. There is now at least a generation of these individuals, and I fear that they will find no reason to use these schools or contribute financially to their endowments.

Because I have no ties to those schools and because my wife's three degrees are from Harvard, I think it is rather unlikely that our own children will consider a historically black college for their own future. Even though I recognize the quality and value of these institutions, my own personal choice would make me an unconvincing supporter.

When we look at the future of the elite black colleges, some interesting patterns are evident. For one, the schools at the top—Howard, Spelman, and Morehouse—will continue to attract the smartest children from elite families. Because they are schools that are well respected in the white corporate and professional community, the graduates of these schools will continue to have access to good jobs and graduate schools. For these reasons, members of the black elite will continue to embrace these three schools for their children.

Unfortunately, the schools that will become less important to the black upper class—particularly those living outside the southeastern United States, where most of the schools are located—are those that are not considered the high-profile, top-tier black colleges. These schools will have a difficult time attracting contributions from black professionals who don't see the schools' relevance, and they may also have a tough time offering unique curricula or activities now that many predominately white universities already have a department devoted to African American studies, a dean of minority affairs, and local chapters of black fraternities and sororities.

The challenge facing many black colleges today is being able to convince black professional parents that their mainstream career-driven children will have as much access to top-paying careers as their white counterparts whether they attend a black college or the prestigious, historically white schools. Even though most of the nation's black colleges will do a great job preparing students for the academic challenges of graduate school, or the professional challenges of the workplace, it may be a school's image and the parents' general impression that contribute to the future success of some of the lesser-known black colleges.

The Right Fraternities and Sororities

I could hear my aunt's anxiety level rising through the phone line.

"So, is she an AKA?"

"No, I don't think so," I answered.

"You don't think so? Either she is, or she isn't."

I rolled my eyes as I looked over at Pamela, the woman I had just gotten engaged to. We were on the telephone from New York, calling relatives to give them the news of our wedding, which would take place in eighteen months. Most of my father's relatives had already met Pamela. My Aunt Phyllis and Uncle Earl were my mother's side of the family and knew few of the details.

Aunt Phyllis already knew that my fiancée was from Detroit and had graduated from Harvard-Radcliffe, Harvard Law School, and Harvard Business School. But foremost on her mind were Pamela's ties to Greek life.

"Well, no," I finally responded. "She's not an AKA."

"Oh, a Delta?" she asked, sounding a little crestfallen.

My uncle interrupted on the line. "Now, Phyllis. Just because you're AKA doesn't mean everybody has to be."

"So, she's a Delta?" she asked again.

I looked over at my fiancée and started to feel a little embarrassed. "Aunt Phyllis, she's actually quite nice."

"I'm sure she is, Lawrence. So she's a Delta?"

"Please, Phyllis."

My uncle and aunt were inextricably tied to black college life. He was an Omega who sat on the alumni board of Ohio's Central State University. Aunt Phyllis, a teacher with two master's degrees, had pledged AKA forty-six years ago at Wilberforce, the country's second-oldest black college. She's the kind of sorority member who finds it hard to believe that not every accomplished black woman would want to be in Alpha Kappa Alpha.

"No, I'm sorry, Aunt Phyllis, she's not a Delta either."

"Oh, I see." As I looked over at Pamela, a long awkward silence fell on the other end of the phone line.

I wasn't sure what my aunt was thinking, but it was probably one of three things:

His fiancée is white.
His fiancée is in one of those lesser sororities.
His fiancée is not a Greek at all.

In the world of the black elite—where race, class, and black fraternity life are intertwined—I'm not sure which of the three assumptions about my fiancée would have been more shocking at the moment. It is an experience I encounter whenever blacks discover that I was not in a black college fraternity. What was certain was that my fiancée won a reprieve when it was revealed that her mother was an AKA and her father was an Alpha.

While there are clearly old-guard blacks who would limit their circle to just those who went to the "right" black colleges (e.g., Howard, Spelman, Morehouse, Fisk), there are also many others who would expand the circle to include those blacks who went to good white colleges, so long as their "good white college" experiences also included membership in one of the black establishment fraternities or sororities.

To understand this mentality is to understand the difference between white fraternity life and black fraternity life. While the former is mostly limited to a three- or four-year college experience, the black fraternity experience begins in college but is an activity that has even greater importance after graduation. It goes far beyond the well-choreographed campus musical "step shows" that the student members present on stage as competitions to show which fraternity can "outstep" the other with synchronized music and dance steps, and the unique traditions they follow in their attire and speech and actions while "on line."

For my aunt, my uncle, and many other blacks, their sororities and fraternities are a lasting identity, a circle of lifetime friends, a base for future political and civic activism. Continuing throughout their adulthood, membership means lifetime subscriptions to publications like the *Sphinx*, the *Ivy Leaf*, or the *Oracle*. It means regulated funeral programs with unique fraternity services that are specifically outlined for surviving fraternity members in attendance. Having attended a college that permitted neither black nor white fraternities, I have long felt alien to—and envious of—the experience that my friends received at other colleges. For many of them, these black Greek-letter organizations provided a forum, postcollege, through which some of the best-educated blacks in America can discuss an agenda to fight racism and improve conditions for other less-advantaged blacks.

Even though blacks have attended southern black universities since the 1860s and northern white universities since the early 1800s, black Greek-letter organizations were not established until the early 1900s—more than 125 years after the white fraternities began. These black college fraternal groups began as small elite social groups that eventually made scholarly discussion and social activism a part of their agenda. When Alpha Phi Alpha, the first fraternity, was begun by a group of black students at Cornell in 1906, it was an important bond between the seven black men who belonged, but was virtually invisible to the rest of the mostly white Ithaca, New York, campus.

But as the presence of Alpha Phi Alpha and the other seven black fraternities grew on black campuses during the early 1900s, they were each known for building their popularity by seeking out certain desirable student candidates (e.g., smart, popular, accomplished, affluent, athletic, good-looking) and turning down others. Even as some campus chapters developed tough standards for minimum grade point averages and other criteria, black Greek life was highly sought after because of the parties and other social gatherings that were offered by fraternity and sorority membership.

Throughout most of the twentieth century, blacks who attended white colleges were not allowed admission into the white fraternities operating on their campuses—just as they were not allowed to live in some of the "white housing" located on or around their college campuses. Because of this, although five of the first eight black fraternities were founded on black campuses, the black fraternities saw their fastest and widest expansion taking place on white college campuses where black students had no housing and where they were facing extreme discrimination and isolation. Clearly, a need for black fraternities had been expressed at both black and white schools.

Most blacks who attended historically black colleges had hopes of joining one of the black fraternities because that was one of the surest ways to become accepted among the campus elite. In the early 1900s, the groups were small, intellectually elite, and rather secretive in their activities. By the 1930s and 1940s, the fraternities and sororities had become more dominant on campus, offering large social gatherings and serving as a magnet for not just the intellectual elite but also the economic elite, who looked at the groups as a way to distinguish themselves from nonmembers who could not afford the membership fees or pay for the kinds of clothes, parties, and automobiles that were de rigueur for members. By the early 1950s, many of the fraternity alumni who stayed active in their graduate chapters had launched important civic programs to respond to the black community and its problems.

While some argued that there was too great a dichotomy between young fraternity members who focused on socializing and the older fraternity alumni who were using their efforts to advance the black agenda, both groups were often attacked as supporting elite organizations that further divided the larger black community into the privileged and the working classes.

By the late 1950s, when many of the fraternity members—students as well as alumni—had gotten involved in the southern civil rights movement, a greater solidarity developed between members and nonmembers. They had a common cause to fight for, whether they were operating on mostly black or mostly white campuses.

A new issue developed for black students who began entering white colleges in the late 1950s and 1960s. Because such bigotry was exhibited by the white Greek-letter organizations on so many campuses where well-to-do, integrated blacks were now students, the black Greeks had an even greater role at the white colleges—except at schools like my own Princeton, where fraternities were not permitted for any racial group.

Barbara Collier Delany's experience at Cornell in 1956 underscored the problems waiting for black students who faced the white fraternities and sororities operating on white college campuses. Delany made national headlines in 1956 when, as a student at the Ivy League campus, she was offered membership in the white sorority of Sigma Kappa. She remembers being one of only a handful of blacks at the college at the time. "I was the first black ever to be offered membership in a white sorority," says Delany, who had grown up in a family of privilege. She belonged to Jack and Jill, debuted with the Girl Friends, and graduated from the elite all-girl Hunter High School in Manhattan. "The girls in the sorority were very nice to me, but the officials at the national headquarters were furious, and they told the students that they had better reject me or headquarters would shut down the sorority's chapter at Cornell," says Delany, who still corresponds with some of those classmates. "When the white students refused to kick me out, headquarters shut down the sorority."

Even though I was a generation after Delany, I was also a Jack and Jiller from New York who would miss both the black college scene and the black fraternity scene. I was slightly envious of a couple of my Jack and Jill friends who attended white colleges that had black campus fraternities. They were able to continue living among whites while replicating the kind of black experiences and interaction we had grown up with through our Jack and Jill weekends. I, on the other hand, was at Princeton, a school that frowned upon fraternities of any type—black or white. Instead, our social activities were built around the eating clubs—a unique Princeton institution that was focused on thirteen stone mansions that lined Prospect, a residential street at the east end of the campus.

"You're just the kind of black that would fit in at Ivy Club," suggested one of my white sophomore classmates, as we considered which eating club would most likely fit our personalities and backgrounds. While most fraternity buildings on college campuses served as places for members to live, eat, and study, the eighty-year old mansions that served as eating clubs at Princeton performed one function: they served as elegant dining halls. They had formal living rooms, parlors, libraries, and—in a few cases—featured waiters, large lawns, patios, and staircases, all of which served as the backdrop for twice-a-year formals.

Although there were thirteen eating clubs—each with at least one hundred members—only a handful of blacks joined the groups. Instead, most chose to eat their meals in the more egalitarian, university-owned dining halls. Perhaps this is why when whites come upon a white Princeton alumnus they ask the question, "Which club were you in?" whereas black alumni are asked, "Were you in a club?" The presumption is obvious, and the two lonely years I spent in my 99-percent-white Princeton eating club bear that out.

Having grown up in a family that clearly valued a strong racial identity, I have, on occasion, wondered why neither my own parents nor my wife's parents—all products of segregated southern schools—encouraged us to enter fraternity life. My parents certainly had nothing to do with my decision to join an eating club; I did that because I was determined to enjoy what I thought was the quintessential Princeton experience. But I have wondered why they didn't insist that I travel to another campus and join a black fraternity. I am fairly certain they would have insisted upon it if I had attended a black college or even a large, predominately white state university.

But I believe that my own parents' reticence was grounded in ignorance—an ignorance that most blacks of my parents' generation would have had regarding life in an almost uniformly white 250-year-old Ivy League university. They had no idea of the rules—for blacks or whites. Like many of the upwardly mobile Jewish, Italian, or Asian parents who were sending the first generation of their families into such bastions of WASP culture, my parents had no idea what the rules would be. Yes, we had cousins and Jack and Jill friends who had attended the schools in recent years, but they had done so only after graduating from the top boarding and private day schools. None of us knew how much or how little of our ethnic culture should be shown or celebrated on these campuses. Maybe we were supposed to latch onto the other black kids. Maybe we were supposed to emulate the dominant culture's way of dress, speech, and thinking. Maybe we were supposed to tiptoe around campus, trying to make the white professors, administrators, and students forget we were black or forget we were even there.

"Now, don't go getting involved in a whole lot of racial protests and sit-ins," said my father as we unpacked the car at Princeton upon our arrival and I remarked on a flier that talked about the school's investments in companies that supported apartheid in South Africa.

Thinking back, I now realize that my parents were somewhat intimidated by Princeton—not because they had not interacted with well-to-do whites before, but because they couldn't advise me on how to go about gaining social comfort on that campus. They weren't going to tell me to assimilate completely into the white WASP culture, but they also weren't going to encourage me to segregate myself—as some black students did—into an all-black existence. Just as they wouldn't have been able to tell me the secrets of surviving at a Wall Street bank, they couldn't give me the inside track on finding social success at Princeton. It was a school that had eating clubs rather than fraternities, few blacks, and few examples of interracial dating or interracial rooming arrangements. And hardest of all, it had no groups analogous to Jack and Jill, no safe harbor like the one the children's group had always provided me after weeklong stretches with little or no black cultural discussions or interaction. Given my parents' inexperience with such a university, it is no surprise that black fraternity life was never discussed as a consideration during my tenure at Princeton.

Today, the National Panhellenic Council, the ruling body of black college fraternities and sororities, has eight organizations. For men, there are Alpha Phi Alpha, Kappa Alpha Psi, Omega Psi Phi, and Phi Beta Sigma. For women, there are Alpha Kappa Alpha, Delta Sigma Theta, Sigma Gamma Rho, and Zeta Phi Beta. A great deal of what has determined the prestige of specific fraternities or sororities depends on the age of the organization, its size, and the wealth and prominence of its members. In fact, many among the old-guard black elite would argue that only three of the fraternities—the Alphas, the Kappas, and the Omegas—and two of the sororities—AKAs and the Deltas—actually fit the "society" profile.

"What made all of these fraternities and sororities so prestigious," says second-generation Alpha member Boyd Johnson, "was the fact that our membership requirements were so much more challenging than in the nonblack Greek-letter groups. High grade point averages and extensive community service are what distinguish us."

In fact, over the years, there have been many white college administrators who express surprise that the black Greek officials insist on such high GPA standards, thus creating a barrier for large numbers of potential members.

Another barrier that stood between the black fraternities and potential members was certain reluctant school administrators who banned the organizations from campus. Not surprisingly, the ultrasocial all-women's Bennett College was one of them. "I really resented the fact that President Jones looked down on sororities and refused to allow them on campus," says Beatrice Moore Smith-Talley, who had to wait until she graduated to pledge AKA. "It would have been nice to have pledged while I was in college, because it's a much different social experience when you go through it as an eighteen-year-old than when you do it in a graduate chapter. A lot more bonding takes place."

Atlanta resident Ella Yates remembers wanting to join the Deltas when she was a student at Spelman in the late 1940s. "But our president, Florence Read, felt sororities would distract us from our work and create divisions and cliques," says Yates, who joined one of the sorority's graduate chapters several years after college.

Although the leaders of Bennett and Spelman finally relented in later years and allowed the Greek organizations to establish themselves on their campuses, there remained some critics. Many of these were white university administrators who simply didn't encourage the creation of fraternities—as I found at my own college. Other critics were specifically critical of *black* fraternities. The most prominent one was Howard University sociology Professor E. Franklin Frazier, who, in his book *Black Bourgeoisie,* criticized the extravagant spending he saw among the black fraternities, who spent outrageous sums during their annual meetings and conventions. He noted, "During the Christmas holidays in 1952, the Greek letter societies meeting in four cities spent $2,225,000."

Because the organizations have grown larger, with richer members, that number would be considerably higher today. True, these five- or six-day annual gatherings feature lavish social activities, cruises, and parties, but many would argue that all of the fraternities use their money to establish and support important and unique programs throughout the nation and throughout the African diaspora.

"These groups can enhance every aspect of our lives," says Dr. Eva Evans, the AKA Grand Basileus, a title used for the highest-ranking officer in the sorority. When Evans says this, I know for sure that she would appreciate my Aunt Phyllis's dedication. As head of the country's oldest black sorority, Dr. Evans also seems like someone who can't imagine why not every accomplished young black woman would want to be shrouded in the pink and green colors of the AKA sorority.

"If you visited us on the World Wide Web or if you looked at some of the Web pages that our AKAs have at Harvard, Wellesley, and elsewhere, you'd be very impressed with what our sorors are doing," she says when asked how active the ninety-year-old sorority is on predominately white campuses. After all, the old AKA was once epitomized by the stereotypical black doctor's or judge's daughter who went to Howard or Spelman before marrying well and settling into a life of quiet volunteerism. But Evans doesn't play into the stereotype.

"As a college sorority, we've always advanced an educational agenda. We always had high GPA requirements. And more than ever, we're pushing the importance of math and science for our girls. We need more black women in those fields," adds the breathless Evans, who lives in Lansing, Michigan, and holds a Ph.D. in administrative higher education. "Of course that's not to say that we don't have some top AKA sorors in those areas—like the astronaut Mae Jemison and the former secretary of energy Hazel O'Leary."

"How about in government?" I ask Evans, testing her almost encyclopedic memory of where every prominent soror is at every moment.

"Oh, please," she responds, as if I were some sort of an amateur. "Some of the most powerful women in government are AKAs. We've got plenty of state officials and mayors like Sharon Pratt Kelly in Washington, Congresswoman Cardiss Collins, Congresswoman Eddie Bernice Johnson, Congresswoman Sheila Jackson Lee—but you know," Evans adds after pausing briefly, "we don't have a U.S. senator yet. Somehow Carol Moseley-Braun got snatched up by the Deltas."

To understand the history of the fraternities and sororities of the black elite, one must both recognize the reason for their existence and understand the comparisons that are often made between them. Even though officers like Evans have a daily agenda that focuses on developing programs, monitoring legislation, and raising funds to benefit members and their international causes, there is an obvious undercurrent among members that causes these students and past graduates constantly to measure one group against another. Unlike whites, who can choose between more than fifty national fraternities and sororities, with only a vague sense of how one group differs from the others, the black elite has a clear sense of which black frats and sororities among the National Panhellenic Council have a strong presence, and which ones don't.

Alpha Phi Alpha is the oldest of all the black Greek-letter college organizations, and it is the one to which most of my friends belong. Founded in 1906, it was the only black fraternity or sorority to have been started at an Ivy League school: Cornell University. Even though there were fewer than a dozen black students at the rural Ithaca, New York, campus at the time, seven Cornell men—now referred to as the Alpha's "Seven Jewels"—formed the organization.

Quickly identifying themselves with programs that emphasized scholarship rather than mere social interaction, the Alphas launched, in 1919, a national "Go to High School, Go to College" campaign to combat the eighth-grade dropout rate of 90 percent among black children. The Alphas also contributed resources and manpower in 1935 to assist in the racial discrimination suit by black Amherst College graduate Donald Murray, who had been rejected by the University of Maryland Law School because he was not white. Not only did the fraternity pay his school expenses, but the group also provided his attorneys, who were Alphas and well-known civil rights attorneys Thurgood Marshall and Charles Hamilton Houston. The fraternity also gave support to a case that successfully challenged the practices of segregation at the University of Texas Law School when Alpha brother Herman Sweatt applied for admission.

Today, the group has over 150,000 members, with 750 chapters spread out over the United States, Africa, Asia, Europe, and the Caribbean. Working in conjunction with the Kellogg Foundation, the fraternity operates mentoring centers in fifteen major cities around the United States that serve as after-school sites for inner-city teenagers. Some say that the stereotypical Alpha is professionally ambitious, bookish, not overly gregarious, and "safe." My father-in-law, who pledged Alpha in 1947, says, "He's the *mensch*—the nice guy—that everyone wants his sister to date."

A veritable "who's who in black America," the membership has included such people as Supreme Court justice Thurgood Marshall; Dr. Martin Luther King; Atlanta mayors Andrew Young and Maynard Jackson; scholar W. E. B. Du Bois; former U.S. senator Edward Brooke; former congressman William Gray, who now heads the United Negro College Fund; Olympic gold medalist Jesse Owens; Canaan Baptist Church head Reverend Wyatt Tee Walker; *Ebony* magazine founder John Johnson; New Orleans mayor Marc Morial; Seattle mayor Norman Rice; Parks Sausage founder Henry Parks; National Urban League directors Hugh Price, Lester Granger, and Whitney Young; Detroit mayor Dennis Archer; San Francisco mayor Willie Brown; New York congressman Charles Rangel; U.S. cabinet secretaries William Coleman, Samuel Pierce, and Louis Sullivan; Congressman Adam Clayton Powell Jr.; and New York City mayor David Dinkins.

"We all knew about the famous Alphas who were out in the real world," says my father-in-law, Albert Thomas, who joined while attending South Carolina State. "The Alpha brothers were always reminding us. Since they had a more socially conservative and academic agenda, I knew that's where I would join."

So carefully does Alpha Phi Alpha monitor its image and reputation that its president, Milton Davis, received an apology and retraction from NBC after the network's show *Saturday Night Live* aired a controversial skit in 1995 that portrayed drunken white college students who were wearing the fraternity's copyrighted logos and mocking blacks who participated in the historic Million Man March. In a letter that was later duplicated and distributed to all the Alpha members, NBC offered a carefully worded apology that insisted the program's intent had been to contrast unruly white frat brothers with the civilized demeanor of blacks who participated in the march. They claimed that the black-and-gold Alpha Phi Alpha outfits had coincidentally landed in the show's wardrobe department.

"That whole branding tradition is not as gruesome as it sounds," says Dr. James Norris, as he defends a popular practice in many of the chapters of Omega Psi Phi. The New York physician joined the Omegas in 1954 while at Hampton University. "It's not as painful or as permanent as a tattoo. It's really just a hot iron with the Omega symbol on it."

Omega Psi Phi is often thought of as the fraternity with the most personality and the most gregarious members. The Omegas' tradition of branding, though shared by other fraternities, is mostly associated with them. Usually referred to as "Q's," many new members of the Omegas have been branded on the arm with a hot iron displaying the fraternity's letters. While the other frats sometimes followed this practice, the Omegas were most famous for it because it played into the group's more macho reputation. Further highlighting that image is the Omegas' secret rallying cry: a barking sound that young frat brothers will often make at large Omega gatherings.

Like most of the other black Greek-letter organizations, Omega Psi Phi was founded at Howard University. In 1911, three students formed the fraternity with the help of biology professor Dr. Ernest Just, a black Dartmouth graduate who also earned a doctorate from the University of Chicago. Most Omegas will quickly tell you that Just is the only fraternity founder to appear on a U.S. postage stamp. Consisting of a large number of physicians and dentists, the group has grown to a membership of approximately 130,000, with 717 chapters around the world.

Among the Omega members are such people as Virginia governor L. Douglas Wilder; Atlanta mayor Bill Campbell; Urban League head Vernon Jordan; Secretary of the Army Clifford Alexander; Howard professor Charles Drew; former NAACP heads Roy Wilkins and Benjamin Hooks; actor Bill Cosby; Rev. Jesse Jackson; scientist Dr. Percy Julian; *Black Enterprise* founder Earl Graves; poet Langston Hughes; former secretary of HUD Robert Weaver; and Howard University presidents James Nabrit and H. Patrick Swygert.

Although equally respected, the fraternity that is least identified with a particular stereotype is Kappa Alpha Psi. It is also the smallest of the three old-guard fraternities. Founded in 1911 at Indiana University, Kappa Alpha Psi now consists of approximately 110,000 members in 660 chapters. While the Alphas and Omegas often find themselves paired off with AKAs and Deltas, the Kappas are not often associated with one of the sororities.

"When I joined the Kappas at Virginia State, very few of my classmates had a sense of what the different fraternities were doing beyond the boundaries of our campus," says Bill Richardson, who pledged Kappa in his sophomore year and recently retired from Schieffelin Somerset and Co., the national distributor of distilled spirits, where he was an executive. "In fact," adds Richardson, who runs into many of his former Kappa fraternity brothers at gatherings sponsored by the Boulé and the Prince Hall Masons, "when I was in college, many students picked the fraternity because of an impression they had of who would fit in there. At Virginia State, it seemed that the premeds and prelaw crowd were Alphas, the athletes were Kappas, and the gregarious social types were Omegas. Since I was on the football team, I chose the Kappas."

Although they have annual conventions, the Kappas also gather every two years at conclaves to elect a new Grand Polemarch, the highest ranking Kappa officer. Among the well-known Kappas are Federal Reserve Board member Andrew Brimmer, New York attorney and businessman Percy Sutton, Congressman John Conyers, Congressman Mervyn Dymally, Congressman Walter Fauntroy, Soft Sheen Products president Edward Gardner, TLC Beatrice founder Reginald Lewis, former Urban League head John Jacob, Congressman Louis Stokes, and former Los Angeles mayor Tom Bradley.

The smallest and youngest of the fraternities is Phi Beta Sigma, which was founded in 1914 at Howard University. With 650 chapters, the

Washington-based fraternity often partners with its sister organization, Zeta Phi Beta, on civic projects such as voter registration among blacks and Project SATAP, a program aimed at reducing teen pregnancy. Although its membership has included Dr. George Washington Carver, Atlanta builder Herman Russell, Chicago mayor Harold Washington, and Congressman John Lewis, the group has never enjoyed the same prestige as the Alphas, Omegas, and Kappas.

Although Ersa Poston is one of Alpha Kappa Alpha's "Golden Girls"—a label the sorority uses for its fifty-year members—she remembers that she and her girlfriends in college had all originally planned to pledge Delta. "Of course it all worked out in the end," says the Washington resident, who served under Governor Nelson Rockefeller and then under President Carter as a commissioner of the U.S. Civil Service Commission, "but in my generation there was quite a bit of blackballing that went on at some campuses. No matter how smart or loyal you were, it was possible for one girl—it required just one—to ruin your chances of getting into a particular sorority. That happened to me when one of the Delta girls found out that a boy she liked also liked me. That was the end of my chances of ever becoming a Delta."

Since she went on to break numerous barriers, including becoming the first black cabinet officer in New York state government, Poston acknowledges that she's done well by the AKAs.

Many say that among the old-guard women of the black elite, you're either a Delta or an AKA—or you're not in a sorority at all. Among this crowd, there are not many choices, and at the undergraduate level, the two sororities are often competing for the same candidates. AKA, the oldest of the sororities, was established in 1908 by a group of women at Howard. Since then, the AKAs have grown to more than 140,000 members participating in 860 chapters.

The competition that exists between the AKAs and the Deltas is so widely acknowledged that it is unusual to find a well-educated black woman who remains neutral on the issue. In fact, the best things I recall my AKA relatives saying about the Deltas when I was growing up were backhanded compliments like, "The Deltas are a great second choice for a girl who can't get into AKA." Since their inception, the two sororities have attempted to distinguish themselves by comparing the grade point averages and other accomplishments of their student members. Even in its first five years of existence, in an act of sheer public relations genius,

AKA staked its claim on superior scholarship by establishing an AKA award at Howard University for that female student who achieves the highest grade point average. Just because of a long-standing family bias toward AKAs, I always presumed that the AKAs were the premier group of the two. Today, I realize that both of them place a great emphasis on achievement—perhaps an emphasis that is not matched by any of their male counterparts.

My aunt, Phyllis Walker, recalls that when she was a college sophomore, "long before the step shows came into existence, we had to learn a lot more about the historic beginnings of the AKAs, and we did it by writing long letters of application to the Ivy Leaf Pledge Club—the senior wing of the sorority that regulated the admissions process—and then attending monthly meetings where the older students tutored us on the history."

Like other sororities, the AKAs would have a "pledge week." If you survive being "on line"—a time when your grades are closely scrutinized—as well as the pledge week, then you are admitted.

"Even though I was presented at the AKA cotillion in Pittsburgh, I always knew that my sister and I would grow up to be Deltas," says Paquita Harris Attaway of Washington, as she recalls her mother's staunch support of the group. "My mom loved being a Delta and ended up serving as vice chairman of the National Panhellenic Council." A tradition of more mother-daughter memberships may be a particular attraction for some pledges who select the Deltas over the slightly more famous AKA group.

Although it is the second oldest of the sororities, the Deltas are the largest of the black sororities, as well as the largest black women's organization in the United States. With just under 200,000 members in 850 chapters, they have been a powerful force in politics, as well as civic and social affairs.

Included in the Delta membership roster through its history have been Patricia Roberts Harris, who served as President Johnson's secretary of HEW and ambassador to Luxembourg; Senator Carol Moseley-Braun of Illinois; Congresswoman Shirley Chisholm; Lena Horne, actress and singer; Mary McLeod Bethune, college founder and adviser to Eleanor Roosevelt; Congresswoman Barbara Jordan, who was the first black to deliver a keynote address at the Democratic National Convention; journalist Charlayne Hunter-Gault; and Dr. Betty Shabazz, wife of Malcolm X.

Delta was founded in 1913 at Howard University, and the Deltas have chapters throughout the United States as well as in Haiti, Germany, Japan, the Republic of Korea, the Virgin Islands, and various countries in Africa.

Vashti Turley Murphy, the daughter of John Murphy, the publisher of the powerful newspaper *Afro-American* in Baltimore, was a student at Howard when she helped start the group. In recent years, the Deltas have gained some additional popularity through the publication of *In Search of Sisterhood,* a book about the Deltas by Paula Giddings, a well-respected Howard graduate and Delta member.

In addition to organizing their own annual national and regional conferences, the Deltas send delegates to international conferences that address human rights issues. For example, the Deltas sent a team of sorors, including eighteenth national president Hortense Canady, to Beijing, China, for the controversial World Conference on Women, at which they presented workshops on research and educational issues. The group has also made presentations at the Congressional Black Caucus's annual legislative weekend. Among their past national presidents are two of black American history's most prominent women: Dorothy Height and Sadie Alexander. Height, who served as president from 1947 to 1958, holds honorary degrees from Tuskegee Institute, Harvard, and many other universities, and is distinguished by her leadership of the National Council of Negro Women and her board membership at the American Red Cross. The group's first national president, Sadie T. M. Alexander, was a Philadelphia attorney, a Ph.D., and the first black woman to graduate from the University of Pennsylvania Law School—in 1927.

"Some of the most accomplished women in American history are Deltas," adds Delta member Sharon Mackel of Cleveland as she prepares to leave a recent Jack and Jill conference to attend a Delta convention. "They are the most inspiring group of people." In many cities, the Deltas, along with the AKAs and the Links, are considered among the most important hosts of the old guard's prestigious debutante cotillions.

Although they perform a great deal of public service and fund-raising to support diabetes research and projects sponsored by the March of Dimes and the National Council of Negro Women, Zeta Phi Beta and Sigma Gamma Rho are not nearly as popular among the old guard.

Founded in 1920 by five classmates at Howard, Zeta Phi Beta has eighty-five thousand members with six hundred chapters in the United States, Africa, Germany, Italy, and the Bahamas. The female Zetas are often partnered with the male Sigmas, and their membership has included writer Zora Neale Hurston, singer Dionne Warwick, actress Esther Rolle, and former National Bar Association president Algenita Scott Davis.

Youngest of all the large sororities, Sigma Gamma Rho was founded in 1922 at Butler University in Indianapolis and has about 72,000 members. Although it has not attracted quite the same number of high-status alumni, Sigma has included Dr. Lorraine Hale, director of Hale House; Congresswoman Corrine Brown of Florida; and Academy Award–winning actress Hattie McDaniel among its members.

As a virtual outsider to the entire college fraternity experience, there have been many social and business gatherings where I have found myself at a disadvantage for not having belonged to a black fraternity. And not surprisingly, to some blacks above the age of fifty, I am "suspect." I am seen as being less connected and less committed to the black culture and the "black struggle." In conversations at New York business networking events or at Martha's Vineyard summer cookouts, there is inevitably a moment when a host or guest asks the question, "So, are you an Alpha?"

And the answer they are seeking is not *whether* I belong to a fraternity. The presumption is that since we are in the presence of other black people, and since we consider ourselves to be authentically black, I must, of course, have ties to one of the great black Greek college groups.

To respond that Princeton had no fraternities is not a satisfactory answer because I could have gotten on a bus and joined at another campus that did have them. To respond that I belong to the Boulé, the prestigious fraternity for adult black men, is also not satisfactory because it does not account for how I spent my four college years. The only response is to admit my guilt, and then somehow spend the next fifteen minutes trying to win back my place among a group that doesn't take kindly to outsiders.

Regardless of how people like me explain our decision not to join black fraternities in college, there are many real challenges to the growth of black fraternities and sororities.

White colleges that have been unsupportive of black fraternities are just one of the challenges that threaten the black Greek system today. This lack of support makes the organizations appear unstable and unpopular, as the groups see their chapters activated, temporarily deactivated, and then reestablished. For example, the Rho Chapter of Delta Sigma

Theta was formed at Columbia University in 1923, then deactivated in the 1960s, and then reactivated again in the 1980s.

"Another one of our challenges," says a Kappa who graduated from the University of Virginia, "is that black frats are not nearly as rich as the white frats. When I was in school, the white frats had generations of rich alumni sending them money to build incredible mansions for their frat houses. We didn't have that kind of support because there were so few black alumni."

But there are problems facing some of the black fraternities that money can't solve: the recent "animal-house" type of hazing incidents that, until recently, were more typically associated with white Greek-letter organizations. In recent years, there have been a rising number of violent incidents during the admissions or initiation stages at several campus chapters. In 1994, for example, at Southeast Missouri State University in Cape Girardeau, Missouri, a student pledge of Kappa Alpha Psi died of injuries inflicted by two frat brothers who later pleaded guilty to involuntary manslaughter. And in 1995, a University of Maryland student sued Omega Psi Phi, and was later awarded $375,000, after he suffered injuries following initiation beatings that lasted over a four-week period. During that time, the student, Joseph Snell, said that frat brothers regularly hit him with a hammer and a horsewhip, and that they forced an electric space heater near his face in order to darken his skin because they felt he wasn't "black" enough.

As the Omega national office said in the Maryland case, such offensive initiations are not sanctioned. The Omega's former Grand Basileus Dr. C. Tyrone Gilmore is frustrated by the aggressive nature of some of the fraternity brothers. "We are seeing an angry, mob mentality among some students that never existed in my father's or my own generation." He adds that the whole intent behind these organizations is to increase camaraderie among members and that all the fraternities and sororities work hard to remove rogue frat brothers who violate the organizations' codes of ethics.

Even though the hazing incidents have created sad moments, they involved a very small percentage of the groups' members. Nonetheless, the incidents have caused some older members to adopt more conservative attitudes. "These new students have it easy," says Delta member Ella Yates. "In my generation you had to walk a fine line to both get into and stay inside one of these groups. Some people say we ought to loosen the academic requirements. I say we should do just the opposite."

Many of the old guard agree with Yates. Many of them recognize that what makes these organizations important and special is the high standards and the prominent members who belong to them. When I attend gatherings of frat brothers or sorority sisters—people who may not have known each other in college, but who share an identity that was formed there—I find myself recognizing that there are few institutions that are able to unify so many people for such lasting relationships. For this reason, the black fraternities and sororities will endure regardless of how admission requirements might change.

The Links and the Girl Friends: For Black Women Who Govern Society

"What about the ones hanging on that rack?"

"Or the ones on that table?"

The saleswoman could see that neither my father, my brother, nor I knew anything about buying evening handbags. We'd been to Bloomingdale's, Bergdorf's and Saks Fifth Avenue, finally arriving at Neiman Marcus by the late afternoon.

"Your mother's taking this stuff seriously," my father remarked. "So, it's got to be something good."

As the slender saleswoman opened up a light blue cloth bag and pulled out a small purse covered with glittering crystals, she smiled broadly. "This is a Judith Leiber," she said as we shrugged our shoulders at a name that meant absolutely nothing to us. "She only uses fine Austrian crystals, and it's a brand that says you've made it."

In an uncharacteristic gesture of conspicuous consumption, my normally conservative father nodded anxiously. "Then it's exactly what we're looking for."

After he handed over something in the neighborhood of nine hundred dollars, the woman put the bag and its blue cloth protector into a matching blue box and the three of us looked at each other with the assurance that the upcoming event justified spending such an exorbitant amount of money for a handbag that was not even large enough to hold a wallet.

"Is this a wedding anniversary?" the woman asked.

"No," he answered. "My wife was just accepted into the Links."

The woman offered us a blank stare and the three of us gave each other a knowing nod. The woman was white. There was no reason for her to understand.

As my mother and every other woman in her crowd would have told you, getting accepted into the Links was a big deal, and it was not something you'd ever need to explain if you were in the company of the right kind of people. In this case, some would say the right kind of people didn't include whites or blue-collar blacks.

Later that week, my mother was initiated into upper-class black America's most elite organization for women: the ten-thousand-woman-strong Links Incorporated. For fifty years, membership in the invitation-only national organization has meant that your social background, lifestyle, physical appearance, and family's academic and professional accomplishments passed muster with a fiercely competitive group of women who—while forming a rather cohesive sisterhood—were nonetheless constantly under each other's scrutiny. Each of the 267 local chapters brings together no more than fifty-five women, most of them either professionals, socialites, volunteer fund-raisers, educators, or upper-class matrons, and is added to only when a current member dies or moves to another city. Along with her longtime friend Betty Shabazz (widow of civil rights leader Malcolm X), Anne Walker (wife of Harlem minister Wyatt Tee Walker), and several other high-profile professional women, my mother was preparing for an elaborate weekend of black-tie festivities in New York, where she and the other new recruits would be feted by friends, relatives, and Links members who had flown in from different parts of the country.

Although not as old as other elite black women's groups like the Girl Friends or the National Smart Set, the Links is by far the largest and the most influential. Founded in 1946 by seven well-to-do black women, it contributes millions of dollars to organizations like the United Negro College Fund and the NAACP Legal Defense Fund and supports hundreds of local charities and scholarship programs in the United States and abroad. A social group with local chapters that meet monthly in members' homes or at restaurants and private clubs, the Links gets its members to donate more than one million volunteer hours a year and has donated more than $15 million to a wide range of charities and programs in the United States and abroad.

On top of all of the volunteer work demanded of its members, the Links also manages to dominate the social calendars of the black elite with its formal parties, annual White Rose balls, debutante cotillions, boat cruises, art auctions, and fashion show luncheons. The stylish women often arrive at its larger affairs in mink, diamonds, and pearls with husbands helping them from limousines. The semiannual conventions are de rigueur for anyone who wants to combine elaborate black-tie dinners with scheduled workshops and debate sessions on such issues as affirmative action, voter registration, national health, or economic development. Following these sessions are golf outings, boat rides, and other leisure activities that allow for networking and socializing among members, spouses, and children from around the country.

"I try to make as many Links events as possible so I can keep up with friends and stay abreast of the current civic issues black people and others are talking about," says Atlanta Link Portia Scott, whose family has been a pivotal part of black Atlanta since it founded the Atlanta newspaper *Daily World* in 1928. "This is a group that does everything from raise money for student scholarships to sponsor cotillions to finance the building of water wells in Africa. How could you *not* want to be a part of it?"

"It took me about twelve years of strategizing, party-giving, and brownnosing to get into this group," admits a Links member who lives in Washington, D.C., and never misses her chapter's meetings or the annual regional and national conventions. "My mother didn't have the connections to get in when she was trying thirty years ago, and she never got over being left out," says the woman. "It's not that I need the validation, but I can definitely say that getting in was worth the eleven or twelve years of anxiety. When you get accepted, the rewards are more than just increased status and a larger number of party invitations. I ended up with a built-in national network of friends who are just like me."

When Philadelphia natives Sarah Strickland Scott and Margaret Roselle Hawkins conceived of the Links, the plan was to create programs where educated black women could focus their attention on civic, educational, and cultural issues. The two women turned to Frances Atkinson,

Katie Green, Marion Minton, Lillian Stanford, Myrtle Manigault Stratton, Lillian Wall, and Dorothy Wright—all women from old-guard Philadelphia families. Their résumés defined the city's black elite.

Scott, who eventually served as the group's first national president, epitomized the well-connected Links member. Born in 1901 and a graduate of the University of Pennsylvania and Columbia University Graduate School, she was a guidance counselor and teacher at schools in Philadelphia and Wilmington, Delaware. Scott's husband, father, and brother were all physicians. Like Scott, cofounder Hawkins was also a teacher who had graduated from Philadelphia schools. Both had children who were active in the then-young but growing elite children's group, Jack and Jill.

Whether by coincidence or design, the Links was to attract future members from the most stellar backgrounds—or at least members who shared a similar blueprint of social and professional credentials.

The current national president, Patricia Russell McCloud, an Atlanta attorney, admits that the organization has always attracted well-to-do women, but says that its mission has never included focusing on social status. "For many years, people wanted to characterize Links members as rich ladies who wore white gloves and sponsored teas and quiet socials, but we are an activist group that takes on important domestic and international projects that assist blacks, children, and others." McCloud acknowledges that there is a preponderance of well-to-do and professional women in the membership, but she insists that the group should be judged by its history of volunteerism and fund-raising and its ability to effect change in the communities it serves.

Noting the occasional articles that have been written about the Links in the style pages or society columns of papers like the *New York Times* or *Washington Post*, McCloud recognizes that the original Links like Sarah Scott were all well-to-do, "but what matters is that they were purposeful."

Of the seven other founding members besides Scott and Hawkins, six were married to doctors and the seventh, Dorothy Wright, was married to Emanuel Crogman Wright, the president of the Citizens and Southern Bank and Trust Company, a black-owned Philadelphia bank. Among the nine Philadelphia women, the most represented church was the extremely patrician St. Thomas Episcopal Church. The women had degrees from such schools as the University of Pennsylvania, Howard, Hampton, and Temple, and had been active in such groups or institutions as Jack and Jill, Alpha Kappa Alpha, the NAACP, the League of Women Voters, the Main Line Charity League, the Bryn Mawr School, and the Philadelphia Grand Opera Company.

"When my mother and her group of friends were forming the Links," says Gwynne Wright, daughter of Links charter member Dorothy Wright, "none of us had any idea how important and influential the group would one day become. I had grown up in an activist home. My parents had hosted Mary McLeod Bethune at our home back when my mother was very active in the National Council of Negro Women. The other women were all rather affluent, but their interests were in improving the welfare of blacks, children, and families, so it was inevitable that the Links would become an activist group."

After its founding, other Links chapters quickly sprung up in Atlantic City, Washington, St. Louis, Baltimore, and New York. There were fifty-six chapters by 1952. Although each new chapter and its proposed membership are closely scrutinized by the national office and other Links members, there are now chapters in such far-flung locales as Beverly Hills, Albuquerque, the Bahamas, and Frankfurt, Germany.

Among the Links' members are some of the most prominent black women in politics, business, education, medicine, and the social world. A look through its membership directory reveals Congresswomen Sheila Jackson Lee and Eddie Bernice Jackson of Texas; Children's Defense Fund founder Marian Wright Edelman; Spelman College president Johnnetta Cole; former Washington, D.C., mayor Sharon Pratt Kelly; the recently deceased Betty Shabazz; former secretary of energy Hazel O'Leary; NAACP Legal Defense Fund head Elaine Jones; and numerous philanthropists, college presidents, judges, physicians, bankers, attorneys, corporate executives, and educators, plus the wives of such high-profile figures as Congressman Charles Rangel (Alma), Vernon Jordan (Ann), American Express president Kenneth Chenault (Cathy), and Harvard psychiatrist Alvin Poussaint (Tina).

"You can generally be sure that the most important, best-connected, most affluent, and most socially acceptable black women in any city belong to the local Links chapter," says a San Francisco Links member who rarely socializes with women outside the circle. "Maybe it sounds a little pretentious, but I simply can't waste time getting to know women who aren't Links. It's an automatic screen that lets me know if this person comes from the right background and has the same values. I'm almost fifty and I live a busy life. I don't have time for people who don't have the right stuff. Rich, educated white women don't hang around with middle-class college dropouts, so why should I?"

As they sit at an elegant mahogany dining-room table twenty stories above Manhattan, Dr. Marcella Maxwell and Mrs. Audrey Thorne compare their experiences as president of one of the oldest and most talked-about Links chapters in the country. The two women refuse to admit it, but their chapter of forty-nine women is ranked by many as one of the most successful because of its high-profile members and because of an elaborate, well-attended annual fund-raiser the group puts on every Easter at the Waldorf-Astoria.

"Nobody misses the Greater New York Chapter's Fashion Show," says a woman who belongs to a Links chapter in New Jersey and who remembers going to her first fashion show when Thorne was president in the 1960s. "I remember reading about those New York Links and seeing their pictures in the *New York Times* and the *Daily News* back in the fifties and sixties when you didn't see black ladies in the white newspapers. I would go to the fashion show not so much for the clothes and the models, but more to see these New York Links women. They'd be walking around the ballroom at the Waldorf or the Plaza talking to friends and smiling for the cameras with so much confidence. It scared me to death, but boy, was it invigorating."

The annual Easter luncheon and fashion show—known as just "The Fashion Show"—attracts nearly one thousand women each year to the main ballroom of either the Waldorf or the Plaza Hotel. The attendees—often attired in gloves and hats—are as high-fashion as the models who parade down the ballroom runway. Several publications, including the *New York Times* and *Daily News,* covered the event when my wife was honored at the group's forty-ninth luncheon.

"I joined the Links in 1954, and we had our first fashion show in the Empire Room at the Waldorf-Astoria," says Audrey Thorne, who served as president of the chapter from 1965 to 1967. "I think we were the first black group allowed to hold a function at the Waldorf."

Marcella Maxwell, the chapter's immediate past president, nods as she looks over prior photos taken at the fashion show. Models can be seen walking down runways, showing off Chanel suits, Bill Blass gowns, alligator handbags, mink or ermine flings, and dresses designed by a host of white and black designers.

"Most of the white hotels were closed to us back then," says Maxwell, who has held many New York state political appointments and is a longtime resident of Brooklyn. "In Harlem, there was the Hotel Theresa, but if we wanted a Midtown hotel, we were usually resigned to the Belmont on Lexington Avenue. That was the only place that allowed black functions."

Thorne, whose husband, Dr. Duncan Thorne, was one of the first black orthodontists in New York, acknowledges that the Waldorf name helped add to the cachet of the event, but also admits that it wasn't easy getting into the Waldorf during those times. "Two of our members, Dorothy Reed and Gertrude Thomas, went around to the larger midtown hotels to see if they would let us in," says Thorne, a former labor union executive, "but most of them simply wanted nothing to do with blacks—no matter how much money we had or what kind of affair was being presented. Of course the white management had no idea who the Links were, so we had to get more aggressive with our approach."

Because many of the Links members were light-skinned and because the banquet manager was favorably impressed with the social stature of the women who bombarded him with requests, the Waldorf finally agreed to rent them space. "It seems ridiculous today," explains Thorne, "but back then, the white banquet managers would actually insist on visiting the host group at one of their meetings before agreeing to rent. Bigotry wasn't so subtle back then, and they wanted to be sure they were letting in the 'right' kind of blacks. It was an insult."

"Although we all take our mandate from the national office and sponsor similar social events and projects in each city," says Links member Patsy Campbell Petway of Nashville, "there are some chapters that have taken on some traditions that are very different from others. For example, the Chicago chapter is known for the debutante ball it always sponsors. My sister's Los Angeles chapter also gives a big cotillion. Atlanta has a jazz brunch, and San Antonio does a lot with artists and photography because Texas Link Aaronetta Pierce is a major collector and art expert."

What is clearly required of all Links members is the adoption of core program initiatives around the areas of education, health, domestic legislation, international welfare, services for youth, and the arts. "Anybody who joins that group just for the social benefits is making a big mistake," says a former Minneapolis-area Links member who says she dropped out several years ago because she didn't have time for all the volunteer hours that the group requests. The organization's book of program initiatives outlines how chapters and members are to implement certain projects. Here's a sample:

- Join forces with the American Cancer Society's National Cancer Initiative mammogram mobile unit and plan mass mammography and extended examination opportunities for local communities.
- Encourage Links members to present papers, seminars, and workshops at international conferences with a focus on improving health, education, and housing.
- Sponsor voter registration drives, get-out-the-vote carpools, phone circles.
- Provide reading materials to students in Rwanda.
- Monitor actions of local, state, and national officials and produce a Legislative Alert Newsletter for inclusion in chapter mailings.

"I like the activism of this organization," says St. Louis Links member Anita Lyons Bond. "We are smart women with money and clout, and we should use it to help blacks who need our connections and our mentoring."

Toni Fay, a New York Link who is also a high-ranking executive at Time Warner, agrees that the group should remain activist. After I met up with her on a plane as she was leaving a recent Links convention in New Orleans, she remarked, "I remember first seeing the Links when I was in high school and I thought of them as a high-society group of ladies, but I've seen them transform themselves from a group of women married to influential men to a group of women who became influential themselves. I'm eager to see this organization go even farther—maybe even start a PAC."

Fay, who is also active with the U.S. committee for UNICEF, the National Council of Negro Women, the board of the Congressional Black Caucus Foundation, the Roosevelt Presidential Library, and the International Women's Forum, is an example of the new politically driven Links member. Many of the younger members like Fay are more interested in the group's ability to help those in need, while some of the old guard would rather focus on the closer-to-home social activities.

"There is a clear conflict between the old guard and the new guard," says a Cleveland Link who joined in the 1970s. "And you really see them butt heads when new members are being nominated. Some of the older women are so insular and so parochial that they really see their chapter as a fiefdom over which they can rule and determine who is acceptable and who isn't. Of course this is just a handful, but when you've got a handful of people doing this in a few chapters, it gets to be a real opportunity for vindictiveness."

Because each chapter is limited to just fifty-five members and because new women are often taken in only once a current member vacates her position by death, by a move to another city, or through resignation (the latter being extremely rare), admission into the Links is extremely competitive. Unlike sororities, members do not join until well after they have completed college or graduate school. Most join in their forties and fifties and stay until they die. The admissions process is a rather confidential one that involves no formal application at the initial stages. There is no one to call if you're interested in joining.

"The only way to get in," explains a Kansas City Link, "is to get one of your friends to nominate you for membership. The main problem is that if the other members don't know you or don't think that you come from their background, your chances of being admitted are nil. The ones who get in are usually those who know at least half of the chapter's membership and who attend Links functions, like the dinners, fund-raisers, or arts shows. Of course it also helps if your social, professional, or economic background will add to the prestige of the chapter—in other words, if you are rich and important."

The Kansas City member points out that prospective candidates are sponsored by a member who knows them well enough to secretly complete an application form detailing personal, academic, and professional characteristics. A big part of the application is being able to list family members—related by blood or marriage—who are connected to what they call "Linkdom." "Once the application is passed around to the membership," she explains, "we have a breakfast or a luncheon where we sort of audition the women. They think they're attending a regular Links activity because we have many other guests there as well, but the truth of it is that we are actually looking them over, giving them a dry run, and seeing how they fit in and interact with the membership. Do they carry themselves well? Are they smart, gracious, interesting? Each of the sponsors is told not to inform the prospective member of what's actually going on, but some of these sponsors have a hard time staying quiet. You know how women can be."

After weeks of reviewing applications and looking over the candidates, the membership votes and either accepts or rejects the candidate. "If the rules are followed properly," says the member, "a 'rejected' candidate never knows she has been rejected because she actually never knew she was being considered."

During a trip I took with my parents to a recent Links convention in New Orleans, I joined two Links members in a restaurant for a conversation they were having with a young woman who was hoping to gain admission to the group (the names and some identifying details have been changed) back in her hometown of Chicago. They allowed me the opportunity to listen in and gather my own conclusions on the do's and don'ts for Links candidates. Lucille, in her late fifties, was from Atlanta, while Charlotte, in her early sixties, resided in Los Angeles. It was Charlotte's niece, Kelly, a young accountant, who was seeking the advice from the two women.

"First of all," explained Lucille, as she gestured to the thirty-two-year-old Kelly, "if you're really serious about this, you should be laying the groundwork long before anyone nominates you. You should be buying tickets to as many Links events as possible. Go to everything."

Charlotte agreed. "That's right. Get in their faces and make them get used to seeing you in their circles. When they start seeing you all the time, they'll start to think you're already a member."

"And when they realize I'm not?" asked Kelly with a lilting laugh.

"Then somebody will say, 'Well, she *should* be because she obviously fits in!'" Lucille concluded, as if the answer were obvious.

Charlotte jotted a couple of names on a sheet of green-and-white Links stationery. "I can think of some ladies that got in after getting their names out on the charity circuit," she added. "Maybe you could start hosting some fund-raisers."

Kelly rolled her eyes slightly. "I think it would take a while to build that kind of recognition."

"Then think about what you've already got working in your favor," said Lucille as she peeked over at Charlotte's list. "What do you have that distinguishes you from other candidates? Do you have a big house? A tennis court? A pool? A vacation home where you can entertain members?"

"Well, Jim and I aren't rich."

Charlotte nodded. "Yes dear, but you've got that lake house you never use."

"Perfect." Lucille underscored her point with an "okay" sign.

"And you drive a BMW," added the proud aunt.

Kelly smiled. "Actually, it's a Mercedes."

"Even better." Lucille jotted down a few words on Charlotte's paper. "Now what about your husband? Is his company the kind of place that will buy tickets at fund-raisers or take out ads in a Links journal? That counts for a lot too."

"That's right, Kelly," added Charlotte. "It's all part of the mix, girl. You've got to play that up. Work it."

"But whatever you do, don't go around telling people you hope to get in," added Lucille with a broad gesture of her opened hand. "Nobody likes to take in somebody who seems desperate. Just be cool about it and play it off like it's no big deal."

"But remember that it *is* a big deal," added the older relative with utter seriousness in her voice.

Kelly nodded at me to demonstrate her agreement.

Lucille picked up the pen again and jotted down some words. "And sweetheart, don't take this the wrong way, but you might go and get yourself some more education—like a master's degree."

"But I'm already a CPA," Kelly snapped back.

"I know you are, but a graduate degree will help your chances," added Lucille, "especially if you're going to be going after one of the older chapters. They really want to see as many degrees as possible."

"Well, she might do better with one of the newer chapters," said Charlotte. "One of those suburban chapters might be easier for you—especially since you're not an Heir-o'-Link and don't have any celebrity ties."

Because competition for Links membership can be so stiff, even the best advice doesn't guarantee admission. Sometimes chapters have become divided over a decision to admit or reject a proposed member. A few years ago, a relatively young chapter on the East Coast underwent a bruising battle over a proposed candidate's eventual rejection.

"She isn't coming into this club *sideways* if I can help it," snapped Ada Evans.

"Well, she seems pretty nice to me, and we need younger, more accomplished women," remarked one of the other club members.

Ada and several members of her chapter (the individuals' names and identifying details have been changed) created a minor scandal that sent waves through the group when they blackballed the candidate in order to get revenge on the young woman's family.

"I don't care how nice she is," explained Ada to the faction of women she had persuaded. "I have my reasons, and I'm going to keep her out." Ada was a Links member of long standing, and she knew how valued each spot was.

Normally, she maintained a neutral position regarding candidates who had been nominated. She acknowledged that she was occasionally swayed by a candidate who had done her or her family a special favor, but today she was influenced by a long-awaited opportunity to settle a score with someone. She had finally found a tenuous connection—and that connection was through Alison Jones-Roberts, a popular young candidate for admission into Ada's Links chapter.

Alison was a Phi Beta Kappa graduate of Yale University, as well as a graduate of Yale Law School and an MBA graduate of Yale's School of Management. Already a managing director at a major investment bank, Alison was smart, soft-spoken, and eager to be considered by the chapter. A native of Chicago, she had been raised in quiet wealth—the daughter of an engineer and a psychologist who had been an Atlanta debutante and a member of AKA. Alison had everything to recommend her: a good family background, a seat on the board of a world-renowned museum, a strong record of volunteerism, an annual salary in excess of $300,000, and a winning personality.

"What is it that you've got against Alison when everybody else likes her?" one of Ada's friends had asked in a private moment. "We're lucky she's not trying to join one of the other chapters," the woman added. "We should automatically let her in."

Ada had nothing against Alison or her family. Her reason for keeping the lovely and popular Alison out was to strike back at Alison's husband, Steve Roberts, who had two years before reneged on getting her own daughter tickets to an inaugural ball in Washington. Although she had met Steve only once, she knew that he had helped out on a presidential campaign. She had heard that he might have access to tickets and had hoped to get her daughter tickets to one of the balls. When the tickets failed to materialize, Ada Evans never forgave Steve. Two years later, the best revenge she could seek was to blackball the woman he eventually married. Alison Jones-Roberts was going to pay the price. Ada launched a campaign to keep Alison out of the chapter. To persuade certain members to go along with her, she promised support for their own candidates and even handed out all-expense-paid weekends to her summer home to those who assisted her in making sure that everyone who was being considered would be admitted except Alison—including women who had never attended the many Links-sponsored events that Alison had supported. Her final blow was to persuade the chapter to accept even an elderly woman who some members believed was showing signs of senility. Although there was a large division in the chapter, Ada was successful in her campaign.

Perhaps more than any other social club, the Links runs itself by a code of conduct and will even dispatch senior officials from its national office to mediate certain disputes. Unfortunately, some of these actions are never discovered by headquarters until it's too late.

"This is a wonderful organization, but we can all run the risk of getting too caught up in a high-society, clublike mind-set," says Anita Lyons Bond, a member of the St. Louis Links. "We need to keep more of our focus on improving the lives of less-advantaged people."

Even though their monthly activities take up a great deal of time and resources, many of the Links women also belong to other groups like the Girl Friends, Drifters, or Northeasterners, or to sororities like Alpha Kappa Alpha or the Deltas. Quite a number are married to men ("Connecting Links") who belong to the Boulé (Sigma Pi Phi) or the Guardsmen, and it is, of course, common for their kids ("Heirs-o'-Links") to be members of—or graduates of—Jack and Jill.

St. Louis Link Anita Lyons Bond, who is a member of the Girl Friends and the Smart Set, is married to a surgeon who belongs to the Guardsmen and the Boulé. Their kids grew up in Jack and Jill. Sharon Mackel of Shaker Heights, Ohio, belongs to the Links and Deltas. She and her husband, a physician, have kids in the Cleveland chapter of Jack and Jill. Hildred Webb, a resident of Houston, belongs to the Links and the Girl Friends. Her husband, John, is a surgeon and an active member of the Boulé. Their daughter, Kim, an attorney, debuted at the Smart Set's cotillion. The Links membership includes hundreds of families like this.

In addition to growing up in a home filled with the women who founded the Links, explains Philadelphia resident Gwynne Wright, "I was also one of the first children in Jack and Jill." Her mother, Dorothy, was Jack and Jill's first national president in 1946. "Once you are a part of one of these groups, you end up knowing many more people in all the other groups too."

As Phyllis Stevenson sat in the living room of her Tudor home in New Rochelle, she flipped through some old photographs of the prominent families that she had grown up with in New York. She stopped to look at an *Amsterdam News* photo that reported on her debut at the Girl Friends cotillion in 1952. The photograph showed her in a white floor-length ball gown surrounded by several young men all dressed in tuxedos. The article detailing the event was written by Gerri Major, the best-known chronicler of black society from the 1920s through the early 1970s.

"Mother, I'm trying to remember the names of the four boys who escorted me at my coming out," said Phyllis as she leaned underneath a light near the piano. "I can see this is Roy, Joseph, and Donald. But who's the other? Is that Sammy?"

Stevenson's mother, Anna Murphy, leaned back and briefly stared out the large picture window into a street lined with sixty-year-old oak trees. "Yes, Phyllis, that's Sammy—he's a radiologist in Boston now, but you have a short memory," she said with a laugh. "You actually had *six* escorts. Two of the boys got cropped out of the picture when the newspaper published the photo." Anna went on to name the other boys and the respective details about them. One had just graduated from Fieldston at the time of the cotillion, one was now an attorney in Philadelphia, four of them were sons of fellow Girl Friends. The yellowing article was just one of hundreds that focused on the activities and families of one of the most stylish and elite organizations established for black women.

This mother and daughter could only be talking about the biggest cotillion tradition of old black society: the Girl Friends' Ball of Roses. For years, in many cities, these Ball of Roses cotillions reaffirmed the importance of the organization along with the families of the girls they chose to debut.

What the Links have in numbers and political clout, the Girl Friends can match in social ties and history. Considerably smaller and two decades older than the Links, the Girl Friends is a league of stylish black women who are accomplished, well-connected, and "Establishment." A high percentage of them are married to physicians.

"We're a lot smaller than the Links because we're a lot more selective than they are," says a Chicago Girl Friends member who notes that the Links organization is no longer as intimate and social as it used to be. "They are all about money, power, and ambition these days. The Links used to be more focused on family background and paid more attention to social status, but now they'll take anybody in if the woman has got money and clout. The Girl Friends aren't into that kind of power."

While it is true that the Girl Friends chapters are less likely than the Links to include large numbers of "power women," many disagree that the Girl Friends is low-key. Some say that the tight-knit group's claim and emphasis on old-guard families are unparalleled.

"Those Girl Friends are almost *too* intimate," says a Los Angeles attorney who was rejected by the group, even with an Ivy League résumé. "Everybody knows everybody's business because all the women in the group seem to be the daughters of members, the granddaughters of members, the nieces of members. It almost feels inbred. While the Links are into power and professional clout, these women are over-the-top with their lineage, long hair, and social status. Even their directory seems clubby and claustrophobic."

While it's hard to discern which criticisms are legitimate, particularly when they are offered by rejected candidates, it is true that the Girl Friends membership directory offers a surprising amount of personal information on each of its thirteen hundred members—perhaps supplying more information than any other club directory. Whereas the Links directory offers the name, address, phone, and spouse's name of each member and the Drifters add the member's birthday, the Girl Friends directory not only presents the name, address, phone, and spouse's name but also offers a photograph, the occupation of both the member and the spouse, the names of the members' children, and information on whether any of the members' daughters are Girl Friends. Many of the members also list the professions of their children.

"Why does the whole organization need to know what my husband and children do for a living?" asks the Los Angeles attorney, who admits she was surprised when she was passed over for membership into the group. "They might as well list where our parents went to college. How is any of this relevant?"

One response to that question is that the families represented by the women in the Girl Friends all place a great value on tradition and position. Others say that such details allow people to understand the historic ties between family members and friends. There is no better representation of this tradition than Phyllis Murphy Stevenson and her mother, Anna Small Murphy. Phyllis was national president of the Girl Friends from 1980 to 1982, and her mother was one of the charter members when the group was founded in Manhattan in 1927. Anna eventually served as president from 1952 to 1954.

Although they are the only mother-daughter team in the forty-chapter organization to hold the title of national president, they are quite similar to the other Girl Friends when one looks at their ties to other institutions of the black elite: summers in Sag Harbor since the early 1940s, when Phyllis was a child, membership in the Deltas, education at Fisk University, childhood ballet lessons with Ada Fisher Jones in Harlem, family membership in the Comus, and friendships with some of the most prominent black families in America.

"When we first decided to create the Girl Friends group," explains Murphy as she relaxes in a living room filled with African art, "most of us lived in Harlem and were just graduating from college or starting graduate school. It was 1927 and I wasn't even married yet. Several of us had grown up together, attended the same schools or summer camp in upstate New York, or taken dance or piano lessons together."

The intent was to establish a club of young women with similar backgrounds and interests who could meet occasionally for social or intellectual purposes. Since they were all quite young, they sought out an adult chaperone to accompany them when they went to public places in New York.

"We asked Bessie Beardon to serve as our chaperone," says Murphy as she points to a piece of artwork hanging near Phyllis's fireplace. From a mere glance at the colorful collage, it quickly becomes obvious to me that "Bessie" is the mother of the renowned artist Romare Beardon. It also soon becomes obvious that Anna and Phyllis's world was populated by many prominent individuals—and some far more famous than the best-known artist to come out of Harlem. I quickly discover that Anna and Phyllis are accustomed to being around people who understand their shorthand.

"One of my fellow Girl Friends was also responsible for giving that piano to Phyllis," explains Murphy with pride. "You know Mildred Johnson. It was Rosamond's." Deciphering the shorthand, I realized that the late-nineteenth-century piano had been owned by composer J. Rosamond Johnson, brother of James Weldon Johnson, who used it to compose the 1901 song "Lift Every Voice and Sing"—now often referred to as the Negro Anthem and performed at black churches, colleges, and organizational gatherings.

After quiet prodding, Anna reluctantly explains to this outsider, "Rosamond Johnson was the father of my closest friend, Mildred Johnson Edwards, and Phyllis grew up as best friends with her daughter, Melanie."

"And I was escorted by Melanie's brother, Donald, for my debut," adds Phyllis as she points him out in a photo that was taken while they were still in high school during the early 1950s.

As the child of a Girl Friend, Phyllis grew up with a network of influential contacts in every major city. "I grew up with two dozen 'aunts' in every city—New York, Washington, Chicago, Atlanta," Phyllis jokes as she thinks about her mother's network of friends from the different Girl Friends chapters. "When my father and mother brought me to the annual conclaves in different cities, it was almost like a family reunion. The mothers, along with their husbands and children, were like an extended family."

Founded in 1927 in New York City, the Girl Friends, Inc., is composed of forty chapters and includes approximately thirteen hundred women in such cities as Washington, Chicago, Pittsburgh, Richmond, Boston, Louisville, Los Angeles, and Atlanta. The club focuses on philanthropic, social, and cultural activities, raising money for local and national charities as well as sponsoring programs for its own members. It was founded by a group of young women who were between high school and college, and who were first led by Eunice Shreeves, who was in college at Cheyney University. Shreeves initially pulled together a group of friends who were living in New York or away at schools like Howard University. Among the group were Murphy and several other women who had come from similar backgrounds. "One of my best friends at the time was Rae Olley Dudley," explains Murphy, "and she along with me and seven others were asked to join."

Like the Links, the members of the Girl Friends come from some of the oldest and best-respected families in their communities. A quick conversation with any of them reveals that they and their husbands are all professionals. An overwhelming number of the members are married to physicians, dentists, and attorneys, with a few college presidents, government officials, big city mayors, and a U.S. Supreme Court justice thrown in for good measure.

"He oughtta get Ersa and Paquita so he can get the song and the political ties," says Murphy as she returns to the verbal shorthand that flies between mother and daughter. Like many women in this crowd, they are reluctant to talk too long to people who don't understand the people or the history of their social world, so I avoid asking questions and exhibiting my lack of information.

But silence doesn't cut it with Murphy and Stevenson.

The mother looks at me. "You *do* know who Ersa and Paquita are?"

I stare back with a blank expression. "Ersa, yes, Ersa—"

"Poston," Murphy adds pointedly.

The name sounds familiar. I've heard my parents or some of their friends mention it. But the only person that comes to mind is a journalist—Ted Poston. "Oh, of course," I respond. "Ted Poston's wife." I recalled that Ted had been the first black journalist ever to be employed by a white daily paper in New York.

"Ted Poston's wife?" Murphy asked. "Some would say *Ted* was Ersa Poston's *husband*. She was our national president from 1982 to 1984. And she was a bigwig in city and state government."

About a beat too late, I remember that Ersa was a Rockefeller appointee and one of the heads of the U.S. Civil Service Commission.

"And Paquita Attaway is in D.C. too," says Stevenson. "She was president from 1992 to 1994."

After leaving Murphy and Stevenson, I later learn that Paquita Attaway's pedigree in the Girl Friends gives the mother-daughter team a run for their money. "My mother, Ethel Ramos Harris, graduated from the New England Conservatory of Music in 1927 and was a charter member of the Pittsburgh Girl Friends where she composed the Girl Friends Hymn," explains Attaway, a retired teacher who was presented at the AKA cotillion in Pittsburgh before earning degrees from Boston University and the University of Pittsburgh. Like many of the Girl Friends, Attaway started life off right. Her father, Chester Harris, was a surgeon who graduated from Tufts Medical School. He was also a member of the Boulé. "My sister, Yolanda, and I were introduced to many children while we grew up in Jack and Jill and attended Camp Atwater in Massachusetts," says Attaway as she recalls the summer camp that was favored by the black elite. Like Phyllis, her sister Yolanda was presented at the New York Girl Friends Ball of Roses Cotillion.

And again like Phyllis, Attaway summers in an old-guard community. "My husband and I bought a house in Martha's Vineyard after my parents had spent years on the Vineyard during the Shearer Cottage days," she explains. Her husband, John, is a member of the Boulé and holds a doctorate in business administration from George Washington University.

With their credentials, Ersa Poston and Paquita Attaway have raised the stakes for the women who are clamoring to join the Girl Friends' Washington chapter.

But, of course, the New York Girl Friends are well aware of their own status. They and their husbands are well-known professionals. Their membership has included the late Vivian "Buster" Marshall, who was married to Thurgood Marshall, the first black U.S. Supreme Court justice. "As a matter of fact," says Murphy, "our national incorporation papers were signed by Thurgood." The New York chapter's history of members also includes Manhattan resident Rae Olley Dudley, whose husband, Edward Dudley Sr., was U.S. ambassador to Liberia under President Truman before joining New York state's supreme court; Jacqueline Denison Russell, whose husband, Harvey, was the first black vice president of a *Fortune* 500 company when he became a leading Pepsi executive; and Laura Mitchell Holland, whose husband, Jerome, was president of Delaware State and Hampton Institute before becoming ambassador to Sweden under President Nixon. Among the newer members is Carolyn Wright Lewis, wife of *Essence* magazine publisher Ed Lewis.

In Detroit, one finds Judge Trudy DunCombe Archer, wife of Detroit mayor Dennis Archer. In New Orleans, there is Dr. Andrea Green Jefferson, wife of U.S. congressman William Jefferson. In Memphis, there is Frances Hooks, wife of former NAACP head Benjamin Hooks.

Erma Lee Laws started the Memphis chapter of the group several years after her friend, Mary Agnes Davis of Detroit, had helped charter the Detroit chapter. Davis, whose husband, Ed, became famous as one of the country's first black auto dealers, had successfully launched a chapter of the Co-Ettes, a smaller elite group. Mary Agnes, an important Detroit socialite and fund-raiser in the 1950s, 1960s, and 1970s, enjoyed the Girl Friends so much that she encouraged Laws to bring the organization to Memphis. "I grew up reading about the Girl Friends in Toki Schalk Johnson's society column of the *Pittsburgh Courier*," says Laws, a former columnist for the Memphis *Tri-State Defender*, "but it was meeting socialites like Mary Agnes and others that got me to help launch my own chapter."

Not to be outdone by their husbands, many of the Girl Friends are educators, attorneys, physicians, professors, high-profile fund-raisers, and government officials. One example is former Clinton cabinet member Hazel O'Leary, who served as U.S. secretary of energy. She became a member of the Minneapolis chapter many years after her mother Mattie Ross Reid helped establish the Newport News, Virginia, chapter in 1938. United States congresswoman Eddie Bernice Johnson belongs to the Dallas chapter.

Unless you are the daughter of a member, getting into the Girl Friends is even more difficult than getting into the Links. Since the chapters are generally smaller (usually twenty to thirty members) than Links chapters, and since no city has more than one chapter, it really does require a wave of resignations or deaths before a woman's chances for acceptance improve. There is no application or membership office to call. You have to be sponsored by at least two members who know you well, and at least two-thirds of the members must vote for your admission. If a woman is turned down, she can never be proposed by that chapter again. "I had a friend who was so devastated by her rejection in the town where she lived that she and her husband bought a vacation home in South Carolina so that she could be reconsidered in another city," says an Atlanta Girl Friend, who sees how candidates have jockeyed for admission. "I felt terrible when she got rejected again. After two years, they sold the house, moved to Washington, and I think she just ended up joining the Hillbillies or something. It was so sad."

In addition to their charity work, for many years the Girl Friends were known for their debutante cotillions, where high-school seniors were presented at the local downtown hotels or country clubs accompanied by their fathers and young escorts dressed in black tie. The Ball of Roses cotillion was adopted by the other chapters but originated with the New York chapter, which holds it around Christmas at the Waldorf-Astoria on Park Avenue. Each debutante is selected through a rigorous process that compares applicants' college plans, career goals, and accomplishments.

While some of the chapters have abandoned cotillions because of a modern tendency to move away from elitist activities, the group has not stopped publishing its annual journal, the *Chatterbox*, which looks and reads like a black version of *Town & Country*. Each chapter's activities of the year are highlighted and illustrated with photographs from formal parties, outdoor picnics, and other gatherings. Members and their families are captured at college graduations, battleship launchings, judicial swearing-in ceremonies, political campaign speeches, White House dinners, and other interesting events. New York artist Romare Beardon designed the cover of the May 1952 *Chatterbox* as a tribute to his mother's ties to the group.

Another popular women's group is The Drifters. Smaller and less driven by family lineage than the Girl Friends, the Drifters consists of thirty chapters, with groups kept to as few as twenty members. With a focus on charitable work like sponsoring scholarships for students going away to college, the Drifters was begun in 1954 in Waco, Texas, and Chicago. Because it represents a generation of founders who were less likely to be housewives, the five-hundred-member group has a high percentage of Ph.D.'s, attorneys, and other professionals.

I first came into contact with the Drifters crowd as a child through my godmother, Dr. Mirian Calhoun Hinds, who has been a member since 1963 and serves as president of the New York chapter. She says that the group is less focused on a husband's status because the members are extremely accomplished in their own right. A former dean at City College with a doctorate from New York University, Hinds cites one of the newest chapters as an example of accomplished women. "Our Memphis chapter is a perfect example of how our group has established itself with intelligent and civic-minded women. Even though it's only a dozen women, eight of them are attorneys who serve on local boards and civic activities."

A former department chairman at City College of New York, Hinds grew up in Orangeburg, South Carolina, the daughter of faculty members at South Carolina State University. "Although I became a member of the Deltas when I got to college," she explains, "I wanted to join an organization of established black women who would help me in my transition from the South to the North. One of my cousins was in the New York chapter, and she told me about the Drifters and the work they were doing in the city. I wasn't looking for a group of society women."

Unlike the Girl Friends, the Drifters are inclined to organize frequent extravagant vacation trips for their members and friends—going to such places as Haiti, St. Martin, Greece, and Africa. Both groups give college scholarships to graduating high school students. Hinds says her chapter has given scholarships and monthly stipends to students who have gone away to Cornell, Spelman, Rutgers, Howard, Fisk, American University in Paris, and other colleges.

Each August, the thirty chapters come together for a national conference and announce the college scholarships they are awarding to the graduating high school students in their respective cities. In addition to their scholarship program, the Drifters have also maintained a non-interest-bearing loan fund for students at targeted colleges since 1969. Among the schools targeted are Bennett College, LeMoyne-Owen

College, Bethune-Cookman College, Howard University School of Nursing, and Virginia Union University.

There's also the Northeasterners, which was founded in the early 1930s. "We have twelve chapters, and most of them include no more than twenty women," says Bebe Drew Price, a native Washingtonian who joined the national group a decade ago.

Cathy Lightbourne Connors, one of the most influential society columnists to write for the New York *Amsterdam News,* has been a big supporter of the Northeasterners, serving as the New York chapter's president. Connors's name can also be found on the membership lists of the ViVants and the Hillbillies, in addition to her long-standing ties to the Links and Alpha Kappa Alpha, which she joined in college.

Connors, whose deceased husband was John Connors, president and founder of American Express Publishing Group, the company that produced *Travel & Leisure, Food & Wine,* and other magazines, believes that black social groups like the Northeasterners serve a valuable role in the country. "Too much credence is given to the black urban project subculture," explains the petite woman as she sits demurely in a blue and ivory linen suit. "When upper-class and middle-class blacks are kept hidden from the community, we fail to produce role models and mentors for young blacks who need to know that there are people like them who still have hope and who have not given up."

But Connors adds that the social groups must recognize that they have to get involved and interact with the people and the local groups they are supposed to be mentoring and helping. In directing the activities of her Northeasterners chapter, she uses certain role models to guide her. One such model is the Logan family, one of the most socially prominent couples in Harlem during the 1940s and 1950s. "Dr. Arthur Logan and his wife, Marian, were affluent and well-connected in the black and white social world," says Connors with conviction in her voice, "and they got their groups and their friends to support important issues. I remember when they gave fund-raisers for Dr. Martin Luther King when Dr. King was a controversial figure. Because of their social credentials, they went with Dr. King to assist in his introduction to Bobby Kennedy and his mother, Rose. This is how we have to use the clout of our social groups—to raise money and to raise awareness about issues in our communities."

When it comes to the social activities, Connors points out, "unlike the Links and other organizations, the Northeasterners do not sell tickets to our formal dances. It's a tight circle of invitations, and the members pay for everything. We have a philosophy that is similar to that of the Guardsmen. We pay for all our events and all of our guests." Originally founded in 1921 with the name The Gay Northeasterners—"when the term meant 'happy' or 'free,'" chuckles Connors—the Northeasterners was started in New Haven, Connecticut, by three sisters. Longtime Harlem resident and Northeasterner member Esther James says the group grew slowly and formed chapters in Pittsburgh, Philadelphia, Washington, Chicago, Indianapolis, and other cities where blacks were concentrated.

Like the Drifters and the Northeasterners, the Smart Set is a moderate-sized women's group with a fairly prestigious roster of female members. "Unlike many of the other groups that were started by adult women, our organization was begun by a group of schoolgirls from the Washington area who were going off to college at Howard," says Ruth White, a Philadelphian who is a longtime member of the National Smart Set, an organization of approximately five hundred women that was founded in 1937.

"Although we perform community service," says Margaret Mayfield Rivers of Memphis, "we are much more like the Girl Friends than we are like the Links. A lot of our focus is on the social events that we sponsor. We generally have three major black-tie social events each year, as well as an outdoor summer activity at someone's home." In addition to spending occasional weekends at a vacation home in the Ozarks, Rivers recalls past years where she entertained Smart Set members in her large home, which is surrounded by a two-and-a-half-acre lawn. She has been in the group for more than thirty years and remembers other early members in the city's chapter like Harriet Ish Walker, the deceased wife of Universal Life Insurance chairman Maceo Walker.

Like her fellow socially prominent Smart Set members, Rivers has all of the right credentials: The daughter of Dr. U. L. Mayfield of Fort Worth, Texas, she pledged AKA at Fisk, then married into a prominent Memphis family after meeting Meharry dentist Frederick Rivers, who was the son of a successful doctor. She is also a member of the city's Sophisticates bridge club and the original Memphis Links chapter. "Like the husbands of many of my fellow Smart Set and Links members, my husband was an Alpha and a member of the Boulé," explains Rivers. "In fact, the overlapping was so common that we began to schedule our meetings to coincide with the same evening as the men's meetings so we would all be out on the same night, and no one would be left at home alone."

Philadelphia's Ruth White comments that since the Smart Set has not grown to the size of the mammoth Links group, the members remain intimate and can keep up with each other's activities through the organization's sometimes fun and gossipy *Smart Set Talk* magazine. "They report on contributions we give to the NAACP as well as the more personal events that happen in our lives. The group is still small enough that we know each other and try to keep up on family stories."

One group that had the potential of being an important national group but resisted spreading beyond its charter city was the Hillbillies. "They just dug their heels in and refused to let people outside their little circle become a part of their group," says a former New Yorker who remembers getting rejected by the social group in the 1940s. Created by a group of well-to-do young women who lived in Sugar Hill—the most exclusive and more northern section of Harlem—in the 1930s, the Hillbillies never became a national organization.

"There was an attempt to start a small chapter in Washington, and they had reciprocal status with the group in Manhattan," explains Anna Small Murphy, one of the charter members of the New York chapter, "but it was never intended to be more than a Harlem-based group." Murphy, who is no longer active in the Hillbillies, went on to be a big name in the Girl Friends as both a charter member and a national president in the 1950s.

Founded in 1956 by a group of women from Philadelphia, Baltimore, Newport News, and Washington, D.C., the Continental Societies, Inc., is another popular women's social club. With chapters in most major American cities and Bermuda, the Continentals—as they are known—provide programs and raise money for young people, families, and institutions that focus on health, employment, recreation, and education. Whether by design or by accident, the group has remained relatively small, adding only six chapters in the last sixteen years, bringing the total to thirty-five chapters. Manhattan resident Dee Matthews serves as the group's national president.

Another rather small group is the Carats, first formed in the mid-1970s. Its three hundred members belong to chapters in thirteen cities including Cleveland, New York, Detroit, Chattanooga, Columbus, and Macon, Georgia. In addition to hosting social functions, the group gives scholarships to college students.

* * *

Some say that the elite women's social clubs are beginning to face new challenges as the number of women in the black upper class increases. Some are watching the good work of the Coalition of One Hundred Black Women and trying to figure out if professional groups like that will satisfy the new, young elite. I've heard the discussions among my family and friends. They argue that the old-guard families will start getting displaced by women with new money and new families that have few connections to what was familiar to the older founders who focused on lineage and black issues.

"It's true that the old families aren't the only ones with money these days," says a Florida Link, "but you'll never see them letting in nouveau wives of athletes and entertainers. They'll hold the line on that kind of member."

Others insist that the growing phenomenon of integration and the increase of interracial coupling will create challenges to the agenda of some of the groups, steering them away from black-focused charity giving. "I can't imagine we'd ever move away from our agenda of supporting black causes," adds the Floridian, "but it's true that an emphasis on one's black lineage may get diluted by these rich young people who live a more integrated life than the rest of us. And of course some of these new biracial kids and adults don't care about blacks at all. But we'll survive them too."

"As educated professional black women, our most important duty is to serve as role models, mentors, and volunteers in our communities," says Evelyn Reid Syphax, a member of the Links, Girl Friends, AKAs, and Coalition of One Hundred Black Women. Although the Arlington, Virginia, resident's famous and well-to-do family can trace its ancestors back to American first lady Martha Washington's grandson, she says family background and social status are not the reasons these women are working together in these organizations. "The kinds of programs and activities that we're involved in should be performed by people who have the resources and the ability to pull them together. If I'm fortunate enough to afford to do these things, it's my duty. These other talented black women feel the same way. Social status doesn't have to be a part of our picture."

While Syphax and other members of old-guard families are convinced that the next generation of young women will join these groups with the same fervor, I am not so sure of it. It is true that the Links, the largest and most aggressive of the black women's groups, has maintained its popularity, but the other groups are challenged by the fact that more women among the black elite are busy with high-profile positions that make it impossible for them to do the things that their nonworking mothers did for these groups a generation ago. Many of my black female friends are saying that they want to join these groups and participate in the social and civic activities, but they are postponing it until they reach their forties. For now, these women are saying, "Let me write a check for a thousand dollars instead of asking me to volunteer my time." For the most part, these groups will wait around for them, but in order to do this, they are going to see the average age of their members increasing. Unless they are willing to wait or accommodate the working professional woman, these elite groups will be populated primarily by retirees and by wealthy socialites who can make the commitment. Since it appears that they are unwilling to change their standards and widen the pool to include less affluent and less accomplished women, these organizations will have to grit their teeth and hold out for the best that the sisterhood has to offer. Unfortunately for them, the "best" may not be as substantial a number as it once was.

The Boulé, the Guardsmen, and Other Groups for Elite Black Men

"If your Boulé friends will back you, people will know you're for real," said my uncle when I told him that I was considering running for political office.

"The Boulé?"

My uncle looked at me in astonishment. "That shouldn't surprise you. Where else are you going to find better mentors? These are the kind of advisers you need."

Although reluctant, within a half hour I was on the phone with some of the most influential men in the country—all of them black, all of them accomplished in their professions, all of them members of a men's group into which I had been inducted three years earlier. The fact that this national organization included virtually every black mayor, congressman, banker, or millionaire was a reality that I should not have forgotten. After all, this was the organization that had already introduced me to E. Thomas Williams.

Many people would say that E. Thomas Williams Jr.—known as E.T. among friends and colleagues—is one of the reasons that there continues to be a demand for elite black men's organizations. Living with a foot in both the black and the white worlds, Williams is a prototype for the integrated black professional who is deeply rooted in old-guard black society yet comfortable at socializing and building a successful career among affluent white businessmen and philanthropists.

His résumé is dazzling. It has included such board memberships as the Museum of Modern Art, the Central Park Conservancy, the Brooklyn Museum, Boys Harbor, the Cathedral Church of St. John the Divine, the Visiting Committee of the Harvard Graduate School of Design, the Eastville Historical Society in Sag Harbor, and the Fiduciary Trust Company. His social memberships include the exclusive University Club and the River Club. In the 1960s, he volunteered in the Peace Corps, where he roomed with Senator Paul Tsongas, and later held senior management positions at the Chase Manhattan Bank before becoming a real estate entrepreneur. For forty years, he has been an important player in the black Sag Harbor community of the Hamptons and has held board memberships with the NAACP Legal Defense Fund, the Schomburg Center library in Harlem, and Atlanta University. As an avid collector of artwork by Romare Beardon and Jacob Lawrence, he received the Patron of the Arts award from the Studio Museum of Harlem.

E.T.'s wife, Auldlyn, as the daughter of a well-regarded Baltimore surgeon, was presented at the Me-De-So Cotillion, then graduated from Westover, Bennington, and Fisk. She was the first black to join Baltimore's Junior League. Their children are graduates of Spence, Andover, and Harvard.

E.T. has ties to the most important men's groups in the world of black society: Sigma Pi Phi Boulé, the Comus Social Club, the Reveille Club, and One Hundred Black Men.

"The reason I joined groups like the Boulé, One Hundred Black Men, and the Comus Club," explains Williams while reclining in a wicker chair at his summer home in a family compound in Sag Harbor, "is that I feel it's important for black people to find places where we can meet, network, and socialize with people that understand our experiences and our concerns."

When you look at the résumés of most men among the black elite, you find fewer different social organizations than are represented by the many women who are active in the social and civic world. Because of this, some members of the black elite insist that women are the primary social organizers. But even if there are not as many different organized men's groups, the number of prominent and accomplished men involved in these select, very elite organizations belies that claim.

The groups discussed here are organizations which are fraternal, professional, or social, but which have no ties to Greek-letter college fraternities. Some of the groups are as informal as the Rainbow Yacht Club, a group of black yacht owners, most of whom are physicians in New York and surrounding areas. The best-known groups are formally incorporated institutions with local chapters throughout the country. The most prestigious of all these groups is the Boulé, a fraternal organization founded in 1904, but many say that all of these groups found their roots in a group that my father and many other family members have participated in: the oldest surviving black fraternal men's organization, the Prince Hall Masons.

Founded in 1776 by Prince Hall, a black Methodist minister who had been born on the island of Barbados and emigrated to Massachusetts, the Masons spun off from an already-established white organization for men. Before later serving a nine-month term in the Revolutionary War, Hall and fourteen fellow blacks living in Boston joined the British Army Lodge of the original Mason organization. When he and the other blacks felt they were not received as well as they had hoped, he asked permission to start an all-black lodge of the same organization. After repeated requests, the British Grand Lodge gave him permission to begin what became a "black version" of the organization. As head of the new organization, he assumed the Mason title, "Most Worshipful Master" of the organization. By the late 1790s, Hall was permitted to form affiliate lodges in Philadelphia and Providence.

While the group never focused on attracting affluent men as members, it has always emphasized the goals of collegiality and intellect. As the oldest ongoing black fraternal group in the United States, it is often seen as a model for the black men's groups that followed. Intensely secretive, the Masons remain a large group with lodges in most major cities and metropolitan areas. When my father became the head of his lodge in Harlem, he found himself instantly linked to a network of black members—they refer to the group as "the Craft"—who are deeply involved in scholarship as well as community projects throughout the country.

Founded in 1904, the Boulé was the first elite national black men's club. Although its official name, Sigma Pi Phi, suggests that it might be just another Greek-letter fraternity, nothing could be farther from the truth.

The Boulé, as it is better known, is older than all of the black college fraternities, and as it is the quintessential organization for professional black men, members are not even considered until they are well beyond college and graduate school. It is considered by many *the* elite men's club, and its membership has included the most accomplished, affluent, and influential black men in every city for the last ninety years.

Distinguishing itself from other men's clubs that might operate only on a local level, the Boulé selects its national membership strictly on the basis of professional accomplishments rather than popularity among a certain local social group. Conducting all of their official activities and social gatherings in black-tie attire with formal ceremonies, Boulé members are men who are attracted to the fraternity because of its intellectual discussions and its interest in promoting scholarship among a group of accomplished black professional men. Although the group no longer conducts its activities in secret settings, as it did during the first sixty years of its existence, it still has strict rules regarding which Boulé events can include nonmembers and which events will exclude unmarried sons and daughters of members. There are even rules regarding the burial or cremation of a deceased member of the Boulé.

"In many ways," a Washington member explained while recalling some of the members that he had met during his participation over the last thirty-five years, "the Boulé is a more select, male version of the Links group. We have attorneys, activists, intellectuals, entrepreneurs, socialites, executives, physicians, ambassadors, judges, and politicians—all of them highly accomplished and highly educated."

The group began when Dr. Henry Minton asked five colleagues—all doctors working in Philadelphia—to join him in forming a social organization that would, as he explained, "bring together a selected group of men with a minimum degree of superior education and culture—men who were congenial, tolerant, and hospitable." Minton, a graduate of Phillips Exeter, the Philadelphia College of Pharmacy, and Jefferson Medical School, was a prototype of the kind of black men that the Boulé would both seek and attract. Not only had he been well educated and married into the powerful Wormley family of Washington, but he was also an achiever who was dedicated to improving the lives of other blacks.

"Dr. Minton was responsible for opening the first two hospitals for blacks in the city—both Douglass and Mercy," says Philadelphia-born Boyd Carney Johnson, who belongs to the Boulé in Westchester County, New York. "The Mintons who came after him were all dedicated to confronting racial discrimination in Philadelphia. When I was growing up, you heard about them constantly because they were opening doors that had been closed to black people previously."

An intensely intellectual group of men, these six doctors chose to model their group after ancient Greek organizations, in terms of structure and nomenclature. The word "boulé" was used because it meant "council of noblemen" or "senate." The members were each referred to as "archons," with the president known as "sire archon" and other officers taking Greek titles such as Grammateus, Thesauristes, Rhetoricus, and so forth.

When I joined the group at the unusually young age of thirty-three, I remember it being an exciting but unsettling experience—not because members made it difficult, but because of the pressure I felt after reading the group's history and hearing friends and family members speak of its prominence. Several weeks before my initiation into the chapter that had nominated me, I got hold of the 450-page Boulé manual and history book. It didn't exactly prepare me for what I was in for.

"So, Mr. Du Bois," said a gravelly voice from behind me as I walked up the steep front steps of the Williams Club—a brownstone on the East Side of New York.

I turned to him and shook my head, "Oh, no, I'm—"

"I know who you are, Archon Graham," the older man responded while extending his hand, complimenting me on my new Calvin Klein tuxedo and introducing himself as Dr. somebody. I was too uneasy to remember his name. I only noticed that he appeared to be a few years older than my parents.

As we walked into the club, we were greeted by a white maître d' who directed us toward an ancient elevator that seemed only large enough to carry three average-sized people. Nevertheless I was squeezed in with four other formally dressed men aged somewhere between fifty-five and seventy years old. "So, how are your parents?"

"Very well, thank you." These pleasant men looked familiar, but my mind was drawing a blank on each one of them. I had no idea what their names were yet felt encouraged that they obviously knew me. A couple were people I'd seen in the New York newspapers—the other two were faces I had met at various times during my childhood. I somehow got the sense that they knew I didn't know their names—and that they enjoyed putting me at that disadvantage.

The four men indicated that they knew my résumé quite well—even my brother's, as one of them remarked that he had referred a patient to Richard's practice. As the tiny elevator car screeched slowly up its path, I could think of nothing intelligent to say. All eight eyes were on me, but all I could do was stare alternately at the starched white collars peeking above these men's conservatively cut tuxedo jackets. I pasted a smile on my face and immediately wondered if my Calvin Klein tux made me seem too nouveau for this group.

"Say something conservative," I told myself. "No, something clever. No—say something racial. No, tell them how grateful you are to be here—no, pretend it's no big deal. No, maybe you're not even admitted yet. Jesus Christ, why doesn't somebody break the ice here?"

The men continued to stare.

And it was getting hotter.

"When is this elevator going to stop? This building is only four stories!" I could hear myself screaming inside, shouting every thought between my ears. As the elevator passed a second floor, I wondered if this was part of the test. Yeah, maybe getting through a three-minute elevator ride in the Williams Club with the Waspiest black men in New York City was the test. Just don't fuck this up, Larry. Just don't fuck it up.

"This is possibly slower than the elevator at the Yale Club," I finally said, shooting my wad at this reticent group.

One of the men cracked a smile without showing his teeth. At least it looked like a smile.

"We'll be going over to the Yale Club with the wives later tonight," said one with glasses. "I hope you and your wife will be joining us."

The door squeaked open and as we were led single file down a hallway, I wondered if the Yale Club invitation was an actual offer—or if it was conditional. "I hope you will be joining us? I hope you will join us." There's a difference between those two sentences, I thought. I'm still not accepted yet!

At the end of the hall, the men went into one room, and then I was directed to wait in a narrow holding room at the other end. That's when I learned that I was not the only new recruit.

I was quickly told by a young member—a Morgan Stanley banker—that three of us were being "brought up" that evening at a private session to start momentarily.

What do you mean, *brought up*? I asked myself. What does *brought up* mean? Is that like *being proposed,* or *being initiated*? There's a big difference.

The tall, slender banker also indicated that there was to be a formal dinner with speeches later that evening with the entire membership and all of our spouses in a downstairs dining room.

Still not sure if I was being initiated or merely voted on, I introduced myself to the other two guys who stood silently against the hallway wall. They were both several years older and seemed amazingly calm. One of them looked black and the other looked white—completely white. The black-looking one was a managing director at a Wall Street investment bank (three years later he was to be profiled with my wife in a *Fortune* cover story on high-powered blacks in corporate America), and the other was an attorney at a major media company. We had all gone to Harvard Law School, and, in fact, there were five Harvard degrees between us.

"So, are we being voted on now, or was this decided before?" I asked the two, in a tone that suggested I wasn't that worried.

"I think we've already been accepted," said the media lawyer with an air of quiet confidence.

Before I could express my gratitude at receiving this information, the door opened and each of our sponsors escorted us into a room of about twenty-two members. A few of them—faces that were familiar from my childhood—gave me a strange handshake. And then the door closed for an initiation ceremony that was to remain confidential.

While the 450-page Boulé history book is distributed only to members, some of what it reveals is information that is already open to the public. But most of the information is held as confidential. "Years ago, the Boulé was an extremely secretive organization," says Grand Sire Archon Anthony Hall Jr., a Houston attorney who presides over the national organization, "and during the period when W. E. B. Du Bois was a member, they worked hard to maintain the club's code of secrecy and to avoid attention or scrutiny."

The gold insignia pins that Boulé members wear on the left sides of their crisp white shirts, for example, are given out at a secret ceremony. But the one-inch triangle with Greek letters on it is visible to any guest who attends a Boulé formal or a Boulé funeral where a deceased member might be buried or cremated with it in accordance to the group's constitution. Nonmember guests have a general sense that members greet one another with an unusual handshake that is not obvious to others or that they display unusual hand gestures when conducting private meetings, but all of this has remained confidential among the membership. The decision to keep these traditions private was made by the small and insular first chapter of the group—long before the Boulé became a national organization.

Because the original Philadelphia chapter was very slow to expand its membership, and because the following chapters in Chicago, Baltimore, Memphis, and Washington were reluctant to open their membership to men who worked outside of the higher-education, medical, dental, and legal professions, the group was quickly accused of being elitist. Whether it was because they didn't want to share their secret discussions or just that they didn't want to dilute their significance by welcoming lesser-known candidates, the members were unapologetic about their decision to exclude. Both then and now.

"People might use the term 'elitist,' but it is accurate to say that this was, and is, an elite group of people," explains Boulé member Harold Doley, who was the first black to own an individual seat on the New York Stock Exchange. "It's a little absurd for black people to apologize when they are educated, accomplished, and successful, and choose to belong to organizations populated with other blacks like them."

Like Doley, most of the members in the Boulé history have been contributors to, and champions of, black causes. But they do not see their membership in the Boulé as their primary means of contributing to charities or institutions. Doley, who purchased and restored the twenty-thousand-square-foot mansion of late-nineteenth-century black entrepreneur Madam C. J. Walker, has contributed generously to, and sat on the boards of, numerous black colleges, including Shaw University and Clark-Atlanta University.

"While all of us have community activities that we support," says Grand Sire Archon Hall, "Sigma Pi Phi was not created with a social action agenda. It was not designed with community service projects in mind. And this is one reason why the public might know a lot more about the Links or the fraternities than they know about us. My wife is a Link, and she and her organization are accustomed to hosting public functions like fund-raisers, cotillions, and civic projects. That's not us."

The membership, which now numbers around thirty-seven hundred, is organized into 105 local chapters, called "Subordinate Boulés," representing all of the major cities and metropolitan areas where the black elite can be found. In addition to including such esteemed people as the scholar W. E. B. Du Bois, who earned a Harvard Ph.D. in 1895 and taught at Atlanta University and the University of Pennsylvania, the Boulé membership has also included Dr. Martin Luther King Jr.; historian Carter Woodson, who earned a Harvard Ph.D. and served as dean at

Howard; Harlem physician Dr. Louis T. Wright; Howard University president James Nabrit; philosopher and Rhodes Scholar Alain Locke, who earned a Harvard Ph.D. in 1918 and chaired Howard's philosophy department; and numerous judges and college presidents.

Among the more recent members are virtually every current or former black mayor of a major city including New York's David Dinkins, Atlanta's Andrew Young and Maynard Jackson, Memphis's Willie Herenton, Baltimore's Kurt Schmoke, Detroit's Dennis Archer, New Orleans's Ernest Morial, Charlotte's Harvey Gantt, and Seattle's Norm Rice. Other members have included American Express president Kenneth Chenault, Washington lawyer Vernon Jordan, *Ebony* magazine founder John Johnson, *Black Enterprise* founder Earl Graves, Essence founder Ed Lewis, Virginia governor L. Douglas Wilder, and National Urban League heads Whitney Young, Hugh Price, and John Jacob, as well as such presidential cabinet secretaries as Louis Sullivan, Robert Weaver, Michael Espy, and Ron Brown.

Also in the group are millionaire Wall Street bankers, investors, money managers, and brokers like Harold Doley, Maceo Sloan, John Procope, Bruce Llewellyn, and members of the Hudson and Houston families in Los Angeles.

And probably more typical than any other group is the doctor crowd that populates the membership. Dr. Melvin Jackson Chisum belongs to the original chapter, in Philadelphia, and remembers when he was first sponsored for membership in the mid-1960s. "Most of the members were physicians who had ties to Mercy-Douglass Hospital," says Chisum, who graduated from the University of Pennsylvania School of Medicine before specializing in internal medicine and rheumatology. "All of my mentors were physicians and they said they wanted me to join. And of course Raymond Pace Alexander was there too. He was Mr. Boulé back then."

Even though some members of the old guard insist that admission to the Boulé has become too easy, most would argue that there is no black men's group—fraternal, social, or professional—that scrutinizes its candidates more closely. To begin with, many of the chapters have a preference for not considering men who are below the age of forty-five.

"When I first joined in 1977," says Grand Sire Archon Anthony Hall Jr., who presides over the organization, "the average age of incoming members was close to fifty-two years, and that was primarily because the Boulé has always been an organization that accepted men who had already distinguished themselves in their fields."

Hall, a Houston attorney, was admitted at the unusually young age of thirty-two, but he had already become a recognizable figure in Houston by that time. A graduate of Howard, he had already been named to the board of a savings and loan bank and elected to the state legislature by the time he was twenty-six.

While many of the members are lawyers, physicians, dentists, professors, politicians, or business owners, there is no professional career requirement that needs to be satisfied in order to be considered for membership. "We only require that the candidate holds a bachelor's degree or an honorary doctorate," explains Hall, who notes that out of the three thousand members only six or seven have used the honorary doctorate method of qualifying for admission.

The selection of members begins as a very quiet nomination process in which the candidate does not even know he is being nominated. A sponsor quietly collects information about the candidate. "This keeps anyone from being embarrassed," says Dr. James "Rump" Jones, an oral surgeon who joined the group in 1964. "This has always been a very staid organization that respects people's stature in a community. It would never want to embarrass someone because he was turned down."

"But as with most of these groups, there is no one you can call for an application to the Boulé," explains E. T. Williams. "People will join because someone has asked them. Most members have friends and relatives in it already, and they simply want to continue socializing with them."

Many members join because their fathers or grandfathers belonged. "When I joined in 1966, I already knew a lot about the group because my father had been a member of the Boulé in Louisville, Kentucky," says Harvey C. Russell, who remembers that his father, Harvey senior, had joined in 1938. Although he had helped found the national group One Hundred Black Men in 1963, Russell says he has enjoyed the fraternal and intellectual atmosphere that the Boulé presents. He served as grand sire archon for the organization after federal judge Leon Higginbotham stepped down in the late 1970s.

Russell is not alone in joining because his father preceded him in the group. In fact, there are many families who have seen multiple generations represented in the Boulé. Highly conscious of history and lineage, the Boulé manual and history book lists dozens of father-son memberships, son-in-law memberships, and family trees displaying generation after generation of blood relations and marital relations that have made the prestigious organization a common thread in a family's history.

"I realized I'd never get into the Boulé as soon as I began reading about the history of my local chapter," says a successful entrepreneur who had hoped to join the Nashville chapter almost thirty years ago. "One of the members—a friend of mine—showed me a page of the book that listed the original Nashville members. Next to each name was his graduate degree and the school he attended," adds the businessman. "I can still remember the schools: Fisk, Columbia, Meharry Medical School, Harvard, Yale. And all the degrees were Ph.D.'s and M.D.'s. I said 'Man, you can forget this right now.' At first I didn't recognize the names, but then I landed on 'Horace Mann Bond, Ph.D., University of Chicago,' and I remembered he'd gone there back in the 1920s. I was clearly in way over my head. Hell, he'd been president of two universities, and all I had was a bachelor's degree and a twelve-person business. I didn't have lineage or big-time degrees."

"I was attracted to the Boulé because the people they invite to join are genuine leaders in their communities who are making lasting contributions," says Los Angeles Boulé member Thomas Shropshire, who compares the group to the Links, where his wife, Jacqulyn, has served as a national officer. "Both of these groups are organizations that seek out leaders who are committed to helping less-advantaged people, and we want to encourage that kind of activity."

When I first joined the Boulé, I was astounded by the willingness of members to spend so much time with an organization that placed so much emphasis on formality. In a world that is becoming increasingly casual at work and play, the Boulé seems anachronistic in its demand for structure. But that is actually part of its appeal.

Whether it's at my Boulé chapter's Christmas party at the Waldorf-Astoria or Valentine's Day dinner at the Harvard Club, my wife and other guests comment on the courtly and sophisticated hosts. The fact that these guys are exchanging gifts from Tiffany's, waltzing with their wives while also discussing politics and football scores makes the group unique for men of my generation.

Among the activities I've experienced with the group in addition to the many black-tie dinners and dances are monthly presentations on racial and political issues and weekend retreats to discuss business topics. There are also many social outings which are limited to the members; the members and their spouses (an archon's wife is referred to as an "archousa"); or the members, spouses, and invited guests. The social functions are typically held in the most formal settings. In New York, it's the Harvard Club, the Williams Club, the Yale Club, the Waldorf-Astoria Hotel, or at a member's apartment or country home. A popular annual picnic for the New York area chapters takes place at the suburban Chappaqua mansion of Dr. John Hutchinson, a thoracic surgeon who also has an apartment in Manhattan.

On the morning of a recent Hutchinson picnic, John and his wife, Anne, sat out by the tennis court watching their son and daughter play a match. "We've had the Manhattan, Westchester, and Brooklyn Boulé chapters come up here for a picnic for the last twenty years," says Anne as her husband, a longtime member, looks over the acres of their gently sloping rear yard. "It is great for the parents and the kids to interact at an informal event."

In Memphis, the settings are mostly formal. "We have our social affairs at the Peabody Hotel or at a club along the Mississippi River," says Dr. W. H. Sweet, a native Memphian who grew up with my parents and belongs to that city's Boulé. According to Ivan Houston, former chairman of Golden State Mutual Life Insurance, the Los Angeles chapter has meetings in a variety of places. "We will meet in people's homes, and we've certainly had functions at the Wilshire Country Club, the California Club, and the Los Angeles Athletic Club."

The group's other activities follow a rather formal structure that has been mandated by the office of the grand sire archon. Because it is a fraternal group that emphasizes social and intellectual interaction rather than community fund-raising or political activism, there are no public forums or public breakfasts like one finds among such fraternal groups as the Rotary Club or Kiwanis Club. When money is raised for a charity like the NAACP or the United Negro College Fund, it is collected directly from the members, who will simply write a check for the contribution.

Many of the chapters also provide college scholarship and youth development programs that encourage members to mentor local students.

Each chapter meets once each month at a private club, restaurant, or conference center that can accommodate the typical thirty-or-more-person group.

"When I was growing up, my parents used to host a lot of Boulé meetings and parties at our home," says Chicago native Maudelle Bousfield Evans, whose father, Dr. Midian Bousfield, was a member as well as medical director and president of Supreme Life Insurance during the 1930s and 1940s. "Everything they did was formal and sophisticated."

Fellow Chicago native Truman Gibson Jr. served recently as sire archon of his chapter and remembers that when his father was in the organization, the group was small enough to meet in members' homes. "Now, we're meeting at places like the Union League Club downtown," he adds, "and places where the real business networking is going on."

The membership meetings usually take place on a Thursday, Saturday, or Sunday evening—once a month—and involve a four-course dinner and include a detailed discussion of a current issue in business, medicine, news, law, or politics that has an impact on black people nationally or abroad. Many Boulé chapters send out reading material to members in advance of the dinner meeting and ask a member who is well-informed on the subject to lead the discussion. The monthly meetings, which last three to four hours, are typically for members only, unless an outside speaker is brought in to present an issue.

"We generally don't need outside speakers because we already have the most interesting and best-connected men among our local membership. They are already making the news," says a Los Angeles Boulé member. "If we want to hear about insurance, we've got insurance company founders like Ivan Houston in our group. Our Hudson family members know all about banking. When we wanted to talk about the L.A. riots, we already had Bernard Kinsey, the cohead of Rebuild L.A., in our group. And if it's medicine, most of our guys are leading doctors."

Ivan Houston, retired chairman of Golden State Mutual Life Insurance Company, is a longtime Boulé member, and acknowledges that he was attracted to the Boulé because it provided another opportunity to discuss issues that black people cared about. "I joined in 1970. I was already friends with many of the members because they were from the city and because my family has been in Los Angeles for many years," he recalls, "and a lot of us are concerned about the same issues in this community. Many of the members are doctors, but we talk about a wide range of issues—from business to politics to international issues."

Chicago lawyer Truman K. Gibson Jr. remembers when his chapter invited the powerful white publisher of the *Chicago Tribune* to join a discussion at a Boulé meeting. "Robert R. McCormick was a conservative member of the Chicago political community," explains Gibson, "and some of our members thought we'd be able to persuade him to be more sympathetic in the already racially divided city."

Grand Sire Archon Hall says that some chapters have attempted to support a more activist agenda. "Twenty-five years ago, a social action committee in the Boulé was created so we could outline a social agenda," Hall explains, "and now we have Eddie Williams, grand sire archon-elect, heading a public policy committee with the historian Archon John Hope Franklin." Williams, who founded the Joint Center for Political and Economic Studies, and Franklin, who is heading President Clinton's race relations initiative, illustrate the kind of powerful team that that can be fielded when the Boulé chooses an issue to investigate and support.

Most members agree that there are very few ways in which the group has changed since its founding in 1904. Those who do note any kind of evolution generally point to three areas: the group's concern for secrecy, the members' racial attitudes, and the standards used for admission.

"For years, the Boulé was extremely secretive," says Los Angeles member Thomas Shropshire, who first joined while he was working for Philip Morris in Milwaukee. "So I frankly had never even heard about the organization until I was out of school. Today, members are more willing to talk about it. You don't really see the group in social columns, but it's also not as hidden as it used to be. It is evolving into a more visible body."

As a part of its ongoing evolution, there is one area where members in specific chapters are expressing concern, and that is in the area of racial makeup. The Boulé was, of course, begun as a fraternity for black men when white fraternities were closed to them. There have, however, been a handful of white members who have been asked to join because they were closely associated with some of the high-ranking members, and shared the same liberal views regarding the role and responsibility of black professionals in America. Among the few white members have been Columbia University professor Jack Greenberg, who worked on the 1954 *Brown v. Board of Education* Supreme Court case and who replaced Thurgood Marshall as head of the NAACP Legal Defense Fund after Marshall was named to the federal bench by President Kennedy. Although there have only been a few of them, most of the other white members have been liberal Jews like Greenberg. In this regard, the Boulé is more progressive than the black sororities and fraternities, where one finds virtually no white adults in the membership.

"I'm happy that the Boulé is open-minded about welcoming non-blacks who share our vision and agenda," says a Detroit member, "but what I don't like is when one of these young black members wants to impress his white boss or colleague by nominating him—especially when the white boss knows nothing about the group and couldn't care less about black people. It makes a mockery out of the group."

An archousa who has been married to a member for twenty-two years is disturbed by a pattern she says was obvious to her when visiting some Boulé friends in Seattle. "I am shocked by the number of black men in this organization who are bringing their white girlfriends and wives into the group," the woman remarks. "It's not my business that all these powerful black men have decided to marry and date whites, but it is my business when I heard some of these women remark that the fraternity might gain more prestige if it contributed to causes that were not just associated with blacks."

Another member agrees that certain chapters have a surprising number of members with white girlfriends. "I never thought I'd see this," he remarks. "I know guys who used to be deeply committed, but now their new spouses are pulling their husbands away from a black agenda. When you think about it, there's no reason why these white women would care about supporting a youth development program for black kids or would encourage their husbands to give more money to the NAACP or United Negro College Fund. I don't see this kind of influence occurring in my wife's Links group, or in the local chapter of the Girl Friends. They all seem to agree that the groups were meant to support a black agenda. What's our problem?"

While there are some people who share these two individuals' concerns, for the most part the Boulé membership does not see the interracial issue as a factor, since interracial couples represent a very small portion of the total membership, and most of those who do participate are respectful of the group's history and its focus. What is interesting regarding the race factor is that when members are asked about the racial makeup of the Boulé, most have no sense of how many interracial couples are represented—and most would be surprised to learn that the membership includes any white men.

The only other area in which some say the group is changing is in the standards used for accepting candidates. Many will find this ironic, since the Boulé is probably the most difficult black elite group of all to enter.

"I think the group is lowering its standards," says a retired physician who lives in Atlanta. "A lot of us go to the Grand Boulés or look in the directory these days, and we meet all these young men that never would have gotten in thirty or forty years ago. Some chapters are taking in five or six members at a time just because they're wealthy or famous. Just because someone is a millionaire entrepreneur or a big-city mayor doesn't mean he should be let into the group." The elderly doctor looks at the bottom shelf of the library in his home and pulls out one of the blue-covered *Boulé Journal* magazines. His wife joins the conversation.

"I'd have to agree," she adds. "Over the last few years, we've met some of the most coarse businesspeople and congressmen who have told us they were members. And it's just ridiculous to admit all these people. Like that chapter in D.C. You can't tell me that there are actually one hundred men that really deserve to be in the Boulé in that tiny city."

The fact that the Washington, D.C., chapter includes college presidents, federal judges, cabinet secretaries, renowned physicians, and high-profile lawyers like Vernon Jordan left the woman unconvinced. Many of the Boulé's members argue that the chapters have grown larger because the black elite is significantly larger than it was in previous years.

And as the organization grows, it continues to maintain most of the traditions that endear it to the membership. "We continue to involve the widows of our deceased members because this is also a family organization that values family relationships," says Grand Sire Archon Anthony Hall Jr. As long as a widow hasn't remarried, she continues to receive the *Boulé Journal,* as well as invitations to events and parties. In most cases, the organization pays for the widow's ticket to these events. The Boulé directories even continue to list the address of a widow under her deceased husband's chapter's membership list.

As the most selective group for black men in America, the Boulé continues to be popular among both the old guard and the new elite who have earned the traditional academic and professional credentials that the original founders embraced.

I first learned of the National Association of Guardsmen from my cousin, Robert Morton, a New York surgeon who had joined the group thirty years ago. Founded in Brooklyn in 1933, the Guardsmen is comprised of eighteen chapters with locations throughout the country. Most of the chapters have fewer than the thirty-member maximum allowed, with an emphasis placed on physicians and attorneys. The original group included several Brooklyn residents who worked for a downtown oil company.

They were James Adams, Armstead Cooper, Wilbur Rogers, Ray Nathan, Marcus Grant, Edward Taylor, Edward Hairston, Joseph Sircus, Everett Miller, Malcolm Fulcher, Harold Carter, and Weston Thorne.

The group is thought of as a purely social organization, so its mandate is very different from that of the more intellectually driven Boulé, or the more politically active women's Links group, or even the children and family organization, Jack and Jill. Although many of the Guardsmen have wives in the Links or Girl Friends and children who grew up in Jack and Jill, the group is very clearly an adult men's social club that eschews formal rules and structure.

While the Boulé's goal has been to gradually increase its size, gain more national stature, and focus its discussions on political and intellectual pursuits, the Guardsmen have chosen to remain small and even more hidden from the public.

"Most of the early members were young lawyers, doctors, and dentists who already belonged to fraternities," says Theodore Payne, the group's national president, who lives in Fairfield County, Connecticut. "I have heard some people describe the Guardsmen as a black man's country club, and I think that description is accurate in some ways."

The group does not have formal conventions with plenary sessions, platforms, and speeches in the way that the Boulé and the fraternities have each year. "In fact, you might get kicked out of the organization if you ever suggested formalizing our get-togethers in that manner," says national president Ted Payne with a laugh. "Most of our members already belong to professional associations in medicine, law, or other areas, and already participate in fraternities where formal conferences are the regular practice. That is not our purpose and everyone understands that."

So averse to formalized practices is the organization that it allows each chapter to establish its own rules governing dues, the frequency of meetings, the size of its local chapter, and other issues. The only strict requirements are that each chapter serve as a host for the group's scheduled weekend retreats and that each chapter send at least two delegates to the national gatherings.

Because the group wanted to remain small and elite, it was a long time before the national office would permit the creation of chapters west of the Mississippi. Today, there are chapters in the cities of New York, Boston, Baltimore, Atlanta, Detroit, Philadelphia, Chicago, Los Angeles, St. Louis, and Washington, D.C., as well as in Virginia, Florida, North Carolina, and New Jersey.

The Guardsmen are particularly well known for their lavish "Guardsmen Weekends," where a host chapter offers an all-expense-paid weekend in the host city for every other Guardsmen member and respective spouse. These weekends take place three times each year and cost the host members thousands of dollars, as they schedule black-tie dances, golf outings, sightseeing, fashion shows, and other events. The first weekend takes place in February, March, or April; the second is in May, June, or July; and the third annual weekend occurs in September, October, or November.

The Guardsmen Weekend scheduling is so important to the members that they set the dates and locations as far as twelve years in advance. For example, a recent Guardsmen Newsletter has already announced that the Atlanta chapter will be hosting the March weekend in the year 2012.

Among the lavish weekends are trips to Boca Raton, Atlantic City, the Caribbean, and western ski resorts. Dr. George Lopez and his wife, Mary, remember a recent weekend in Florida. "Our host was a multimillionaire attorney who spared no expense," says Lopez, who is also a member of the Boulé. "Not only did they entertain us in public places but our host threw open his home to us as well." As a society columnist, Mary Gardner Lopez has covered many social affairs. She was one of the founders of the Doll League, a national women's organization, and is a regular on the black elite circuit. "We've been to weekends in Palm Springs, Bermuda, Martha's Vineyard, Newport, Montreal, Hilton Head, you name it," says Mrs. Lopez. "No one moves around as much as the Guardsmen."

Alfonso Orr of New York and Tom Shropshire of Los Angeles say that they have flown cross-country and from business conferences outside the United States in order to return in time for the golf and tennis tournaments taking place at Guardsmen Weekends. Their wives, both members of the Links, also juggle their schedules for these extravagant events.

"You've never seen people spend money until you've been to a Guardsmen Weekend," says a Boston member who recently returned from a Los Angeles retreat. "My wife and I were blown away. The group must have spent a quarter of a million bucks on us during the three days."

Besides the extravagant weekends, each of the chapters has its own local events. While the different cities don't sponsor cotillions as some of the fraternities do, they have theme parties and other gatherings. For example, the Los Angeles chapter gives a big New Year's Eve party.

Each of the members is given a medallion that has a quill, a winged foot, a lion, and a helmeted guardsman embossed on it. The group is extremely selective about beginning new chapters and accepting members into the chapters that are already in existence. Because the Guardsmen requires a financial outlay greater than all the other men's groups, many eligible men take themselves out of the running.

Past and current members include Virginia governor L. Douglas Wilder, former U.S. secretary of commerce Ron Brown, and Hampton University president William Harvey.

You can ask ten different Guardsmen about the method for gaining admission to the club and you'll get ten different stories. This is primarily because each of the chapters is allowed to establish its own rules governing admission.

"To this day, I still don't understand how I got turned down, when I knew and got along with everyone in the group," says a physician who lives in Boston.

"Some chapters have unusual rules for admission," admits national president Ted Payne. "You may encounter a chapter where twenty members want to admit you, but two people don't. If the rules say that two negative votes can keep you out, then this is what can happen."

While there are some father-son memberships reflected in chapters like Atlanta and Philadelphia, the group does not have the number of "legacies" that one finds in the Boulé.

"I was attracted to the group for family reasons," says Thomas Shropshire. "Since my brother had been in it for years in Atlanta, I decided to get a group together to form a chapter out here. Although it was traditionally an East Coast group, the newest chapters are generally west of the Mississippi."

Whether the members come from the East or West Coast, most of them look to the group exclusively for social interactions. They have none of the pretensions that Boulé members have, even though there are some overlaps in their membership rosters.

"Even though there's some overlap between them, the Guardsmen guys have always been a lot more fun-loving than Boulé men," says an attorney who belongs to both. "But just because they know how to party doesn't mean they weren't all ambitious guys with accomplished families."

* * *

There is no black-tie dinner ticket more coveted among the black elite than those for the annual Comus Ball that takes place each year after Christmas in New York.

Phyllis Stevenson fondly recalls the Comus Balls that she and her husband began attending in the 1960s. Frank Stevenson, now deceased, was president of the group during the 1980s. "People used to start calling us around September," Phyllis explains, "all of a sudden renewing old friendships because they knew the formal was approaching and they wanted to get tickets." Phyllis, who had been one of the most celebrated debutantes to come out with the Girl Friends in the early 1950s, is well-schooled in the art of high society diplomacy. "I understood the popularity of the event—I've gone to at least twenty of the annual balls. But I told people that I had no control over distribution of the invitations. They didn't print more just because more people wanted to attend."

"Each member gets only ten invitations," says longtime member John Procope, who was once publisher of the *Amsterdam News* and now coheads the insurance firm E.G. Bowman, on Wall Street, with his wife, Ernesta. "We don't sell tickets because we want to act as true hosts. We pay for each of our invited guests," he explains.

Phyllis Stevenson says that most members have their wives, one or two family members, and their very closest friends at their tables, thus providing virtually the same crowd year after year. As she flips through a stack of photo albums that chronicle past years of the Comus Ball, it becomes obvious that there is one crowd, indeed. Many of the tables look nearly identical Christmas after Christmas.

"My wife and I would love to go to that ball, but they won't sell you a ticket no matter how much you offer. And I'm not going to beg," says a New York physician who has wanted to attend for the last fifteen years. "It's a closed group if you're not related to a member or if you're not his best friend or business partner. At this point, I just want to go out of curiosity."

A popular couple on the New York social scene, Frank and Phyllis Stevenson included on their membership and guest lists the names of the people who had been tied to the Comus group for decades. Phyllis not only is the daughter of Anna Small Murphy, one of the founders of the Girl Friends, but was also president of the exclusive women's group in the early 1980s. Before Frank became an administrator at the New York state supreme court, he and entrepreneur Richard Clarke had opened a successful travel agency, Hallmark Holidays, in Manhattan to serve the corporate sector. Clarke, who was the first black named to the board of the Metropolitan Museum of Art in New York, is a regular at the holiday event.

"Like most members of the Comus Club, my husband was a native New Yorker," says Phyllis. "He went to Stuyvesant High School and Lincoln, so he knew this group when they were all young." Other members at the time were Dr. Hiram Bell, Dr. Cecil Forster, Dr. Leslie Alexander, Dr. John Parker, Judge Herbert Evans, Judge George Fleary, Judge Charles Lawrence, Judge Robert Couze, Dr. Charles Greene, Dr. Robert Somerville, Dr. Oscar Allen, Dr. Thomas Leach, and many other prominent Brooklynites.

Founded in Brooklyn in 1923, the Comus Club was started by men who lived and worked in the city, holding a variety of professional and nonprofessional jobs. Unlike the Boulé, whose membership has always been professional, the Comus evolved into a completely professional crowd only after the first couple of decades of its existence. Unlike the other black elite groups, the Comus has only one chapter. The club owns a brownstone on Decatur Street, which members use as a clubhouse where they host Saturday evening meetings, card games, and other affairs. Throughout the 1940s and 1950s, the Christmas ball was held at the Hotel St. George in Brooklyn, but with crowds of five hundred or more, it is now held in larger and newer facilities like Terrace on the Park in Flushing Meadow Park or New York hotels.

A sought-after credential on any black society family résumé, the group is highly selective about admitting new members. Getting any of the approximately seventy members to reveal anything about the selection process is virtually impossible. In fact, memberships seem to be held by many families who pass them between relatives or across generations. "I was proposed in 1957 by my wife's brother, Dr. Cecil Forster," says John Procope. Forster, who was a Brooklynite like his sister, Ernesta Procope, was a New York psychologist and president of the New York Psychological Association. He went on to become president of the Comus as well.

Comus member E. T. Williams also has ties to this organization's many family connections. "Ernesta Procope's brother is my godfather," explains Williams. He joined eleven years ago. And Earl Arrington continues the chain. "My mother and E.T.'s mother were very close friends in Brooklyn," says Arrington, a real estate investor in New York, "and since Brooklyn was once a very small community, the Comus guys knew each other's families very well."

There are also many members who have memberships with the other groups. E. T. Williams, Dr. John Parker, Dr. Vernal Cave, and Dr. Hobart Jarrett are among those who have belonged to the Boulé as well. John Procope and Judge Herbert Whiteman belong to all three: the Comus, the Boulé, and the Guardsmen.

Although they have never been as numerous as the women's social groups, several other men's social groups have operated locally or regionally within the black old guard. Among them are the Gallivanters, the Fellas, the Westchester Clubmen, and the Reveille Club, all of New York; the Forty Club and the Royal Coterie of Snakes in Chicago; the Rogues of Detroit; the Bachelor-Benedicts in Washington; the Commissioners and the Ramblers of Philadelphia; and the Illinois Club of New Orleans. There are also many men's social groups that are built around specific activities such as boating. New York's Rainbow Yacht Club, which is made up mostly of well-known black physicians like Harry Delany, the chief of surgery at Albert Einstein Medical Center in the Bronx, includes two dozen men who travel in a fleet of yachts during summer weekends. Others gathered to establish the annual Black Summit, an all-black ski vacation in Aspen, Colorado.

Some of these groups sponsor important annual events. For example, New York's sixty-six-year-old Reveille Club, which includes such members as New York Hospital surgeon William Curry, Carver Federal Savings Bank chairman Richard Greene, Judge Fritz Alexander, former mayor David Dinkins, and Harlem real estate investor Ed Meyers, sponsors a "Man of the Year" award. Philadelphia's Ramblers have two annual formals. The Bachelor-Benedicts in Washington remain famous because of the debutante cotillion that introduces daughters of some of the city's oldest families.

Of all the men's groups, One Hundred Black Men is seen more as a professional organization and less as a selective social club. While some of the most prominent black men in America belong, the group does not have the prestige of a social club because its intent is to serve as a professional networking group. Its chapters are often quite large, and membership is based merely on sponsorship and payment of dues.

Considered a leading force in economic development, mentoring, and networking, One Hundred Black Men was begun in 1963 and now has forty-six chapters around the country. In its early years, the group attracted only its host city's old black family names, but today it has expanded to include young corporate executives, bankers, lawyers, physicians, accountants, entrepreneurs, and politicians. Most chapters have monthly meetings on a selected weekday evening, hosting speakers from the worlds of business or politics, where they discuss practical business strategies, personal improvement, local economic development, or legislation of interest to the black community.

Since members represent some of the most affluent black men in their cities, many of the chapters have established scholarship funds for inner-city young people or have "adopted" special charitable causes such as a local hospital, a nursing home, a summer camp, the NAACP, or another group that needs their fund-raising support.

Among the founders is Harvey C. Russell, who was the first black to attain the position of vice president at a *Fortune* 500 company. As a Pepsi executive in the 1960s, Russell served as a mentor for many young black managers and students who were pursuing careers in corporate America. "When we started the group, there were only fourteen or fifteen of us, and most of the members were working in local government," says Russell, while relaxing at his summer home in Martha's Vineyard. "Our intent was to create a local networking group for business and political leaders, and we had no idea it would grow into a national organization."

For years, Harvey and his wife, Jacqueline Denison Russell—a member of the Girl Friends—have hosted fellow One Hundred Black Men members and families at their Oak Bluffs home, which is just steps away from the Martha's Vineyard ferry. Since forming the first One Hundred Black Men chapter, the New York organization has been particularly successful in its high-profile fund-raising. Oprah Winfrey was so impressed with the group's work that she contributed $100,000 at the group's annual black-tie dinner in 1996, hosted by national president Thomas Dortch and New York president Luther Gatling.

"It was impressive to see close to a thousand black male businesspeople all in one room recognizing the need for us to take a leadership role in our community. It was one of the snazziest One Hundred Black Men events I'd ever attended," says a member who serves as a One Hundred Black Men mentor for kids in the local schools, "but I must admit that a few of us were a little annoyed that all twelve hundred blacks were in formal attire, while the mayor and governor of New York showed up in suits. We represent some of the most influential people in America, but some of these white politicians have no respect for us."

Among the members in One Hundred Black Men are Herman Cain, CEO of Godfather's Pizza; Milton Irvin, partner at Salomon Brothers; Ken Chenault, president of American Express; William Campbell, mayor of Atlanta; Bruce Llewellyn, CEO of Philadelphia Coca-Cola Bottling Company; Richard Parsons, president of Time Warner; Clarence Smith, president of Essence Communications; Kweisi Mfume, former congressman and head of the NAACP; Norman Rice, mayor of Seattle; Ronald Kirk, mayor of Dallas; and General Colin Powell.

Many One Hundred Black Men members participate in the group because it gives them the opportunity to participate in the community service projects that the socially focused groups like the Guardsmen do not partake in on a group level. The group allocates 21 percent of its spending for educational programs and another 13 percent for mentoring programs. Many of the chapters also operate conflict-resolution and antiviolence programs.

As a member of the Boulé, the Westchester Clubmen, and One Hundred Black Men, I am an unabashed cheerleader of these groups, and I often remind my peers of the need to continue the traditions that prior generations began. Although I initially joined because I liked their activities and was flattered to be asked, I have come to recognize that this is finally the safe harbor that I had never been able to find in other organizations where people looked at me—the successful black man—as the oddity.

Because many of these elite men's groups were initially focused on social rather than civic activities, some of my friends have not bothered to investigate the value that these organizations are now adding to their communities. Because of this, they often limit their activities to the alumni chapters of their college fraternities. As groups such as the Boulé and the Guardsmen become less secretive and gain the national exposure that women's groups like the Links have had, they will begin to attract more than just the sons and nephews of the older members. Perhaps the exposure will remove some of the prestige and the mystery, but it will serve to maintain a national roster of men who can call themselves the best and the brightest.

Vacation Spots for the Black Elite

"So, Maggie, who's here?" my mother asked.

"I saw Velma yesterday. And Anna had a book signing on the boat Thursday afternoon. And some of the Links from Philadelphia came up last week."

"And what about Donald and Charlene? I heard they were looking at some property over in Vineyard Haven?"

"*Looking at?*" the woman asked while rolling her eyes. "You've obviously been gone one summer too long. That place was built, moved into, and landscaped nine months ago—all three thousand square feet of it."

"Well, we spent last summer in Sag Harbor. You know how much easier the weekend commute is."

Maggie shrugged without empathy. "Well, no one said this stuff was easy. You can't be in both places at one time. You know you really have to pick."

My mother didn't miss a beat. "Well, I picked and I'm back. We'll be at the tennis courts this morning and the Ink Well this afternoon. So call me—and tell everybody I'm back! Tell 'em I'm back!"

That was the rallying cry that I remember hearing at the beginning of each summer that we spent in Oak Bluffs, on the island of Martha's Vineyard in Massachusetts. When my parents spotted their good friend Maggie Alston strolling along Circuit Avenue, the summer had officially begun. This was long before the groups of black college students started coming in the 1980s. It was before President Bill Clinton's highly publicized 1994 Vineyard vacation with Vernon Jordan and the Kennedys. It was before Black Dog Tavern T-shirts had become a cheap cliché—seen on the backs of people who couldn't even find Vineyard Haven on an island map. For three months each year, the three-block stretch of stores on Circuit Avenue and the short strip of sand along Seaview Avenue—which we blacks call the "Ink Well"—was the center of our universe. It was black heaven—a world that few of us could abandon, even for half a summer. Unless, of course, we wanted to get left behind.

Even though we'd been going there since I was two years old, such was the arrogance of black privilege on that island that it never even occurred to me that white people had summer homes on Martha's Vineyard until I was ten or eleven years old. Of course I saw white people at the Flying Horses, at Our Market, and at the tennis courts off South Circuit Avenue in Oak Bluffs. But I assumed they were just passing through as guests of black people who had homes there, or as unrooted tourists. Just people passing through a place that was ours.

But of course Martha's Vineyard had white families. The black neighborhoods of Oak Bluffs were dwarfed by the white sections in the town and by the white population that dominated the rest of Martha's Vineyard. But I was a summer kid who defined the resort by the boundaries of the black neighborhoods and by whole days and evenings spent with our extended black family in our all-black tennis tournaments, all-black yachting trips, all-black art shows, and all-black cookouts, and the white vacationers had no relevance for me.

As I grew older, I saw what my younger and more naive, self-satisfied eyes had missed. As an adolescent, I finally paid notice to the racial lines that long ago had been drawn between blacks and whites on Martha's Vineyard. I eventually even saw the many hierarchies that existed within the groups of blacks who summered there. But in spite of these changed perceptions and my newfound confrontation with reality, the one unalterable impression that remains today is that when vacationing among our own kind, in places that have been embraced by us for so long, there is a comfort—and a sanctity—that makes it almost possible to forget that there is a white power structure touching our lives at all.

Today, America's black elite is closely associated with three historic resort areas that became popular as a result of laws that had kept other vacation spots exclusively white. They are Sag Harbor, Long Island; Oak Bluffs, Martha's Vineyard; and Highland Beach, Maryland. In the past, and to some extent still today, blacks also choose Hillside Inn, a black-owned resort in Pennsylvania's Pocono Mountains; and Idlewild, Michigan, a small town two hundred miles north of Detroit that was a popular escape for the midwestern black elite. In recent years, the elite have built ancillary vacations around the annual Black Summit ski vacation event that brings hundreds of black skiers and their families to resorts in Aspen and Vail, Colorado.

Although it's the best-known resort among the black elite, the tiny, seven-square-mile village of Oak Bluffs, Martha's Vineyard, is actually a mostly white community with only a 5 percent black year-round population. Expanding each year as more black professionals from all over the country continue to buy or rent homes there, Oak Bluffs has been a popular area for elite black families since the 1930s and 1940s.

Located on the quiet, ninety-two-square-mile triangular island of Martha's Vineyard, off Cape Cod, Massachusetts, Oak Bluffs is one of six small island towns on Martha's Vineyard—the others being Vineyard Haven, Edgartown, Chilmark, West Tisbury, and Gay Head. The island, which has a year-round population of thirty-three hundred residents (it swells to eighty thousand during the summer), was originally chartered in 1602 by Bartholomew Gosnold of England, who named the island after his daughter, Martha. Already living there were members of an American Indian tribe, the Wampanoags.

Beginning in the 1890s, blacks who lived in Massachusetts and parts of Rhode Island began to move full-time to—and open businesses on—the island of Martha's Vineyard. At the same time, there were a handful of well-to-do black families that began establishing roots there as a summer vacation spot.

Among the first blacks to locate there were the great-grandparents of Robert W. Jones, a New York City real estate developer and consultant whose family had come from Boston. The well-to-do family owns twenty-five lots on the island today. "Although we have always had a home in Oak Bluffs," explains Jones, "prior generations in my family purchased property in the Edgartown section in the 1890s." Jones's great-grandmother, Phoebe Mosley Ballou, bought the family's original parcels at a time when blacks were not allowed to stay in any of the hotels on the island.

"My family and the family of the writer Dorothy West were the first to own homes and host black summer visitors, since the white inns had a whites-only policy," explains Jones. "Since this is the liberal Northeast, people are often surprised to hear that the segregated hotel rule was actually in place on the Vineyard. But when I was summering there as a child in the 1930s and 1940s, that was the rule all over the island."

An alumnus of Howard, New York University, and New York Law School, Jones grew up in an accomplished family in the Back Bay section of Boston during the 1930s. His background was not rare for the kinds of blacks who could afford to summer at the resort. His father, John, was born in Cambridge, Massachusetts, attended Harvard, and became the first black graduate, in the 1920s, of Suffolk Law School. Today, a portrait of his father hangs at the school in Boston. "My aunt Lois painted the portrait many years ago while she was still an art professor at Howard," says Jones. Aunt Lois is Lois Mailou Jones—a well-known fixture on Martha's Vineyard whom my parents first met at one of her many gallery showings. Black and white homeowners have been snatching up her works since she first joined the Howard art department faculty in 1930. Today, her works are on display at the Metropolitan Museum of Art, the Hirshhorn Museum and Sculpture Garden, the Corcoran Gallery of Art, and the National Portrait Gallery.

Gesturing to the street-corner location where Our Market, a busy Oak Bluffs general store, now sits, Jones recalls, "My family and the West family were next-door neighbors right here on New York Avenue until around 1903, when a fire burned the buildings down. Then they moved to Circuit Avenue. My grandfather, Thomas Vreeland Jones, bought more property in 1912 in Oak Bluffs and then moved to Pacific Avenue, which is where the family home is today."

Like a few of the other "originals," Jones has clear memories of Martha's Vineyard before it became popular with affluent blacks from all over the country. "I spent every summer of my life in Oak Bluffs, but when I was a kid, the only blacks that came here were ones from Boston and Washington. No New Yorkers, no southerners, no midwesterners. The Shearers, the Ashburns, the Dixons—all Boston and D.C. people—were our social circle."

The other kids Jones played with were white—a fact that struck me as odd, since my own childhood experiences in the 1960s and 1970s provided almost no contact with whites on the island.

Like others, Jones points out that it was not until Adam Clayton Powell Jr., the outspoken congressman, and his wife, Isabel, and their son, Preston, started visiting there that Oak Bluffs gained true popularity among the New York black elite. "The Powells started coming to the island in the late 1930s and early 1940s and started to bring these high-profile friends," says Jones, who remembers the congressman and still remains in contact with his widow, Isabel.

But even with all the affluent, high-profile blacks who were coming from Washington, Boston, and New York, the island's inns and hotels—even the 120-year-old Wesley Hotel on Lake Avenue—did not allow blacks to board in their rooms until the height of the civil rights movement.

"And believe it or not, that hotel was owned by the Methodists," remarks a woman who remembers her relatives being turned away from the ornate building when she was spending her childhood summers during the 1950s in Oak Bluffs. Even with the Jones family connections to the prominent members of the Methodist church groups that controlled much of the property and civic affairs in the town—Robert's mother was a member of the Methodist community's historic Tabernacle in Oak Bluffs and raised money for the religious structure's lighted cross—they and other blacks had to entertain their black visitors in their own homes. "The only place for blacks to stay was the Shearer Cottage on Rose Avenue," Jones added, while noting that the owners of the cottage—the Shearer and Ashburn families—were an important link between new black families and island culture.

Some find it ironic that a community with so many religious roots would have supported such bigotry. It was a team of activists in the Methodist Church who first visited Oak Bluffs in the 1830s and established an annual meeting in subsequent years, when they would pitch large tents and establish two- or three-week retreats for their members. After conducting these retreats for several years, the group of Methodists expanded and purchased several acres in the center of town, where they constructed small Gothic-styled houses on the property. The property is called the Methodist Campground. Sitting on the land today, just steps away from the harbor, are dozens of colorful, hundred-year-old, wood-framed Gothic—or "gingerbread"—houses that, according to black residents in town, have always been inhabited by whites.

"I don't ever remember seeing a black family in one of those gingerbread houses either," adds Alelia Nelson, "but I can't imagine why any of us would really be trying to move into one of those tiny houses. They are so close together, and the road is ten feet from your front door. There's no privacy." Nelson, who has lived in one of the largest homes overlooking the Oak Bluffs beach for nearly forty years, believes that the more desirable homes are the ones that have a view of the water—not the ones on the Campground.

Such is the way in which black Vineyarders discuss the early treatment of their kind in the resort community. Like my own parents, most people here eschew the debates on racial conflict and, instead, adopt an air of quiet resignation about the past. The bigots who came before them were typically much less affluent whites, so to dismiss them now as "white trash" is a gesture that—if not equally racist—makes the memories a little less painful.

My parents and many of my friends' parents always seemed to be aware of the island's past attitudes on race, but these people also always seemed to suggest that we had little reason to complain because nowhere else in the world were we going to find blacks like us living such a relatively conflict-free life.

Since Oak Bluffs is often perceived as some kind of oasis in the otherwise white-dominated world, no one in this black community seems to want to get a straight story on the Campground. They just all seem a little uncomfortable with it. And they are satisfied with maintaining that level of discomfort.

"Even though Edgartown and Vineyard Haven were even less welcoming to blacks, Oak Bluffs has its own blemish with the Campground houses," says a New York attorney, who points out that he can't recall any black families that have ever been welcomed into the historic section.

According to one longtime black Oak Bluffs summer resident who says he doesn't want to stir up anything, the Methodist Campground association has to approve the sale of the houses on the property. "Anyone can buy a house on the property, but since the association owns the land, it's my understanding that they can refuse to let you stay there," says the man, who quickly adds that he wouldn't want to live that close to people like that.

Robert Jones's grandfather, Thomas Vreeland Jones, had purchased a Campground house in the early 1900s and had it placed on a flatbed with wheels in order to have it moved out of the white neighborhood and pushed up along School Street to Pacific Avenue, where many of the town's blacks were living at the time.

"Most of us blacks lived on the other side of New York Avenue—closer to East Chop. That's where we were more comfortable and that's where my own grandfather moved when he got here in the thirties," says Doris Pope Jackson, whose grandfather, Charles Shearer, was one of the first black landowners in the community.

Despite the history of the Campground neighborhood and the current makeup of the predominately white area, Oak Bluffs has long been considered the most hospitable area for blacks on the island. Even in the 1960s and 1970s Oak Bluffs was one of the two towns where my parents allowed us to roam unsupervised.

Although there were several black families like Robert Jones's that initially moved to the School Street area of Oak Bluffs, a much larger number came to settle in the area Doris Pope Jackson refers to several blocks north—on the other side of New York Avenue. This second black neighborhood was located near a part of the very white East Chop section of Oak Bluffs, and it is sometimes referred to, by older black residents, as the "Pork Chop" neighborhood. The area, which takes in such streets as Highland Avenue, Rose Avenue, Mountain Avenue, and parts of Munroe Avenue, includes such residents as Isabel Powell—the congressman's widow—and the writer Dorothy West.

This is also the neighborhood where the historic Shearer Cottage can be found. Doris Pope Jackson remarks that her grandfather, Charles Shearer, a Hampton graduate, established the inn on Rose Avenue and made it a stop for many blacks who had avoided the island because of the inability to find innkeepers who would accept nonwhite guests. "We can accommodate up to sixteen visitors at a time," says Jackson, who now owns the inn, along with several other buildings that are operated under the Shearer Cottage name. Jackson was an early member of the island's elite black women's group, The Cottagers, and her family is well known in Oak Bluffs.

Not long after some of these blacks arrived in the area near East Chop, many white residents began to aggressively buy up the undeveloped East Chop beach area during the 1940s and 1950s in order to maintain white dominance over the neighborhood. These white residents also established an exclusive club, the East Chop Tennis and Yacht Club, that black neighbors say did not have a black member for several decades, even though there were many well-to-do black residents living in the area.

Even though the oldest black residents of Martha's Vineyard acknowledge the elitism evident among the white residents that have settled in East Chop, Edgartown, and other neighborhoods, they are not so quick to admit to the healthy dose of snobbery that they, themselves, serve up each summer for the black people who are new to the summer resort.

Even as a child, I was aware, upon our arrival, of the need to make a good impression. In fact, each year before we came up to Martha's Vineyard from New York, we would spend a couple of weeks amassing new beach paraphernalia at Abercrombie and Fitch and new summer clothes at Lord and Taylor's and Saks Fifth Avenue. One year, when I was around six or seven, just hours before we were to leave town, my mother hurriedly drove us over to one of the stores with the demand, "My sons need two or three matching outfits for clamming."

When the salesclerk shook her head in confusion, Mom said, "You know, short pants—somewhere between bathing trunks and long trousers—something stylish and nautical. Something that won't get wet if they're in ankle-deep water." My brother and I knew that it wouldn't do to suggest that we could go hunting for clams or shells by simply rolling up our khakis or by wearing cutoff jeans. We knew that no kid in our group would show up looking so mismatched and "middle class." It did not matter that they *were* middle class. So, instead, we ended up with some horribly effete yellow-and-white suits—I'd call them culottes—that would have been perfect for Little Lord Fauntleroy on a pirate ship.

But we were not alone in the pursuit of making good first impressions in Martha's Vineyard.

From the moment one prepares for the trip to Oak Bluffs, the goal is to be well prepared to show your best. Because my parents used to move us to Oak Bluffs for nearly all of July and August for our summers, we would fill two cars to the brim with our new summer clothes, our newest toys, four bicycles, beach paraphernalia, and boxes of school study guides, *Mad* magazines, and the last several issues of *Ebony* and *Jet*.

My father would spend two hours packing our Pontiac sedan and Buick station wagon in the garage the night before, and then awaken us in pitch-black darkness at 4 A.M. The cars were so overloaded that in my grogginess it felt as though we were stealing away like runaways.

"Keep the blinds closed. We don't want these people seeing us going away," my father would whisper to us as we tiptoed out to the garage with our final baskets of food for the five-hour trip.

I'm sure he didn't mean it that way, but because we were the only blacks in our neighborhood, taking off for two months of solitude among our black friends, my brother and I always imagined that "these people" referred to the white people who lived around us. (Whoever "these people" were, by 1971—the year our house was burglarized—my parents wised up and realized that they'd better start telling the neighbors to watch our empty house.)

So, with no sounds outside except for the hiss of a neighbor's automatic lawn sprinklers and our electric garage door coming up and then down again, I would get into the Buick with my mother, and my brother would get into the Pontiac with my father, and we'd head north for a highway that would begin our three-hundred-mile trip to Martha's Vineyard.

It is at Woods Hole that one gets the first look at the black elite of Martha's Vineyard. Woods Hole is a port community at the southern edge of Cape Cod, and it serves as the docking location for ferries taking the forty-five-minute trip between mainland Massachusetts and the small island.

Even though there are four times as many white families sitting in their cars in the large lot waiting to board the ferry, it was always the black people that we were looking at. They were the faces we were expecting to recognize.

When I was a child, walking the line between the cars in the early morning sunlight gave me an exhilarating, almost cocky feeling as we saw other good-looking black people hop out of station wagons and fashionable sedans.

Today, on that line, blacks give each other a knowing nod as they pass by in their tennis whites and sunglasses, with copies of *Black Enterprise* magazine folded underneath their arms. We don't always know each other, but we'll generally use this opportunity to meet one another if it looks like the person might belong to one of our social or civic groups.

"I put my best face on when we arrive at Woods Hole for the ferry ride," says a forty-year-old Washington, D.C., resident who has boarded the ferry to Oak Bluffs each summer for thirty-three years. "We always car-hop when waiting to drive aboard. You get to see who's coming out, what people are driving, who's giving parties, who's changed jobs, who got divorced, and who died. It's a great prelude."

This prelude of people-watching and "profiling" continues during the forty-five-minute ferry ride until the boat docks at Oak Bluffs. Then people head down to one of two places: the Ink Well or Circuit Avenue.

The Ink Well is the nickname for a section of the Oak Bluffs beach that the black elite have informally taken over as their own. Just a few blocks southeast of downtown, running along Nantucket Sound, the Ink Well is a four-block stretch of beach that dips several yards below Seaview Avenue—the road that borders the beach. Not surprisingly, its name comes from the fact that its chief patrons are black. The closest beach to the town's busy center, the Ink Well also abuts the rocky area where the Steamship Authority ferry loads and unloads. This makes the location bad for solitude but great for getting your name around.

"If you want to quickly get an idea of who is out here on any particular weekend, you should walk the length of the Ink Well around 2:30 or 3 P.M. after the whites have left," says the forty-year-old Washingtonian. People stand up along the sidewalk and wave down to friends and children who are playing, swimming, reading, tanning, building sand castles—and most of all, keeping track of who's around.

"My daughters loved going to that beach," says Dr. Beny Primm, a New York physician who bought an Oak Bluffs house in 1971. "They saw their friends from home, from school, and from Jack and Jill there. It was an extended family."

Jacquelyn and Bill Brown own an 1869 modified Campground-style house that has a view of the beach. "Initially we were going to buy a place in Newport," explains Bill Brown, an MIT graduate, who first visited the island in the late 1960s and stayed at Shearer Cottage, "but what attracted us to Oak Bluffs years ago were unique factors like the Ink Well. We didn't want to be spending summers in places where we were the only blacks. Our kids got enough of that during the school year."

Longtime Vineyarder Alelia Nelson laughs when she hears people talk about the Ink Well. "Originally that term was used privately among blacks; then one day Louis Sullivan, George Bush's cabinet secretary, gave an interview to one of the white newspapers and he mentioned the term 'Ink Well,'" explains Nelson as she sits in the rambling home that wraps around one of the most picturesque corners in Oak Bluffs, "and that's how white people learned about the nickname we'd made up for our section of the beach. And then came that dreadful movie about the Ink Well. Not only was it insulting; it wasn't even filmed here. They did it down in Virginia someplace." Nelson, who is the goddaughter of millionaire entrepreneur Madam C. J. Walker, brought her children to all the beaches on the island—not just the Ink Well. Her daughter, Jill, a former *Washington Post* reporter, has written about the summers she spent there. Her son, Stanley, a filmmaker, has worked on projects focusing on the community. "I was not going to be intimidated by people who said the only beach for us was the Ink Well. So I brought my kids to all the beaches here."

Like the Primms and the Browns, well-to-do families populate this beach throughout the summer. On any given afternoon, there are doctors from Philadelphia, radio station owners from Detroit or Pittsburgh, bankers from New York, attorneys and teachers from Washington, and journalists from Los Angeles. Everyone has one or two books stacked on top of issues of *Essence, Vanity Fair,* the *New York Times,* the *Vineyard Gazette,* and the black newspaper from home. Pen and paper are also de rigueur for copying down phone numbers of old friends one will inevitably see. It's a perpetual networking scene.

"Unfortunately, there's also this coarse element that has started showing up here on the Fourth of July," says a New Jersey dentist as he looks up from the beach to a group of teenagers playing a radio and sitting on a railing overlooking the beach.

"They obviously have no business here," adds the dentist's wife as she adjusts the umbrella behind her and rubs sunblock on her pale, lightly tanned skin. "All these loud, dark-skinned kids coming over here *for the day*. They nearly destroyed Virginia Beach—and now they think they can just plop themselves down here. Just because we're black doesn't mean we have to put up with this."

"They think they're fitting in, but they are clearly not our kind of people," adds another woman as she drops the *Vineyard Gazette* by her side. "But I guess we can all tolerate them two days out of the whole summer. After all, they always just go back to wherever they came from."

A short walk from the Ink Well is Circuit Avenue, the main shopping street in Oak Bluffs, which consists of about two dozen shops, restaurants, and small businesses, and a post office. The sidewalks on the narrow street are jammed with people stopping and gaping at the slow-moving cars, straining to see who of their friends is back on the island. Among the newer institutions on the street is the dance club Atlantic Connection, which attracts young people, who line the street outside on Friday and Saturday summer nights.

With the increasing number of blacks visiting the island, there seem always to be new ways to separate the elite from the ordinary. Ever since I was a teenager, I recall that one popular method of establishing divisions was by asking every new face one question: "Do you rent, or do you own?" It's a question that you hear black people asking up and down Circuit Avenue as they bump into vaguely familiar black faces, or completely new black faces that seem inviting enough to welcome with a few words of introduction.

"If they've owned a house here for twelve years, I'd know 'em," said a gray-haired woman, interrupting a bridge game that I sat in on recently.

"Well it's supposed to be some fabulous place with six bedrooms," added her partner. "Something like five thousand square feet, with a tennis court. And he's a surgeon from Washington, and she's a lawyer."

"I'm telling you, Alma," insisted the older woman. "Surgeon from D.C., five thousand square feet. I'd know 'em if it was true. Never heard of 'em."

A third woman, also a longtime owner in Oak Bluffs, took a sip from her iced tea. "Well, that's *probably* because they're in Chilmark!"

"Ha! I knew it!"

"Well, there you go."

"Okay, then."

I looked around at the women sitting on the screened porch of a home not far from the town's Waban Park.

"What kind of black folks are going to be in Chilmark? Is his wife white?"

"They must be hiding. Might as well be in Chappaquiddick. Or Nantucket."

There was nothing more to say about the Washington surgeon to this group of women. He and his wife were so irrelevant, they were not worth further discussion.

Among the blacks who summer on the island, there are fine distinctions made between those who own in Oak Bluffs or Vineyard Haven, those who *rent* in Oak Bluffs or Vineyard Haven, and those who own *or* rent in places like Edgartown, West Tisbury, or Chilmark. Blacks who live in the latter three towns might as well live on the island of Nantucket. The old-guard blacks dismiss these vacationers as "Incognegroes" who are trying to hide out in faraway white communities.

"Historically, blacks were not welcomed anyplace but in Oak

Bluffs—and later in Vineyard Haven," explains a Chicago attorney who owns an Oak Bluffs home, "so people always question blacks who choose to live outside these two areas. We kind of wonder if you're hiding or trying to separate yourself for some reason."

While any person, black or white, can place themselves on the Ink Well beach section and attempt to blend in, there is no mistaking who the old-guard people are in Oak Bluffs. Among the established summer residents is a small club known as "The Cottagers," a group of one hundred black women who host clambakes, bridge games, and fund-raisers for various island charities. The close-knit, almost clannish group not only owns their own clubhouse on Pequot Avenue but they also give out scholarships to students headed for college.

Hester Boxill is a member of this selective club that has been around for more than forty years. "Our group is limited to one hundred members," she explains, "and while many of our members live in Oak Bluffs, the primary requirement is that they own their own homes somewhere on Martha's Vineyard." Along with Boxill, there are such other members as retired teacher Helene Wareham, Mary Manley, and Agnes Louard, one of my mother's former professors from the Columbia University School of Social Work.

Louard, who has owned three different Oak Bluffs houses since 1964, is a big fan of the Cottagers and their activities. "We have bridge groups here on Monday mornings, as well as dance, exercise, and arts lessons for children," says Louard, who often runs into her fellow Delta sorors at Cottager events. "Although we have meetings at the clubhouse on every other Wednesday, our most popular event is our annual house tour which takes sightseers to the historic houses in Oak Bluffs. We have as many as three hundred guests, and most of them are white visitors who are interested in the community's history."

"My sister, Liz White, and her friend, Maggie Alston, negotiated the purchase of the Cottagers' clubhouse at a time when black organizations rarely owned their own buildings," says Doris Pope Jackson as she points out a building on Pequot that once served as the neighborhood firehouse.

In addition to the Cottagers, there are other institutions and events that continue to link the elite blacks on the island. One of those events is the annual Tucker Tennis Tournament that takes place at the family compound of Judge Herbert and Mary Tucker. The invitation-only gathering begins in early August and brings together some of the most prominent summer residents as observers and participants in a mixed doubles tournament where the women pick their male tennis partners through a drawing.

Dr. Beny Primm, an anesthesiologist and leading medical expert on drug addiction, has been visiting Martha's Vineyard since the 1940s and is a popular player in the Tucker Tournament. His name has been drawn by such neighbors as PBS journalist Charlayne Hunter-Gault, who recently built an Oak Bluffs house with her husband, New York investment banker Ron Gault.

Another popular summer tennis tournament was established at the Oak Bluffs public tennis courts more than twenty years ago for Labor Day, and, while it attracts white residents as well, is a must-attend for the blacks in town.

In addition to the very oldest families such as the Wests, the Joneses, the Shearers, and the Ashburns, there are many other established names on the island.

There are Harvey and Jackie Russell, whose cedar-shingled house people first see when the ferry approaches the Oak Bluffs dock. In addition to being a former sire archon in the Boulé and a charter member of One Hundred Black Men in New York, Harvey became the first black to be named vice president of a *Fortune* 500 company when Pepsi promoted him in the 1960s. His wife, Jacqueline Denison Russell, an avid tennis player and a member of the Girl Friends, is the daughter of Major Franklin A. Denison, the highest-ranking black in the United States Army in World War I.

Others in the New York contingent are Walter Lowe, a retired AT&T executive; Jill Nelson, the former *Washington Post* reporter, and her mother Alelia, who has a prominent house overlooking the water; and Howard Johnson and his son Brad, who have owned such New York restaurants as the Cellar and Memphis, and the trendy Los Angeles eatery Georgia's. Howard recently sold his million-dollar Vineyard Haven home, which overlooked the harbor.

Another group is the Boston crowd. Although they were the first contingent to arrive on Martha's Vineyard, black Bostonians are clearly no longer the dominant voice. "Those New Yorkers now outnumber us three to one," jokes native Bostonian Jack Robinson, who lives in Oak Bluffs and founded the Martha's Vineyard Resort and Racquet Club. In addition to Robinson, over the years prominent Bostonians have included former senator Edward Brooke, who was the first black U.S. senator after Reconstruction. The new Boston crowd includes Harvard professors like James Cash, Charles Ogletree, Randall Kennedy, Chris Edley, Cornel West, and Henry Louis Gates.

Harvard graduates Fletcher and Benaree Pratt Wiley are among the Boston set. "Flash" and "Bennie" bought in the 1970s and are active in many civic groups, including the Links. Flash is an attorney and head of the Boston Chamber of Commerce, while Bennie is the sister of former Washington, D.C., mayor Sharon Pratt Kelly, and runs the Partnership, a nonprofit board of corporate executives who study business and social issues. "I wouldn't have raised our kids anywhere else," says Bennie, who owned a Vineyard Haven toy store, Giocatolli, in the 1980s. "It's so safe, it reminds you of a 1950s community. Everyone looks out for each other's children."

Others vacationing from around the country on the island include Washington power couple Vernon Jordan and Ann Dibble Jordan; former insurance executive Earl Adams and Ezola Adams from Atlanta and New Jersey, whose son is married to Vernon Jordan's daughter, Vickie; former U.S. secretary of health and human services Dr. Louis Sullivan and his wife, Ginger; and Grace Frye of Los Angeles, who owns an estate in Vineyard Haven at the edge of Lake Tashmoo.

And there are also a growing number of black celebrities, like Spike Lee, who have bought property on the island. "But those celebrities aren't a part of our crowd," says one matron who has owned for more than forty years in the Pork Chop section. "They're here today, gone tomorrow—no stability and no background." The woman pauses for a moment, then nods approvingly. "But I guess Spike Lee *did* go to Morehouse."

As old names disappear from the area, new ones are quick to replace them.

While the old guard was sad to see the home of deceased New York physician and *Amsterdam News* owner Dr. C. B. Powell fall into disrepair after it was given to Howard University, they were all thrilled to hear that Harvard professor Henry Louis Gates had purchased the old Overton property overlooking the Ink Well. "We hear that Skip Gates is thinking of establishing a library there for black studies research," says Primm. "That's the kind of place that will help to further enrich black Martha's Vineyard." Mary Gardner Lopez agrees. Although she and her husband, George, now live in the Sengekontacket section, she remembers living two doors down from the Powells. "C.B. and Lena had the largest and most beautiful piece of property," Lopez remembers, "but when they died, nobody had the kind of money to keep up what was the size of three large houses. There's a lot of history and we need an owner who appreciates that."

A critical area in which blacks have failed to succeed in the resort is their ability to establish businesses there. In fact, given the concentration of well-educated, sophisticated black businesspeople who vacation on the island, I am struck by the dearth of black enterprises in the area. It is as puzzling as Harlem's 125th Street—a neighborhood filled with black consumers, yet virtually no black merchants to serve them. Longtime black neighbors I talk to in Oak Bluffs cannot think of even five establishments owned by blacks on Circuit Avenue. "There's Cousin Rose's Art Gallery that Zeta Cousins owns, Summer Breeze Gift Shop that's run by Diane Tarter, and the Carousel Clothing Store that the Foy family owns," says a resident.

"We are the richest blacks in the country and we have virtually no businesses here," says an embarrassed Oak Bluffs resident whose family has summered there for three generations. "It's a disgrace. Part of it is our fault because we get up here and just do a lot of profiling and social climbing. But white people are also at fault because they are not eager to support black businesses."

A few blocks up the hill from Circuit Avenue is the Martha's Vineyard Resort and Racquet Club, which was established by Jack Robinson in 1989. The struggle to open this black-owned club highlights the challenges faced by black merchants who experience the subtle racism in the area.

"The town government gave him a real runaround," says a black resident who lives off New York Avenue near the club's two-acre site. "They knew that black people with money were coming here, but the idea that we would have a country club too made us appear too permanent. It scared the town to death."

"It felt great to open the first black-run country club on the island because the white residents here already had their clubs," explains Robinson, a commodities trader and former Boston NAACP president. "They had the East Chop Club down the road, and they had clubs over in Edgartown, but we were stuck entertaining in our backyards or on the public beach." Robinson's club, which is the only Vineyard resort with clay courts, has a fitness club, sporting goods shop, miniature golf course, jazz club, computers, sun deck, and hotel suites.

"When the white residents, who quietly opposed the club, finally walked in and saw that these black folks had an eighteen-hundred-square-foot lobby in that building," says a New York Avenue neighbor, "they flipped. Now they see both blacks and whites staying at the club."

Doris Stewart Clark founded the only black-owned licensed bed and breakfast on the island several years ago. A charming pink-and-white Dutch colonial built in 1906 in Vineyard Haven, the Twin Oaks Inn has won numerous awards from vacation guides and was host to many of President Bill Clinton's staff during his high-profile 1994 summertime visit. But Clark was not welcomed with open arms when she was ready to establish the inn in the 1980s.

"Not only was it a three-year fight to get a license from the local zoning board," says Clark, whose relatives first came to the island more than sixty years ago, "but one of my white neighbors on the block specifically opposed my inn despite the fact that there was an inn operating on the same street and forty others that the zoning board had already approved prior to mine. The others were white-owned, and this neighbor actually told me that I should consider moving my business plans to Oak Bluffs, where there were a larger number of blacks."

Since Clark has turned Twin Oaks Inn into a popular stop for prominent blacks and whites, the neighbors have tried to forget the quiet dispute. Still, black residents are aware of the thin line of comfort that exists among the growing numbers of blacks in various areas outside of Oak Bluffs. I am particularly aware of this thin line when I am in Edgartown, the whitest and richest town on the Vineyard. Today, I feel quite comfortable perusing the merchandise in Edgartown stores like Bickerton and Ripley or the town's palatial Harbor View Hotel, but as a child, it was the one town that my parents—and many other blacks—had no interest in visiting. Edgartown was considered a haven for older, white, blue-blooded Republican families, and its busy shops and residential streets are still populated by an almost exclusively white group of residents and tourists.

But even with the subtle distinctions between the black and white sections and the nuances of those who own and those who rent, people who come to Oak Bluffs still believe it is the supreme place for the black elite. Longtime resident Robert Jones notes that his closest and oldest friends are people he met either at Howard or while summering as a child in Martha's Vineyard. "From the minute that you line up for the ferry ride from Woods Hole in Cape Cod, you get a sense of the Vineyard's informality and its energy."

I remember how I felt as a child each year on that ferry. We lived a wonderful life in New York, but it was one where race always seemed to be a factor in the things we saw and heard. In Oak Bluffs, we felt in control of our environment and protected against the offenses that were often witnessed by black children and black people living in a predominately white society. As we approached the harbor each June and caught the first view of those beautifully weathered, gray-shingled cottages on the bluff, I felt a serenity that I never felt on the mainland. It is a mood I am still able to capture when I return as an adult today.

While not as well known as Oak Bluffs, a second summer resort community that is highly popular among the black elite is Sag Harbor, a small village on the eastern end of Long Island, New York.

Although blacks have populated Sag Harbor and sections of nearby East Hampton, Long Island, since the early 1800s, in the early years they were primarily year-round laborers who worked in the whaling industry. During the nineteenth century, Sag Harbor and New Bedford, Massachusetts, were two of the northeast's largest whaling ports. According to E. T. Williams, a Sag Harbor resident who has sat on the board of the town's Eastville Historic Society, it wasn't until just before the 1920s that black professionals started to purchase or build their own summer homes.

"The first community of blacks here lived in the Eastville section of Sag Harbor, and they lived amongst a small group of Native Americans," says Williams, a real estate developer who spent his childhood summers in Sag Harbor and who also has a home on Manhattan's Upper East Side. Today, Williams, his wife Auldlyn, and their two daughters live on a four-acre Sag Harbor family compound where they maintain other family structures. He says that in the early 1800s, many of the blacks in the area were fishermen and craftsmen. "In fact, the church in our area, St. David AME Zion, was built in the 1840s and was a stop on the Underground Railroad as slaves were smuggled from the South."

While many black residents in the early 1900s owned houses on Lighthouse Lane, a street off of Route 114, many more were sent to buy land along the water, as most other sections of the village were reserved for white purchasers. The blacks were pushed into the remote east end of the four-square-mile town. It is for this reason that many affluent white Sag Harbor residents and vacationers are virtually unaware of the long-time presence of blacks in the community.

"I started coming to Sag Harbor with my parents and grandparents in the 1940s, and black people were very clearly establishing our own neighborhoods," says Williams, who, along with his friend Bill Pickens and other children of the Sag Harbor elite, grew up in a circle of black boys and girls who spent the summer taking tennis, swimming, and sailing lessons. As they became teenagers, they even formed a social club called the Centurions, which ended up nominating and sponsoring some of the prominent young girls for various New York debutante cotillions.

Although we did not come to Sag Harbor as regularly as we summered in Martha's Vineyard, my family first visited Sag Harbor in the mid-1960s for two- or three-week periods. As much as we liked the people, then and now, when I go back today, two- or three-week periods are still about as much as I can take, given the social intensity that one experiences in this Hamptons environment. Unlike Oak Bluffs, which has traditionally attracted a more diverse, laid-back crowd from the South and Midwest as well as the Northeast, the Sag Harbor black crowd is mostly a fast-paced, sophisticated, and often cynical New York crowd that likes formality and likes to avoid the rustic whenever possible. For example, in Martha's Vineyard, dirt roads and sandy driveways are common, while Sag Harbor's norm is paved streets and designer gravel.

Since the 1940s, the black elite has settled into five specific sections of Sag Harbor—all on the east side of the small village. Each of the neighborhoods is off Route 114 and consists of no more than five or six small roads that had been left unpaved until about twelve years ago. While a handful of black society families live outside these neighborhoods, most own and socialize in these five tight-knit sections. Three of the neighborhoods offer beachfront property along the Sag Harbor bay. The first neighborhood—and the one with the most high-profile residents—is Azurest. Here, in a neighborhood of about one hundred families, the most desirable address is on Terry Drive, which runs along the water. This is where, among other well-to-do black families, Dr. and Mrs. Chester Redhead live.

For three decades, the Redheads have helped to direct the black social scene of Sag Harbor, which is made up of some of the most successful professionals in the northeast.

A prominent New York dentist and trustee of Howard University, Chester says, "We decided to buy in Sag Harbor rather than Martha's Vineyard because of the close proximity to New York. Although we always visit our friends in the Vineyard, it's much easier to live here because of the short drive or plane trip. We have no ferry rides like they have in Martha's Vineyard, so I can see patients on a Friday afternoon in New York and be out here that same evening."

Redhead, whom I have known since I was a child, will either drive his Rolls Royce Silver Shadow from the city or take a forty-minute trip by plane and meet his wife, Gladys, at the East Hampton airport. The two of them look at the resort town as being an easy extension of the comfortable life they have carved out in Manhattan and the tony suburb of Scarsdale, New York.

"This has always been a wonderful place to raise kids in the summer," adds Gladys as she looks out over the deck to the beach and waves up to their son Raymond, who is flying a small red-and-white Cessna 172 single-engine plane a few hundred feet above their contemporary-style home. "Everybody knows everybody here. I never had to worry about our three sons because this is a family community and we all looked out for each other. And that's still true today."

To see the truth of Gladys's point, one only has to walk down their narrow street, which follows the beachfront. Terry Road, which has just nine homes, includes the residence of Ken Chenault, president of American Express; Earl Graves, publisher of *Black Enterprise* magazine and owner of the Pepsi bottling franchise in Washington, D.C.; Cecil Broderick, a New York physician; Arch Whitehead, a New York headhunter; and several other well-known names in the black New York crowd. Three of the nine families on their street have kids who grew up together in Scarsdale, and the others hail from Manhattan, suburban New Rochelle, and other parts of New York and suburban New Jersey.

As a gardener trims away at a pine tree (a sight you'd never see in Oak Bluffs) in the Redheads' front yard, real estate developer E. T. Williams is getting into his blue Mercedes station wagon to run into town. Vascular surgeon Bill Curry—on his way to the Shinnecock Hills Country Club—is thanking Cecil and Mercedes Broderick for a referral, and decorators are arriving next door at Ken and Cathy Chenault's to add finishing touches for a party to be held in honor of Alma Brown, the recently widowed wife of U.S. commerce secretary Ron Brown. Two doors down, caterers are opening cases of soda and laying out furniture on the deck of Earl Graves's stately two-story shingled home. Graves and his wife Barbara are moving back and forth from the kitchen to the deck with their grandkids and daughter-in-law, Roberta—who went to Yale with their son, Butch—as they play host to Bob Holland, former CEO of

Ben and Jerry's; his wife Barbara; and Link Inez Richardson and her husband, Franklyn, a fellow Boulé member, who is a resident of Scarsdale and pastor of Grace Church in Mount Vernon, New York.

"There's a lot going on in Sag Harbor, and I've spent virtually every summer of my life here—including the one when I met my wife," says Bill Pickens, a member of one of the first black families to summer in the community. Pickens is not alone in noting that everybody is somebody in this community.

"Sag Harbor consists of a very New York-based group of families," remarks Gladys Redhead as she walks up from the hot beige sand behind her home. As she recalls a quick trip to the nearby Saks Fifth Avenue and a recent gathering the day before, she says, "In just a few hours on this beach, in town, or at someone's cocktail party, you will run into people from all over Manhattan, Brooklyn, or suburban New York and New Jersey."

Gladys herself is one of those many native New Yorkers who can be found hosting reading groups and art shows for the visitors and residents of black Sag Harbor. Like her husband, Chester, she is a tireless host and supporter of events that fill the calendars of the Sag Harbor crowd. A graduate of Howard and Columbia University Graduate School and a retired schoolteacher from suburban New Rochelle, she grew up in one of those well-connected, Sag Harbor-type families. "When I was growing up in New Rochelle, my father was an established dentist there, so I grew up knowing kids who were vacationing in places like this. Since both of my parents had gone to Columbia and were active on the school board in New Rochelle, my whole life has been New York-focused."

Adrienne Lopez-Dudley agrees that Sag Harbor's influence is a very New York one. The daughter of New York socialites Dr. George Lopez and Mary Gardner Lopez, Lopez-Dudley is an executive in the New York office of Nickelodeon. Having spent summers in both Sag Harbor and Martha's Vineyard, she says, "there are more New Yorkers here than you'll ever find in the Vineyard."

Adrienne, who married the son of Edward and Rae Dudley—another old Sag Harbor family—also lives in Scarsdale with her children. Growing up in a house where her father practiced dentistry with many of the Sag Harbor crowd and a mother who wrote the society column for the *New York Voice,* Lopez knew the inside scoop on who was who in the popular resort community.

Even though the majority of the Redheads' social circle are fifty and older, there is a younger thirty- and forty-something crowd that one spots along the streets of Azurest. And for all of them, there's an interesting story or credential that seems to get whispered about them the moment they're out of earshot. There's the architect Garrison McNeil and his wife Cheryl, who live in an unusual two-story cedar house that seems a bit too much like the Hamptons for this very un-Hamptonslike street. ("They've got two kids—really smart—one at Swarthmore and one at Hunter High, and you know he was one of the architects who redid the Apollo Theater in the eighties. There's talk that he may be designing the new Pathmark too.") And Steve and Barbara Evans Williams. ("Steve's father was a doctor in Harlem, and you know his sister, Saundra—she was one of the first black women law partners. Saundra's husband is Don Cornwell, who used to be at Goldman Sachs and now owns Granite Broadcasting, which owns all those TV stations. And you know Barbara's brother, John Evans, is a dentist in New York—and you remember John's wedding three years ago to Vanessa, that lawyer with the green eyes. I think she's a Delta.") And so forth.

While I often chuckle to myself when I hear talk like this, it reminds me of why I was never able to relax in Sag Harbor. Even when I visit today as an adult with close friends, I get the feeling that people are watching, talking, and keeping track. Because the neighborhoods are less rustic, more developed, and more densely populated by New Yorkers who know each other, new visitors are recorded immediately and become the subject of daily observations and conversations.

Each summer, the Azurest neighborhood association and its dues-paying members sponsor a large beach party for its residents and guests. People get particularly excited when a high-profile Azurest resident like *Essence* editor Susan Taylor and her husband, Khephra Burns, show up. "Now, that's a glamour couple," confirms one of the older residents when Taylor and Burns pass by on one of the Azurest roads.

A second black neighborhood along the water is Sag Harbor Hills, which is just to the east of Azurest. The sought-after street there is Soundview Drive, where there is an uninterrupted view of the beach and bay. Living in the Hills are such people as the restaurateur and TV personality Barbara Smith and her husband Don Gasby, who together run her New York and Washington restaurants, B. Smith's.

"Dana, bring them up to the deck," says the glamorous Barbara Smith as her polite ten-year-old daughter leads three unexpected guests into her gleaming white house—a postmodern structure of glass, steel, and aluminum. "This is where I get away from everything," says Smith, a former model, as she walks through a kitchen that looks like a California television studio. In fact, at the time of my visit, Smith (who is often called the "black Martha Stewart") was just launching her nationally syndicated television talk show, which she hosts for viewers interested in household and lifestyle tips. "Our daughter loves it here," she says. One of the more unusual homes in the Hills section, the house has a roof made of bright steel and aluminum, and a great portion of the outer walls is constructed of tinted glass, thus offering an unbroken view of the Sag Harbor bay.

Also living in Sag Harbor Hills is the Day family. Dr. Thomas Day, a retired dermatologist who divides his time between Sag Harbor and Manhattan, is a die-hard New Yorker who has owned his home for thirty-six years. "I was born in Harlem Hospital, but this is home," says Day, whose great-grandfather was a cabinetmaker who became well known in the early 1800s in North Carolina, where he designed the Day bed. A catalog in his living room reveals that many of his ancestor's furnishings are on display at the North Carolina Museum of History in Raleigh. "There are many people out here who have interesting ties to history," says Dr. Day, who practiced in Harlem for several decades. "You just have to meet a few people and then the stories just open up."

The people that Dr. Day is talking about have been showing up in big numbers for the Hills' annual dance, which has taken place every July at the Salty Dog or the Waterside.

"And this is the last of the three black beach neighborhoods," says Chester Redhead as he steers his large Rolls onto Lincoln Drive in the Ninevah neighborhood. As he drives me down the block, it's obvious that he knows the people and the houses well. At the mere sight of his car, friendly waves are elicited from sundecks and driveways along the road.

Unlike Azurest and Sag Harbor Hills, Ninevah is a very small neighborhood. "We are only about thirty families and, as we joke, all our streets are named after Republican presidents," says Barbara Brannen, who first bought a house in Sag Harbor in 1953. "There's Lincoln, Wilson, Harding, and Taft. My husband and I persuaded a lot of our friends from New York to move here, and many of them are starting to make this their year-round residence."

Although Ninevah is considered more low-key than the other two beachfront neighborhoods, its annual arts festival brings out the neighborhood's old guard and their friends every August in big numbers. In addition to Brannen, who was president of the Brooklyn Girl Friends in the 1970s, president of the Queens chapter of Jack and Jill in the 1960s, and a columnist for the *Amsterdam News,* Ninevah has such other well-placed people as Sylvia Hayes, whose family owned Small's Paradise in Harlem; Gwendolyn Dukette; Dr. Buddy Gibbs; and socialite Hazel Gray.

Just one block away from the three beachfront neighborhoods, on the opposite side of Route 114, are two other black neighborhoods that I am certain I never visited or heard of as a child. The first is Chatfield's Hill. Judy Henriques's family has owned a house on Lighthouse Lane in the Chatfield's Hill section for more than forty years. "My dad bought this house in 1952 at a time when virtually all the houses in Chatfield's Hill were owned by whites." A successful architect, Al Henriques eventually saw other black professionals move onto his street once the lots filled up in the waterfront neighborhoods. Among the active residents in Chatfield's Hill is neighborhood association president George Burnette, who is also chapter president of the American Association of Retired Persons.

"Come on in," says Ernie Hill as he beckons some more unexpected guests up the stairs of his multilevel contemporary home in the fifth neighborhood, Hillcrest Terrace. "Ardeth and Alma are on the deck off the family room."

Passing by a foyer, staircase, and living room lined with paintings by Frank Wimberly ("Everybody out here has at least one piece of Wimberly's art in his house," says Chester Redhead in a quiet, approving tone), one finds Ernie's wife, Ardeth, sitting on an elevated shaded deck with Alma Brown, widow of Ron Brown.

"Did you see Michael on the beach today before he left to golf?" asked Brown, referring to her son. "He and Tammy are over at Ken and Kathy's."

From the deck, you can see the woods in back and the narrow road that continues up a steep hill. As you continue up the road, you pass the lovely homes of people like Gaynelle Spaulding Woods, a big name in the Girl Friends and one of the founders of the Baltimore chapter of Jack and Jill. It is quieter here and you get the distinct impression that this neighborhood is less populated, less pretentious, and less black, compared with the go-go neighborhoods across Route 114.

"The place to be is in the Hills or Azurest," says one resident walking a golden retriever. "It's a real scene over there. Chatfield's and Hillcrest are really not in your face. It's nice here, but it's not what most visitors want when they come to sightsee in the Hamptons. You don't get all the glitz in these two neighborhoods."

"That's because white people have infiltrated it and they're not keeping their property up," says a Ninevah resident who points out a white family's backyard in Hillcrest Terrace, where wet towels and linen are hanging on a rope tied between the house and a pine tree thirty feet away.

"Would you look at that?" says a New York attorney who once owned a house in Sag Harbor. "This is why I sold and moved to Martha's Vineyard. You get too many of these uneducated rednecks deciding to move here as year-round residents, and they have their pickup trucks, loud music, and bad habits."

The people of Chatfield's Hill are not the only ones who think about the racial makeup of the homeowners in the historically black neighborhoods of Sag Harbor.

"When we first came out to Sag Harbor, they didn't want us living on the other side of town, or in the village's downtown area, so we came out here," says a resident of Sag Harbor Hills. "Now, we're losing some of the houses on our own waterfront when some of these children get married to whites and get divorced. The white spouse keeps the house and that's that. At least three houses in our neighborhood have turned over to whites who want nothing to do with us or with the support of the neighborhood association that makes this a nice place to live."

Because the housing lots in Sag Harbor are relatively small and offer little privacy, the black Sag Harbor crowd are highly visible and accessible to each other. This visibility means that everybody knows everybody, everybody talks about everybody, and everybody remembers everybody and everything. For new visitors, that's a bad thing, since they never know how to respond. But for residents and longtime vacationers, it's a good thing, because, for the most part, the people in this tight-knit group think of everyone as part of an extended family. For this reason, the people here have a long collective memory about residents from the past. Reminiscing is a sport for many.

The Sag Harbor families have favorite names that they like to hear discussed at their parties. So aggressively do you hear these same names mentioned over and over that it becomes obvious that such discussions are meant to shut out new arrivals who have no ties to the old Sag Harbor elite or to the old social scene that was once dominated by people like the Hudnells, Dudleys, Bannisters, Taylors, Harts, and Guiniers. If you can tell a story linking yourself or your family to any of these people, you can at least appear to be second-generation Sag Harbor. And that's saying a lot.

Other names that get talked about often are Dr. Stewart and Yvonne Taylor, who started flying in each year from Rochester, Minnesota, two generations ago. The most talked-about individual, however, seems to be Emilie Pickens, who had been president of Jack and Jill in the 1950s and was also founder of the children's parade that takes place in Sag Harbor Hills each summer. Her son, Bill, married Barbara Brannen's daughter Audrey—thus uniting two old-guard families. A couple who seem to be mentioned with great frequency are Julius and Hazel Gray. "During the 1950s and 1960s, Hazel Thomas Gray was the consummate socialite and hostess who always knew who should be on which guest list," says a Sag Harbor regular who always saw the Grays in Sag Harbor and at the extravagant biennial debutante balls that Hazel hosted with her Girl Friends chapter in Brooklyn. "I get a thrill whenever I see her on Ninevah Beach."

Homes in Sag Harbor are for dropping in, unlike those in Martha's Vineyard, where house visits are secondary to the time people spend at the Ink Well or on Circuit Avenue or at the Cottagers' clubhouse. As Chester Redhead tools me around the five pivotal neighborhoods in his dark-green Rolls Royce Silver Shadow (he and his neighbor Earl Graves both have Silver Shadows), the popularity of informal house visits is evident. The front doors are open for a constant flow of familiar faces.

Another growing trend in this community is for parents to buy up additional homes near their own so that their guests or adult children will have nearby quarters. Like Earl Graves and others, Redhead recently purchased another home near his own in order to have his grown sons nearby during the summer months.

As one ventures outside the eastern end of the village, one finds a few black families who have located there in the last fifteen years. Robert and Elizabeth Early have one of the newer homes near Sag Harbor's marina. Although they live outside the five historically black sections, most of their friends live in those areas. Dividing their time between their summer residence and a home in suburban New York City, they give frequent get-togethers for members of the Girl Friends, the Links, and Bob's Boulé chapter.

"A lot of the people who come out here for the summer have ties to the same organizations," comments Phyllis Stevenson, a former national president of the Girl Friends group, as she stood in the Earlys' living room during a recent July Fourth cocktail party. "There's lots of Howard graduates here. And many of us are members of the Girl Friends, like me and my mother. Elizabeth is on the board of Howard with Chester. And there's Ernesta Procope over there. She's in the Girl Friends too. And you know my friend Richard Clarke, who used to be in business with Frank. And there's Al and Rosa Hudgins—I think you know their daughter Kendall—they're over next to Earl Arrington. They are both in the real estate business."

Stevenson, an attractive and outgoing woman who lives in suburban Westchester County, stood next to her mother, Anna Murphy, who had helped found the Girl Friends in the 1930s. They have been coming to Sag Harbor for more than forty years. As new guests entered the room and embraced Phyllis like long-lost family members, she added, "Most of us have known each other for years. Every summer, Sag Harbor brings us back."

The credentials of the people in this Sag Harbor living room are inspiring. For example, Donald Stewart and his wife Isabelle are executive directors, respectively, of the College Board and the forty-year-old organization Girls Incorporated. He is the former president of Spelman College, and she sits on the board of the Museum of Modern Art. Ernesta Procope and her husband John were one of the first blacks on Wall Street with their insurance brokerage firm, E. G. Bowman and Company. She's been on the board of Avon and he's been the owner of the New York *Amsterdam News*. In addition to being a real estate investor, Al Hudgins is a former head of Carver Federal Savings, one of the nation's oldest black banks. Richard Clarke, owner of a personnel firm, sits on the board of the Metropolitan Museum of Art.

"Of course there are many white Sag Harbor people—the media people, for example—who have no idea that we're out here," says a resident of the Ninevah Beach area as she drives her German sports car down Redwood Road, which runs along Sag Harbor Cove. "Like these people that live around here. All white. All lily-white," she says with a wide sweep of her hand. "They think of Sag Harbor as being theirs—all white—because they don't see us out their window or in big numbers along Main Street, the way you see us in Martha's Vineyard."

E. T. Williams says that although the black and white neighborhoods have always been separate, these two groups have always interacted harmoniously. "I remember that even in the 1950s, the white store owners were always respectful of the black children who went in and out of the shops," Williams says as he recalls his early years, as well as the experiences of his daughters. "John Harrington, the chief of police," says Williams, "knew everybody's parents, whether they were black or white."

If there is any discussion of race relations, it has more to do with the fact that the Sag Harbor black elite want very much to preserve their five neighborhoods for their next generation. "We need to keep Sag Harbor for our children," explains Earl Arrington, who lives in a large modern home with a swimming pool. He thinks of his daughter, who is about to receive her doctorate from New York University. "There are important social and business contacts for our kids in this community. They have been here for whites for many generations, and we should try to preserve the same opportunities for generations of blacks."

To continue the legacy, several members of the black elite have taken a leadership role in the resort community's civic affairs. Barbara Brannen of Ninevah has served as president of the village's John Jermain Library, and her husband, James, served as president of the local Knights of Columbus and for eight years as a village trustee. "We are more than just a few black families in this village," says Barbara Brannen proudly. "We have been here for several generations and we have a stake in its future."

Beyond the larger resorts of Oak Bluffs and Sag Harbor, there are smaller, more regional resorts like Highland Beach on the Chesapeake Bay in Maryland, Idlewild in Northern Michigan, and Hillside Inn, a black resort hotel in the Pocono Mountains of Pennsylvania. These more intimate communities do not have the same national reputation as the other two northeastern villages, but they nevertheless deserve mention.

"Oak Bluffs may have the numbers, but we have the history and the lineage," says a retired Howard professor whose family has vacationed at Highland Beach for three generations.

Unlike Sag Harbor and Oak Bluffs, Highland Beach was established specifically for and by blacks who were not allowed to visit the racially segregated communities in the area. The land was first purchased by Frederick Douglass's son, Major Charles Douglass, in the 1890s as a forty-four-acre parcel along the Chesapeake Bay. Douglass built a home on Wayma Avenue known as Twin Oaks.

Among the first residents of Highland Beach were Washingtonians like poet Paul Lawrence Dunbar, Judge Robert and Mary Church Terrell, and other black professionals. While not the most prominent of the black elite vacation spots, Highland Beach is the community that enjoys the most interesting place in black history. It was the first Maryland town to be established and incorporated by blacks and was home to famous black political leaders, intellectuals, and businesspeople who built cottages in the community. Interestingly, it has its own mayor and a board of commissioners to handle common municipal matters and local ordinances.

"There are no more than twenty-five or thirty people living there year-round," says Washington resident ViCurtis Hinton, who started going there as a child, "but everyone comes back in the summer." Considerably smaller than the Oak Bluffs crowd, "everyone" is a relative term because Highland Beach consists of fewer than one hundred homes. I found it to be similar to Oak Bluffs in its informality, yet almost identical to Sag Harbor in its density and its parochial attitudes: After all, it is mostly a Washington, D.C., group that runs the community. Historically, the Highland Beach neighborhood has included such people as the Memphis Church family; the recently deceased Robert Weaver, who served as secretary of HUD under President Johnson; and Patrick Swygert, the president of Howard University. Although its image is much lower-key than that of Oak Bluffs or Sag Harbor, the community now attracts many Baltimore, Richmond, and Philadelphia families in addition to a solid core of well-to-do Washingtonians, who actually seem to be in control. "It's the Nantucket of black resorts," says a Washingtonian whose family has seen four of its generations at Highland Beach. "It's small, private, and intimidating to outsiders. And we like it that way." Less than an hour's drive from the capital and just five miles south of Annapolis, its location makes it a popular summer and weekend getaway.

Traveling along the property line of the Hillside Inn resort is a steel-gray Jaguar. It climbs a steep incline covered with a half-foot of icy snow as three of its passengers try not to look over the side of the road, where a fifty-foot drop lies just inches from the unprotected driveway's edge. "So what's everyone so quiet about?" asks Judge Albert Murray, who has been driving the treacherous snow-covered roads of the Pocono Mountains ever since the early 1950s when he first opened this black-owned resort in the famous Pennsylvania resort area.

The Pocono Mountains in Pennsylvania have never achieved the popularity among the black elite of Martha's Vineyard, Sag Harbor, or Highland Beach, but blacks in the Northeast know about the history of Hillside Inn, a small resort hotel owned by retired New York City Judge Albert Murray and his wife Odetta. "My wife had always said that she wanted a place in this resort area where blacks could come for a vacation with the same amenities that whites had," explains Murray, who had become a wealthy landowner by the late 1950s after graduating from Brooklyn Law School. "We were both very disturbed that the resorts in this area did not allow blacks to stay in their hotels. There was only Archerd Cottage."

While there were many top resorts during the 1950s and 1960s, like the Shawnee Inn, which hosted President Eisenhower, the Murrays wanted to offer a luxury resort for blacks as well. Before they transformed the inn into a hotel, it had previously been a boarding house with roots going back to the late 1800s. "Since the Murrays are not society types," says a returning guest from Philadelphia, "they don't create the kind of social pressure you feel in Sag Harbor or Oak Bluffs. This is a quiet place that lets you disappear from the gossip scene."

"People don't talk about Idlewild as much as they used to when there were clubs and major entertainers visiting during the summer," says Detroit attorney Joseph Brown, who has been visiting the Idlewild resort in northern Michigan since he was four years old. "But there is still a loyal group of people who come from Detroit, Chicago, Cleveland, Milwaukee, and St. Louis. In fact each of those cities has an Idlewilders Club that arranges annual parties during the second week of August."

A lakeside resort area approximately two hundred miles north of Detroit, Idlewild was a haven for the black elite between the mid-1920s and the early 1960s. During its peak, Motown performers and others like Dinah Washington came to the Phil Giles Hotel, Paradise Club, Flamingo Club, and El Morocco Club to perform. "When integration arrived and opened up resorts and hotels in other areas, Idlewild lost a lot of its popularity," says a Chicago physician whose family owned a house on Baldwin Road for many years.

"Today, a lot of us go there for horseback riding, fishing, and a peaceful escape," says Brown, who has a sixty-year-old cottage on one of the lakes. Brown, who is married to Detroit attorney C. Beth DunCombe, the sister-in-law of Detroit mayor Dennis Archer, remembers when the Idlewild Clubhouse was still in use. "When I was growing up, before we midwesterners started going east to Martha's Vineyard, the clubhouse was filled from Memorial Day to Labor Day. Now the most loyal followers are the retirees who came back to live and the children whose parents had owned cottages around the three lakes."

I am the first to admit that the old guard are somewhat reluctant to see the new black professional families and the new nonprofessional blacks or whites intrude upon the resorts that they carved out on their own three generations ago. Whether it's because these new visitors are disrespectful of the traditions that preceded them, or because they are new faces and names that fail to share the same college and social club affiliations, the black elite holds on tightly to its history in these vacation communities. Since I grew up with many of these families, I recognize yet disagree with their fear that their extended families are being diluted by people who mock the culture of black society. When I return to Martha's Vineyard or Sag Harbor today, as I did as a youngster, I do so with a feeling that black people have a history and a stake in our surroundings. It's a security that I will want my children to experience during their lives as well. But I think that it is possible, and necessary, that diverse groups be able to coexist in these settings.

New York resident Phyllis Murphy Stevenson says that all of these resorts—the small ones like Idlewild and the big ones like Oak Bluffs—play a role in bringing together the black elite. "I remember riding out to Sag Harbor from the city with my parents in the backseat of our Cadillac in the late 1940s and early 1950s," says Stevenson, whose parents had a house in Sag Harbor Hills next door to people her mother had known as a child. "It was an exhausting five-hour trip across bumpy, dusty roads, but there was something very special about arriving in a welcoming community that was uniquely our own. When you grow up in a place surrounded by people who feel like an extended family—people who are well-educated and accomplished, people who care about your history and your future—you feel inspired. When a black child spends summer after summer in an environment like this, she grows up feeling as though she can accomplish anything."

Black Elite in Chicago

"All this was founded by a black man. Jean DuSable."

I trudged quietly up South Parkway in the hot sun, staring at the back of my father's highly polished black wingtip dress shoes as they whipped back and forth ahead of me along the sidewalk.

"Of course they don't give him much credit for it—but DuSable was black. That was your mother's high school—DuSable High School, over on Forty-ninth Street. Named for a black man."

The first time I visited Chicago, I was in my early adolescence. It was the early 1970s, and my parents were on one of our week-long "see-the-country-and-embrace-your-black-heritage" vacations. These annual six- or seven-day excursions were usually squeezed into a spring or winter school break, or at the end of a leisurely summer spent in Martha's Vineyard. Supplemented by readings that my mother and father picked out of the eight-volume, tan-colored *Negro Heritage Library* books that sat in our den at home, these "black heritage excursions," or "working vacations," as I now call them, were deemed a necessity because Mom and Dad feared that my overwhelmingly white neighborhood and school experiences in suburban New York were doing little to inform my racial identity or enhance my black self-image. So, they did exactly what the parents of our other black friends were doing: increased our appreciation of black America by taking us out of our white neighborhoods and bringing us to where the larger black experience was taking place. So, while our white friends were jetting off to the Rocky Mountains, Niagara Falls, or Disneyland, my parents took me and my brother on trips to historic black schools, museums, and communities. From the time I was four or five years old until the time I left for college, we went on over a dozen weeklong excursions to different cities and locales that had special meaning to black America. The South Side of Chicago was one of those destinations.

Before we had packed and left for the airport, my brother and I had been coached on just about everything "black" about Chicago. We'd been told that this was where *Ebony* was published; where Afro Sheen was made; where the first black U.S. congressman had been elected; and where Lorraine Hansberry, author of *A Raisin the Sun,* had grown up and where she based her famous play. And in recent weeks, we'd heard from friends and relatives that this was where "Dr. Odom and Dr. Claiborne—people like us" were being thrown in jail for daring to drive through all-white or mostly white neighborhoods.

As we walked up South Parkway (now known as Martin Luther King Drive), we took snapshot photos of buildings and storefronts that my parents pointed out. Like most black urban areas of the 1970s, the South Side of Chicago had lost much of the aesthetic polish that it had when the black elite was still living there in numbers. Even though Mom and Dad were telling me about the black professionals who conducted business, entertained, and lived along these streets in the 1920s, 1930s, 1940s, and 1950s, I was not impressed. By now most had moved to the North Side office buildings and to the more southern suburban Chicago neighborhoods, leaving what looked like an urban, working-class ghetto.

My lackadaisical mood must have fully captured the attitude of the middle class integrated child who was uninterested in the black struggle that preceded him. I was hearing great stories about important entrepreneurs, socialites, and political figures who came out of the South Side, but all I was seeing were working-class and out-of-work black people standing along Wabash, South Michigan, and Indiana Avenues. A crumbling, all-black neighborhood to someone with my very "white" and very elitist perspective undermined any chance for me to believe that there were once people like me living here. And therein lies the difficulty in teaching middle-class, integrated black children about their people's past.

For the most part, the integrated suburban black child equates black neighborhoods like the South Side with poverty and other alien circumstances. Because we know just enough well-to-do black friends in sparkling white city neighborhoods or suburbs, we assume that this is where and how accomplished blacks always lived. Because we are the black children of black professionals living across the street from the white children of parents working at similar jobs and earning similar incomes, we find it incomprehensible that it wasn't always this way. We divide first by class, and later by race, and conclude that a community that looks worn and urban today would never have included people like us among its residents. It was just illogical that we could have lived with, shopped with, and gone to school with people who had so little in common with us socially and economically: No place exemplified this better than Chicago's South Side, and because I was one of those post-1960s integrated suburban black kids, I was slow to understand what my parents were telling me.

We had just left a service at one of the city's black society churches and I was already beginning to sweat in the new blue suit that my father had picked out for me in the "Paddock Shop" section at Barney's for this occasion. As we dodged some chips of broken glass along the sidewalk, I thought about the recent incident that had precipitated our trip. Because of my elitist attitude toward this all-black South Side experience, that was where my interest stood. And we had just been talking about it in church. I don't remember all that was said at the time, but I clearly recall the point of the discussion.

For years, members of the black elite in Chicago and elsewhere had heard accounts of police brutality against innocent blacks in the inner city, but we too often reacted in the way that some white people react: "That's too bad about *those* people," we would say. We felt detached and uninvolved because "those" people were another group—another class. "Poor blacks are not like us," we wrongly told ourselves. "They don't relate to us and we don't relate to them." In fact, some upper class blacks even expressed some cynicism, convincing themselves that perhaps these urban blacks had done something to justify brutal police treatment.

In retrospect, I understood all that the minister and others in the service had been implying. Because of our own class-based narrow-mindedness, segments of the black population often didn't react to or care about racially biased abuses until one of our own—a black member of our socioeconomic class—had been victimized. Every city's black elite had one of these pivotal moments when an abuse reached into its own group and shook up an otherwise complacent community of professionals. In the spring of 1972, shortly before our visit there, Chicago's elite community had experienced its incident.

For years, ever since the ghettoization of blacks occurred on the South Side around the second decade of the 1900s, white Chicago police officers had been stopping and harassing blacks who dared venture east of Cottage Grove Avenue or north of Twenty-sixth Street, the early boundaries of the newly outlined black belt. However, it wasn't until two prominent black dentists, in two completely unrelated incidents, were jailed and harassed by the police that the black elite in the city (and elsewhere) became outraged.

In the spring of 1972, Dr. Herbert Odom was arrested after police noticed that one of the rear lights on his Cadillac was not working. The second arrest took place when an even more prominent dentist, Dr. Daniel Claiborne, suffered a severe stroke while driving along a South Side street. Dismissing his unconsciousness as intoxication by alcohol, Chicago police officers arrested Dr. Claiborne despite his obvious need for immediate medical attention. Instead of taking him to a hospital, they put him in jail, and he died shortly after the incident.

As the black elite read about these kinds of incidents in their local black weekly papers and in *Jet,* a weekly magazine that reports such incidents faithfully, they saw their quiet lives being disrupted even more as they ventured into affluent and peaceful white neighborhoods. My own parents saw this in the late 1960s one quiet afternoon when a police squad car attempted to take me and my eight-year-old brother off our residential street; the officer presumed that the red wagon my brother was pulling me in must have been stolen from a house in our white neighborhood.

"Rich blacks don't start seeing the light about bigotry and police abuse until it starts happening to their own. It's not real to them when poor blacks are getting beaten up," says a South Side physician who knew Dr. Claiborne and recalled how the incident mobilized many of the professional blacks in Chicago. "A lot of us knew that poor blacks in the Ida B. Wells projects, Cabrini Green, and Robert Taylor Homes were being beaten up, but now Mayor Daley and his racist cops were hitting too close to home. There was a reason why—years after integration—the

U.S. Civil Rights Commission called Chicago the most residentially segregated city in America."

Middle-class and upper-class blacks all over Chicago and elsewhere were talking about what happened to the black doctors in Chicago. It was another issue that justified our visit.

"This is where your Uncle Telfer's office was in the 1930s and 1940s."

My brother and I stood on the corner of Forty-eighth Street and Wabash and stared up at a row of nondescript buildings that made little impression on me. "Uncle T," as my relatives called him, was one of Chicago's first black bail bondsmen. Along with the attorney and real estate broker who rented offices in the small corner building, he was a part of the group of black South Side businessmen who were quickly outnumbering those in other northern cities like New York and Philadelphia. A longtime Republican, he was a fan and supporter in the 1930s of his Chicago neighbor, Congressman Oscar DePriest, the first black to be elected to the U.S. House of Representatives following Reconstruction. Since Uncle Telfer had migrated north, ahead of many of my other relatives—some of whom followed in the early 1940s, with the rest of the black migration—he warned them that Chicago racism could be as insidious as southern bigotry, but here he could at least use his political influence to give our Memphis relatives access to jobs in the city or admission to the University of Chicago. These were the trade-offs and compromises for us when we came north.

When he was profiled by *Ebony* in the late 1940s, toward the latter part of his career, Uncle T owned two apartment buildings and a successful business that put him in touch with a wide variety of people around the "black belt" of Chicago. For years, the article about him and his colleagues—yellowed and faded—remained taped to our refrigerator door, reminding us that getting out of the South and moving to the northern cities made good sense.

Today, when I return to Chicago, I spend most of my time on the North Side—in office buildings, law firms, banks, hotels, and department stores that seem to reflect barely much more integration than existed during my uncle's time. But when I go there today, I have few of the personal ties to the black community that my parents have, and almost no connection to the white: I go there as an outsider. I feel outside of the almost uniformly white North Side and feel outside of what is still a uniformly black South Side. As an outsider looking at the lack of integration in both neighborhoods, I would say that the situation hasn't changed much at all.

Truman Gibson Jr. and Maudelle Bousfield Evans, two insiders from the black belt, insist that the city changed a great deal during the decades they spent there. As black South Side residents, they saw Chicago in the 1930s become a leader in producing black congressmen. They saw Chicago in the 1940s become a community of black elites who fought amongst each other because of political party labels and patronage. And they saw Chicago after the 1950s become a town of more unified blacks who became equally outraged by the uniform mistreatment they faced under the Daley machine. In the 1960s and 1970s, they saw a Chicago that reluctantly opened up opportunities to upwardly mobile blacks. And in the 1980s, they saw the election of one of their own race and class to the position of mayor.

Although their fathers both grew up in the late 1800s, graduated from Harvard and Northwestern Medical School, and then became well-to-do businessmen while running the most powerful black-owned business in Chicago between the 1920s and 1950s, Gibson and Evans remember their privileged youth against a backdrop of a midwestern city that remained racially polarized longer than any other. It was a city that was run by a white elite that used restrictive covenants and other bizarre discriminatory laws to keep blacks in their place—on the South Side.

As adults, Gibson and Evans have seen much of the city's old bigotry erode. Always accomplished and successful, they and their colleagues have pushed farther north into neighborhoods and institutions their fathers never would have imagined. Gibson, a successful attorney and member of the Boulé, laughs when his all-black chapter meets at the Union League Club downtown. "When I was a student going to the University of Chicago, both the school and the members of this club were part of the conservative leadership that worked to enforce restrictive covenants against blacks and keep us out of good neighborhoods and important networking clubs like this," explains Gibson, "but now we've gotten to see blacks head a lot of major institutions, including the Union League."

For Evans, a founding member of the Chicago Girl Friends who holds a master's degree in biology, the pivotal point was probably the week she and her husband, a retired magazine publisher who was also the nation's highest-ranking black advertising executive, moved into a duplex apartment on the sixty-fourth floor of the John Hancock

Building, a downtown luxury commercial and residential building that remains a centerpiece in the city's skyline. "Most people who knew Chicago's history with blacks were surprised that we were able to get that apartment, because for years we had politicians and real estate brokers who were unapologetic about their support of segregation," says Evans, "but it was 1969, and we had more choices by then. We were no longer required to live on the South Side. My husband, Leonard, wanted to be nearer to the office where he published *Tuesday* magazine, and I wanted to be closer to the office where I worked with the United Negro College Fund. We hadn't yet elected a black mayor, but blacks were gaining more clout."

Since Chicago's city neighborhoods were more segregated than the neighborhoods of other cities, many wonder why so many black-owned enterprises succeeded there. In fact, many New Yorkers ask why New York's black population of the 1940s and 1950s—a larger and less confined community than Chicago's black South Side—was not producing black-owned businesses at a rate equal to that of Chicago. The explanation that many offer is surprising. Many black businessmen reason that Chicago's extreme segregation made it possible—or better yet, necessary—for blacks to build businesses that attracted undiluted black support. Black Harlemites in New York, on the other hand, were spending their money in the white-owned establishments that populated most of Harlem and in other nonblack neighborhoods where blacks lived and shopped.

And—using New York for a further comparison—because its black population of the 1930s and 1940s, for instance, was less ghettoized and was dispersed in many noncontiguous Manhattan, Brooklyn, and Bronx neighborhoods—Harlem, San Juan Hill, Greenwich Village, Brownsville, Ocean Hill, and the South Bronx, for example—black economic power was diluted. Chicago's blacks were ghettoized into one relatively easily defined area, and because of such boundaries—south of Twenty-sixth Street—black-owned and black-run businesses were able to fuel the community and flourish.

Dempsey Travis, a real estate developer and major landholder on Chicago's South Side, agrees. "For the last sixty-five years, it's been very easy to draw the boundary lines between whites and blacks in this city," he says. "Almost regardless of black people's wealth, we have still remained primarily on the South Side. And that's so ironic because this city was settled by a black man. In fact, going back before the twentieth century—in the 1890s—blacks were actually living in every neighborhood of the city."

First founded and settled by the black explorer Jean-Baptiste DuSable of Santo Domingo, Haiti, in 1773, Chicago was begun as a thirty-acre land parcel. DuSable, working as a fur trapper and trading-post operator, eventually owned in excess of four hundred acres. He and his new Native American wife remained in the area until 1800, when he moved to Missouri.

With an early black population that was much smaller than those of southern cities like Washington, Memphis, Atlanta, and Richmond, Chicago had a small black elite in the mid- and late 1800s—it consisted of only a few families. Most of them lived very integrated lives: They interacted while working together with liberal whites who had been abolitionists when the Underground Railroad moved black southern slaves into the North. The black elite of the period included people like physician Daniel Williams, Pullman Train Company executive Julius Avendorph, caterer Charles Smiley, and attorney Laing Williams. They were all educated people who lived, worked, and socialized among whites. "In fact," says Travis, who also wrote the book *Autobiography of Black Chicago*, "at that time, there were blacks living throughout the North Side and elsewhere. Though we were small in numbers, we were represented in every census tract."

Travis points out, however, that the total black population was still under fifteen thousand people. It was not until around World War I, the time of a major black migration from the South to the North, that a substantial black population arrived in the city. Most of these black southerners came—about seventy thousand of them between 1900 and 1920—as a result of the *Chicago Defender*, a black newspaper that was read in the South by educated blacks eager to escape their more rural environment. When these blacks arrived in town, the old-guard black families and their social clubs immediately decided who was "in" and who was not. Truman Gibson's parents and Maudelle Bousfield Evans's parents were clearly "in" as far as the black old guard was concerned. Interestingly, as old-guard blacks were busy trying to separate the "society blacks" like themselves from the new working-class arrivals, whites were making plans to ghettoize both groups together on the South Side. And they quickly did so by establishing restrictive covenants that moved blacks out of white areas.

In fact, the white community responded quite aggressively to black mobility during the early years of World War I. In the working-class and middle-class white neighborhoods that saw blacks moving in, white residents simply bombed the houses or set them afire. In more upscale neighborhoods like Hyde Park, which surrounds the University of Chicago, white residents organized a full-blown plan to preempt any sales to upwardly mobile blacks who might be able to afford homes in the well-to-do community. My Uncle Telfer, who died before the upscale neighborhood allowed blacks to buy homes there, had saved a copy of Hyde Park's neighborhood newspaper, published in 1920, which read, ". . . Every colored man who moves into the Hyde Park neighborhood knows that he is damaging his white neighbor's property. Consequently . . . he forfeits his right to be employed by the white man. . . . Employers should adopt a rule of refusing to employ Negroes who persist in residing in Hyde Park."

Soon after that time, restrictive covenants making it illegal to sell homes to blacks, regardless of their wealth, were strictly enforced.

But regardless of how violently whites reacted to the influx of poor and upwardly mobile blacks, the old-guard blacks of Chicago had their own dismal way of responding to their fellow blacks in this northern city. They were not happy to see them arriving.

"Not surprisingly, elitism was quite evident. But the rules governing black society in Chicago were always slightly different from the rules that were used in the southern cities," explains former *Chicago Defender* society columnist Theresa Fambro Hooks. "In the South, black society was determined by the years your family had lived in a particular city and by their ties to one or more of the nearby black colleges like Howard or Fisk or Spelman. But the rules were different in Chicago because almost everybody was new—almost all of them had migrated from the South. There were very few old families and there were no old local black universities to be tied to."

So the standard for black society in Chicago became, instead, financial success and, to a lesser extent, family ties to a few of the northern white universities. In both regards, the Gibson and Bousfield families were at the top. Acceptance by the right schools, the right churches, and the right clubs proved that. In fact, Truman Gibson Sr. and Midian Bousfield were a part of a triumvirate of powerful black men in Chicago.

A look at their credentials and the business they built reveals why: "My father was a 1909 graduate of Harvard and was responsible for forming Supreme Life Insurance Company of Chicago," says Gibson, who sat on the Supreme board for several years, "but Dad had actually worked for two other insurance companies before. He'd been friends with Mrs. Alonzo Herndon, whose husband was the millionaire owner of Atlanta Life Insurance, and it was Mrs. Herndon and W. E. B. Du Bois who encouraged my father to go to Harvard after finishing Atlanta University. So after Harvard, he came back and became an executive at Atlanta Life."

Having trained under one of the richest black businessmen of the South, Truman senior went north to Columbus, Ohio, where he started Supreme Life of Ohio. "In 1929, my father merged Supreme Life with two other insurance companies and moved my mother, brother, and me to Chicago," says Gibson, who was starting his first year at the University of Chicago at the time.

That same year, Chicago became the first congressional district in the nation to elect a black to the U.S. House of Representatives since the late 1800s, when a handful of blacks had been elected during the temporary black political gains of the Reconstruction period. "With us having Oscar DePriest, the only black congressman in the country," says Gibson, "you would have presumed that this town was some bastion of racially liberal open-mindedness. Believe me, it wasn't."

A member of the Chicago Boulé, Gibson's father, Truman senior, formed a relationship with Maudelle's father, Midian, and with attorney Earl B. Dickerson. Dickerson was a 1920 graduate of the University of Chicago and later became Chicago's assistant corporation counsel and, in 1939, a city councilman. The third member of the team, Dr. Midian O. Bousfield, was a 1909 graduate of Northwestern Medical School.

"My father started practicing medicine in Chicago in 1914, and he'd become president of a company called Liberty Life Insurance," recalls Maudelle Bousfield Evans, who spent most of her life in Chicago. "My dad, Truman's father, and Harry Pace, who headed Northwestern Life of Newark, New Jersey, then merged all three companies in order to create Supreme Life. My father later became medical director of Supreme as well as president of the National Medical Association."

With business leaders of the sophistication of Gibson, Bousfield, and Dickerson, black Chicago had done a much better job of creating a coalition of black talent than had cities like New York, whose black community was divided by the Harlem and Brooklyn cliques and was somewhat late in building strong black businesses and electing national black political leaders.

"The Gibsons, Bousfields, and Dickersons were like Chicago royalty when I was growing up," says Ronne Rone Hartfield, who remembers reading about Earl Dickerson's daughter, Diane, in the city's black society columns. "The black community in Chicago was concentrated enough that everybody knew who those families were. And among my own group of young friends, we all followed Diane's fairy-tale life in the papers. Because of her dad's position in business and politics, she had access to places and people that we didn't know. Even her coming-out party got reported in the papers," says Hartfield, who attended the all-black Wendell Phillips High School and the University of Chicago for both college and graduate school before becoming executive director of education at the Art Institute of Chicago.

But even with a legacy of such well-heeled black businessmen and even with its history of serving as the hometown for the first three blacks—Oscar DePriest, Arthur Mitchell, and William Dawson—to serve in a post-Reconstruction U.S. Congress, Chicago and its black elite remain a conundrum when students from other cities with large black populations analyze Chicago's inability to elect local black leaders with any consistency. Unlike Atlanta, Washington, Detroit, New Orleans, and other cities with a long history of a black elite, Chicago has managed to elect only one black mayor. It has done no better than communities like Minneapolis, Seattle, or Denver—cities that have elected black mayors with newer and smaller black populations.

"The problem that always plagued upscale blacks in Chicago," says a retired city employee, "is that they were always jealous of each other. Everybody thinks there can only be one HNIC—one Head Nigger In Charge—to represent the black community. And the white people here have been cynical enough to play us off against each other and keep us divided. The old mayor Daley perfected the game of dividing the black elite by utilizing tokenism as a means to keep blacks angry at each other. Elevate one black and demote the other. Keep each one of us looking over our shoulder."

"Just look at the history of the most talented blacks in this city," says a retired black Cook County employee. "We're always selling each other out or talking down the rising black stars. People did it to John Johnson after he became a millionaire by publishing *Ebony*. They tore him down because he grew up poor. They did it to George Johnson when he turned Afro Sheen and his Johnson Products company into household names. They said he and his wife, Joan, were too "nouveau." When Jesse Jackson was running Operation PUSH here, we said he was too young and anti-establishment. We said Jewel Lafontant was too pro-establishment. When we got Harold Washington elected as mayor, we joined in with whites when they questioned his sex life. Early on, even the *Chicago Defender*, our own black newspaper, went after some of the liberal black leaders. Washington and Atlanta have their skin color fixation. New York has its American-born and West Indian-born conflict. New Orleans has its obsession with French family lineage. And we've got the HNIC problem that came out of Daley's machine politics of rewarding and anointing Uncle Toms."

Some of the older black Chicagoans agree that their community would have been better off if some of the elements among the black elite had worked together sooner. For one, the very first black congressman from Chicago, Oscar DePriest, was a staunch Republican who was elected by a mostly black, working-class community. A major supporter of President Herbert Hoover, he disappointed many of his black constituents when he spoke out against the liberal New Deal programs that Franklin D. Roosevelt had put in place. The very people that stood to gain from these programs had elected a representative who chose the party line over his community's needs. Similarly, the *Chicago Defender*, the nation's most influential black newspaper, was published by the wealthy Abbott family, also supporters of the Republican establishment. Although the paper later came to support liberal programs that aided blacks, the patrician publisher Robert Abbott, who lived in a brick South Parkway mansion, infuriated many black Chicagoans by his occasional tendency to turn a blind eye to the needs of his readers and neighbors who had so much less than he.

Another person who became a divisive element in the black community was the well-to-do black attorney Arthur Mitchell, who got himself elected to Congress in 1934 after having moved from Washington only six years earlier. Even though he was elected by the mostly black First Congressional District, he very quickly distanced himself from other Chicago blacks by giving a famous speech in which he declared that he did "not represent Negro people in any way." Educated at Tuskegee Institute, Columbia, and Harvard Universities, Mitchell was later credited with filing a Supreme Court case that led to the desegregation of passenger trains. "But he only filed that suit because he, himself, got pulled out of a first-class seat on his way to Arkansas while he was in Congress," says an unconvinced Chicagoan who had lived in Mitchell's congressional district in the early 1940s.

"And interestingly," says Dempsey Travis, former president of the Chicago NAACP, "we've always had a visible group of black Republicans." A further division between elite blacks in Chicago has, indeed, been the presence of many of them in the Republican Party. My own Uncle Telfer had been one of those individuals until he was won over by the results of Roosevelt's problack programs. But there were many who remained Republican.

"Like that Jewel Stradford Lafontant," says a retired *Chicago Defender* reporter as he recalls Lafontant, a lifelong well-to-do black Republican and one of the city's most accomplished professional women. "There she was, sitting on the boards of all these *Fortune* 500 companies—like Mobil and TWA—and supporting all these white Republicans. I was aghast when I saw her standing up on the podium at the Republican Convention in 1960—seconding the nomination of Richard Nixon. Of all the people she could have tied herself to. That one picture captured all the problems with the black elite in Chicago. We were in the middle of the civil rights movement, with black people getting our butts kicked in every part of Chicago, on every city's six o'clock news report, and this brilliant black woman—from the University of Chicago Law School—stands up there in pearls, in front of an all-white convention, and nominates someone like Nixon. We've always been a small percentage of this city, but as time goes on we keep trying so hard to be accepted by the white majority that we just sell our souls. I don't think her father or grandfather would have done what she did."

Although the remark made me uncomfortable, I was well aware of the very Republican and very accomplished Jewel Stradford Lafontant, since her son, John, had been a couple of years ahead of me at Princeton. Her father had been a high-profile Republican lawyer in Chicago during the 1930s and 1940s, and my mother had lived a few doors down from the Stradfords on Chicago's Washington Park Court at a time when Jewel's father, Francis Stradford, gained notoriety for arguing against restrictive covenants in housing. A hero for many of the black South Siders, he supported the family of writer Lorraine Hansberry when, in 1936, the Hansberrys were refused an apartment on Sixtieth Street because restrictive covenants stated that only whites could rent in the building.

While the characterization of Lafontant is not entirely a fair one—because she did, indeed, support many black organizations like the Chicago Urban League—the perception that some of the wealthier blacks failed to work together politically with the working-class blacks is closer to reality. Today, the city is 40 percent black, and the white Irish, Polish, and German communities are much better able to unify and keep the mayoral seat in the hands of the white electorate and, particularly, in the hands of the Daley family.

Adding to the *Chicago Defender* journalist's view of Lafontant, there are many more accurate stories of elite black families who at one time produced older, problack activists who had helped the community but later raised children who, in one way or another, did not live up to the larger black community's expectations. Along with Francis Stradford and his daughter, Jewel, people often talk about Earl Dickerson and his daughter, Diane, in this group. Earl was the high-profile attorney who had been so outspoken for the black South Side during the 1920s, 1930s, and 1940s. A city alderman, an assistant corporation counsel, a corporate executive, a well-to-do attorney, a Democratic committee member, and a member of the black community's elite, Earl was long heralded as one of the leaders of black Chicago. His only child, Diane, however, eventually became an outsider because of what some blacks say she did *not* do for the black community.

"Diane was something else. I can still remember that white convertible she drove around in after she graduated from Francis Parker," remarks a friend of my mother's who welcomed me into her Hyde Park home during a recent visit. "She was like black royalty—wealthy, good-looking, sophisticated. But it just saddened me to see how little interest she took in black people."

"Francis Parker?" another friend asks when the story is repeated. "So, that explains it."

Explains what?

It is obvious that when outsiders ask what happened to Diane Dickerson and all the money, good looks, and sophistication she inherited, few of the old-timers are willing to answer. Some are willing to point out that she was typical among the black elite's children: less concerned with and less tied to the black community than their parents. Others remain a little annoyed and share the facts that they are reluctant to forget.

"Oh, she married into the Brown family—you know, Sidney Brown, who was the first black on the school board."

Another Chicagoan chimes in. "That's right, she married his son, Nelson, the lawyer. Handsome couple—and what a wedding!"

Then everyone falls silent.

"So what happened to her?" I ask with obvious curiosity as I glance through some of the aerial photographs taken above the extravagant 1950 garden party that was given for Diane's debut.

"Well, she died two years ago," someone says. "But even before that she just kind of disappeared."

"Disappeared" sounds like code for something, so I press on. I ask why I never saw her name in any of the directories in any of the black organizations, why she wasn't listed in my mother's Links roster, why her kids aren't pictured in my family's collection of Jack and Jill *Up the Hill* annual yearbooks, and why her name wasn't appearing in the columns of the *Chicago Defender*. Given the professional and social stature of her father in Chicago, it seems that she would be well ensconced in the city's, as well as the country's, black society.

"Well, Diane was a little different from her parents. *They* belonged to the black world," says one woman who was Diane's age and had attended both her debut and her wedding. "Her mother was in the Northeasterners and her father was in lots of black groups, but—" Her voice trails off before the thought is completed.

Finally, one of the men in the group continues the sentence: "—you're not going to find Diane in our groups. After her folks sent her to that Francis Parker School, that was the end of her life with black people."

"The rest of us all were going to DuSable High School," said a svelte, middle-aged Chicagoan who had known the Dickersons and understood the man's point. The woman had graduated a year ahead of my own mother. "DuSable was where you went if you were black. And the Dickersons were high-toned people with money, so they sent her to this all-white private school on the North Side. Next thing we knew, she went to some white college out west."

"And then," the DuSable High woman continued, "she married Nelson Brown—this good-looking lawyer from a good family—when she came back. But then with all her good background and history, she divorces Nelson and then ups and marries some white man that nobody ever heard of."

"Somebody named Cohen—"

The group shrugs while two of the women roll their eyes in disgust. Like many blacks of my parents' generation, they have little understanding of interracial marriage and are particularly perturbed when they perceive that it's the best-educated and most-respected blacks who are abandoning the black community the fastest.

"—and then her *third* husband," one of the women adds, "was this *other* white man—some senior vice president from the National Bank of Chicago. So, after that, there was no way she was going to be interested in our clubs and things. She just totally went, if you'll pardon the expression, *Incognegro*."

"I heard she wasn't even buried here," one of the men remarks, with a sad expression. "Even though her family defined black Chicago, I heard they buried her with her new white family's relations down south."

"That's Francis Parker for you," adds one of the men who knew the Dickerson family from St. Edmund's Episcopal Church. "You can't send rich black kids to a school like that and expect them to come back and stay on the South Side with black folks."

This group of men and women knew that Diane left the family church and converted to Judaism when she married her second husband and that she never joined a black sorority after attending all-white Mills College in California. Rather narrow-minded and judgmental, they looked at the young woman's personal decisions without putting them into the context of the upper-class, mostly white-oriented childhood that her parents had created for her. And although none of them knew her personally, they all assumed that Diane had disappeared into a white world that was unaware of her family's role in both black and white Chicago.

Not surprisingly, this insular group had labeled Diane incorrectly. My conversation with Diane's last husband, Charles Montgomery, revealed an independent woman who was looking for an identity outside the control of her powerful father's circle.

"Diane's father wanted her to attend one of the eastern colleges like Wellesley or Smith or Barnard because he had lots of business dealings on the East Coast and could have visited her often," said Montgomery, while vacationing at his summer home in Maine, "and for this reason, Diane decided to go to college in California. She loved him but wanted her own identity." Montgomery, a well-established member of Chicago's white upper class, is a tremendous fan of Earl B. Dickerson and knows all the details of his contributions to the black citizens of Chicago. "Earl faced a lot of discrimination, but he changed Chicago."

This group of men and women who sized up the popular daughter of Earl B. Dickerson now launch into a discussion of other prominent black South Siders who sent their kids to white North Side private schools—like the Latin School—or northern suburban schools like New Trier

High, which drew from wealthy Wilmette, Winnetka, and Kenilworth—and then saw them marry outside the race and never return.

"You know what we mean, don't you?" one of the women asks.

Actually, I knew quite well what she meant. In fact, I had recalled my mother telling me that while she was attending the all-black DuSable High in the late 1940s, she had dated one of the two or three black students who were attending the suburban New Trier. The boy used to pick her up in her black South Side neighborhood and ride north on his motorcycle through Chicago, past the all-white but somewhat liberal suburb of Evanston, into the rich white conservative towns of Wilmette and Winnetka. I remember her telling me that their dating ended as she came to realize that growing up as one of the only blacks in his town and school had made him angry, confused, and embarrassed by his racial identity. When he announced that he would prefer to date a white woman or a black woman who could pass for white, his problems put a quick end to the teenagers' friendship.

Several years later, when Truman Gibson Jr. sent his own daughter, Karen, from the South Side of Chicago to the North Side to school, he also picked the expensive, private, and uniformly white Francis Parker School. Like the Dickersons, he was also Chicago royalty. "Earl Dickerson's daughter, Diane, had actually been the first black at the school a few years before, so we really weren't blazing any trails," says Gibson, an attorney who represented the boxer Joe Louis and who worked on many high-profile cases, including the 1940 Supreme Court case, *Hansberry v. Lee*, which helped outlaw race-based restrictive covenants.

"Since the 1920s, we'd been told we had to live, work, and go to school on the South Side, but I was not going to have my family limited by that," Gibson says as he explains that his daughter eventually went off to Sarah Lawrence College. "I just wanted a top school for my daughter."

And when the old South Siders proudly announce that Karen Gibson had been raised in Jack and Jill and now lives in Harlem, they add that she was one of the few blacks who "survived the North Side experience."

"Many of us who raise our kids in white neighborhoods and send them to white schools are wise enough to teach them to be proud of their black identity," adds a black South Sider whose son attends the mostly white Latin School. "It's a daily activity to teach our kids these messages. They get no reinforcement on this at school, so we teach it at home."

When I listen to the way some people speak of Jewel Lafontant and Diane Dickerson, children of an old black Chicago elite, the tone and words sound familiar. My brother and I grew up hearing similar comments from black strangers or distant relatives who feared that we were being raised in environments that were too white to ever allow us to later identify with the black community.

"With that white neighborhood, and all those white schools and white friends," my Aunt Earlene had said to me on the evening of my wedding as we stood in a Fifth Avenue mansion on New York's Upper East Side, "we were sure that you would lose interest in black culture and, no doubt, end up marrying a white woman." Even though Peter Duchin's distinctly nonblack orchestra was busy playing George Gershwin at our reception, filled with many of my white friends, what mattered most to her—and to many of my relatives and black friends at that moment—was that I had "remained at home" by marrying a member of our race.

The fact that I married a black woman from a similarly integrated background and remained active in the black community is a scenario that pleases, yet still surprises, black strangers and relatives who raised their own kids in all-black settings. It is an unfortunate fact that the racial loyalty of integrated blacks who come from the middle or upper classes is constantly called into question. And what is also unfortunate is that so many of us measure race loyalty by noting who "marries black" and who "marries white." I once focused on that issue and I now realize that it was both wrong and bigoted to do so.

"I knew my three children were going to be among a handful of blacks in the Latin School of Chicago," says Eleanor Chatman, who raised her two daughters and son in the elite Pill Hill neighborhood, "and I knew that this was a school that was known for having wealthy white children—the kids of Marshall Field and Adlai Stevenson. But my husband and I knew there was a way to still help them maintain their black identity. Ironically, all three of my children became president of the school's black student organization. They spent Saturdays volunteering for Jesse Jackson's Operation Breadbasket, and my two oldest daughters even went to the march on Washington. So they obviously were able to maintain that balance of black identity within their white academic surroundings at the Latin School and Northfield Mount Hermon."

Just as some blacks may have, years ago, unfairly accused second-generation Chicagoans like Jewel Lafontant and Diane Dickerson of selling out, I know that many elite blacks speak about me, my wife, and many other integrated black offspring in these terms. It is just the reality for the children of black parents who attempt to move outside of the black mainstream. But not surprisingly, just as members of the old guard will criticize those blacks whose money and status have given them *too* much access to the white upper class, they also continue to thumb their noses at lives that are "too black" and "too old-world."

For instance, blacks who lived on the West Side of Chicago were considered "country" and unsophisticated by old-guard South Side blacks. And even though there were many poor blacks living on the South Side as well, South Siders insist there were no professional blacks and no cultural or business institutions worth noting west of State Street.

"Nobody wanted to be caught living on the West Side—nobody. Some people even said it was *hotter* on the West Side," remarks Theresa Fambro Hooks as she explains how old-guard blacks viewed different sections of Chicago from the 1940s through the 1970s. "These days, the South Side is basically anything south of Madison Avenue. But back then, most people said the line was a couple of dozen blocks below that—at around Twenty-fourth Street. The border on the east was Cottage Grove, and the border on the west was State or Wentworth," explains Hooks, who covered black society for more than twenty-five years in her gossipy social column "Teesee's Town" in the *Chicago Defender*. "Where you always found white people," she explains, "was on the North Side and in the neighborhoods near the lake."

"When I was growing up, you didn't end up in the black social columns if you lived outside of Hyde Park, Kenwood, the Highlands, or South Shore," says a retired teacher who belongs to the Links, "and you didn't even know there were such things as social columns if you lived on the *West* Side. That's where the 'country' blacks were. They were our old-world equivalent."

People on the South Side were the ones who belonged to the right black social clubs like Jack and Jill, the Boulé, the Links, and the Girl Friends. "We have also always had an inordinate number of elite men's social groups," says South Side resident Vivian Patton Durham, a friend of my parents, "like the Royal Snakes, the Druids, the Assembly, the Original Forty Club, and the Capricorns, which always included professional men who lived on the South Side." Married to Charles Durham, an attorney and former municipal judge, Vivian is a member of the Carousels, an elite black woman's club that includes many South Side women.

"It was the South Side men who controlled black society in the 1930s,

1940s, and 1950s," explained Theresa Fambro Hooks, a longtime friend of the Durhams. "In fact, if you were a girl from a good family on the South Side, it was the men's clubs who would sponsor your coming-out party. Most of the groups gave formal dinner dances, but it was the Royal Coterie of Snakes—a group of black doctors, lawyers, and businessmen—who gave a formal debutante cotillion every Christmas." As Hooks and Durham point out, the women's groups—the Links, Drifters, Royalites, Girl Friends, and Carousels—took over the social scene after the men's clubs moved in other directions. Most of the members today come out of the South Side.

In addition to the right black social groups, the South Side community had the stores, restaurants, churches, and activities that black elites cared about in Chicago. Of course they didn't have Marshall Field's, Sears, or Mandel's as the whites did on State Street, but they did have The Parkway Ballroom, the Regal Theater, the Tiki Room, the Savoy, and night clubs like the DeLisa. These were all destinations for black celebrities who performed in or visited the city. They also had smaller, black-owned establishments—law offices, insurance firms, medical and dental offices, beauty salons, record shops, and pharmacies. And Provident Hospital, the South Side's black hospital at Fifty-first Street, was a popular employer for old-guard black physicians who eschewed ties to Cook County Hospital, where the poor blacks were more likely to be treated.

Further differentiating the South Side from the West Side was the fact that South Side blacks with means had access to activities that were particularly favored by, and more associated with, wealthy whites. Tennis courts, indoor swimming, and riding stables were an example of this. During her high-school years, my mother was one of the young South Side blacks who kept a horse at a black-owned stable located near Fiftieth Street and Forrestville, where a handful of whites also boarded their horses. Although the groups of young people socialized at different parks and beaches—the blacks at Jackson Park and the whites on virtually every other beach north and south of Jackson—one could see both races riding through trails near the University of Chicago.

With these opportunities on the South Side, it was rarely necessary to venture to the West Side. In fact, whenever my mother did head in that direction, during the 1940s, the experiences were not pleasant ones. For a short time, her parents sent her to the mostly white Englewood High School, which, at that time, made it clear to West Side and South Side blacks that they need not try to join Englewood's school activities or more challenging classes because they would not be welcomed.

Looking specifically at the South Side, black professionals had a few favorite neighborhoods. Although several of them later owned homes in the elite and mostly white Hyde Park–Kenwood section, "there was a core group of professionals living in the Julius Rosenwald building on Forty-sixth and Michigan in the 1940s and 1950s," remarks Fern Jarrett, a Chicago resident who has belonged to many Chicago groups including Jack and Jill, the Links, and the Girl Friends. Jarrett's husband, Vernon, remembers that the Rosenwald was almost self-contained, making it almost unnecessary for its black residents to have to leave the neighborhood. The Rosenwald was built for blacks and was named for one of the founders of Sears and Roebuck. "Julius Rosenwald was an important name in this city and he was a major philanthropist in black education," explains Vernon Jarrett, who was one of the first black columnists for the *Chicago Sun Times* and *Chicago Tribune*. "The Rosenwald complex even had its own nursery school, activity center, and areas where the Girl Scout and Boy Scout troops could meet. That's where the upwardly mobile families wanted to be in the 1940s and 1950s."

Several years later, another important address for this group was established when Building 601 opened in the Lake Meadows complex at Thirty-first Street and Martin Luther King Drive. "If you lived at 601, you were in the midst of the whole Boulé and Links crowd," explains a former resident who now lives in Evanston. "You got all that, plus the park and the water across the street."

Even today's young black professionals recall how the neighborhoods were divided during the 1960s and 1970s—the last period before affluent blacks finally moved downtown or out to the suburbs altogether. "One of the reasons this city had such a strong black professional class when I was growing up is that segregation pushed us into just a few areas and we all ended up supporting each other's businesses—whether we were lawyers, doctors, or entrepreneurs," says Eric Chatman, a Chicago banker who grew up in a well-to-do South Side neighborhood referred to as "Pill Hill" because of the concentration of physicians living there. Pill Hill, which was populated originally by Jewish physicians and then by black physicians who were affiliated with South Chicago Community Hospital, lies roughly between Eighty-ninth and Ninety-fifth Streets with Stoney Avenue and Jeffrey bordering, respectively, on the west and east. The son of an obstetrician and a travel consultant, Chatman remem-

*B*y the late 1890s, through her hair care products company and chain of beauty schools, Madam C. J. Walker had become the first self-made woman millionaire in America.

*I*n 1906, Walker had this twenty-thousand-square-foot mansion built on her estate overlooking New York's Hudson River in Westchester County. Her employees and guests pose around the rear terrace and pool in a 1920s photo.

Madam Walker's daughter, A'Lelia, poses in the living room of the Walker mansion in the 1920s. A'Lelia was a Harlem socialite who owned two town houses where she entertained and raised funds for writers during the Harlem Renaissance.

Mae Walker Perry was the millionaire granddaughter of C. J. Walker. After attending Spelman College, she was married in this 1923 wedding, which cost in excess of $60,000. The ceremony took place at Harlem's old-guard St. Philip's Episcopal Church.

Madam Walker stands here with educator Booker T. Washington and other businessmen in 1913, when she gave thousands to the YMCA, the NAACP, and other groups who were combating the lynching of blacks in the South.

Alonzo Herndon, Atlanta's first black millionaire, poses with his wife, Adrienne, and son, Norris, in 1908.

Alonzo Herndon built this mansion in Atlanta in 1910, five years after he founded Atlanta Life Insurance Company. Today, the black-owned firm has over $220 million in assets.

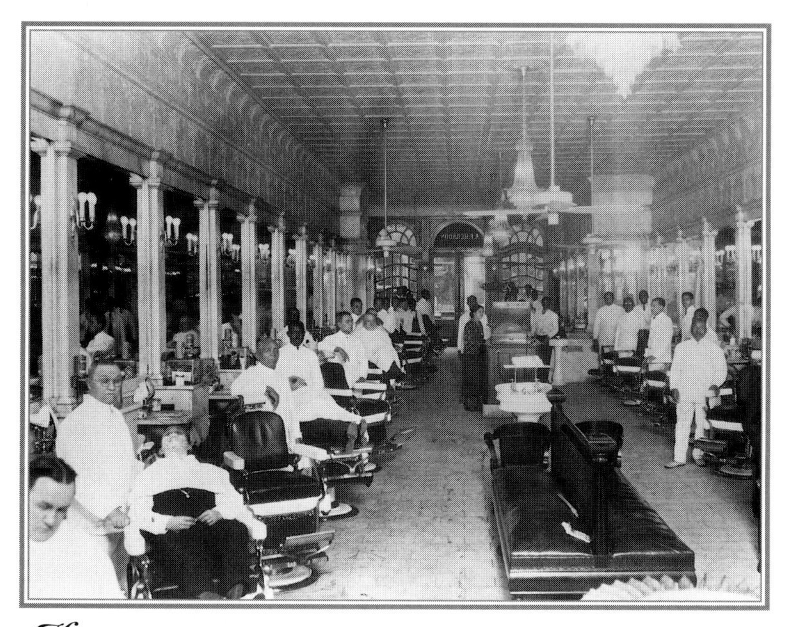

Herndon's other Atlanta businesses included a twenty-five-seat men's barber salon that employed only black barbers and served only white businessmen. Located on Peachtree Street, it featured leather benches, marble floors, and crystal chandeliers.

Supreme Life Insurance of Chicago was run by the city's black elite. Seated in the boardroom in the early 1930s are *(left to right)* attorney Earl B. Dickerson, Harry Pace, A. P. Bentley, Dr. Midian Bousfield, W. Ellis Stewart, and attorney Truman Gibson Sr. Bousfield and Gibson were in the Chicago Boulé.

Since 1938, Jack and Jill has been the premiere by-invitation-only social group for black elite children. Here, Jack and Jill mothers pose at a 1951 convention. There are now 220 chapters throughout the United States.

Jack and Jill president Dr. Nellie Gordon Roulhac of Philadelphia *(center)* is flanked by four other national presidents at the 1955 convention. They are *(far left)* Alberta Turner, Emilie Pickens, Roulhac, Edna Seay, and Dorothy Wright. Wright's family founded Philadelphia's oldest black bank.

The Girl Friends is one of the most selective black women's groups in the nation. Charter members gather in a New York home in the early 1930s. Anna Small Murphy *(back row, fourth from left)* launched the club's famous cotillions when she was president in 1951.

Phyllis Murphy Stevenson *(far left)* was the most celebrated debutante in 1952, when she came out at the Ball of Roses cotillion in New York. Here, at seventeen, she joins her friends and escorts. Years later, Phyllis was president of the Girl Friends and her husband headed the all-male Comus Club.

A Girl Friends Ball of Roses cotillion formation being presented in the early 1960s at the Waldorf-Astoria Hotel.

Oak Bluffs, Martha's Vineyard, has been a popular summer resort for the black elite since the 1930s. Here, writer Dorothy West *(third from right)* joins friends on the island in the 1920s.

Clifford Alexander was named secretary of the Army under President Carter. He married Adele Logan, the daughter of successful Harlem physician Warren Logan. Clifford and Adele both attended the Fieldston School and Harvard before Clifford went to Yale Law School.

Jewel Stradford Lafontant was just one generation in a long Chicago dynasty. Her grandfather and father practiced law in the city before her. Here, in 1973, she serves as deputy solicitor general of the United States under President Nixon. A powerful Republican who made waves when she seconded the nomination of President Nixon, Jewel later sat on the boards of the Mobil Corporation, Revlon, and TWA.

An appointee of Governor Nelson Rockefeller, Ersa Poston held positions in New York State government and then in Washington when Rockefeller became vice president. Married to Ted Poston, the first black journalist to work for a mainstream newspaper, Ersa is a member of the Links, the Girl Friends, and AKA.

U.S. senator Edward Brooke *(far left)* was elected by Massachusetts residents in 1966 and was the first black senator since Reconstruction. Here, he joined Detroit debutante Mary-Agnes Simmons and Edward Davis, the first black to own an auto dealership. Davis's wife is prominent fund-raiser Mary-Agnes Miller Davis.

Detroit socialite and philanthropist Mary-Agnes Miller Davis *(far right)* chats with First Lady Jacqueline Kennedy and Lady Bird Johnson during a 1961 press conference. A founder of the Co-Ettes, a national civic organization for black girls, Mary-Agnes has been at the center of Detroit society since the 1940s.

As publisher of *Black Enterprise* magazine and owner of a Pepsi-Cola bottling franchise, Earl Graves Sr. is an important figure in New York's black elite. He raised three Ivy League–educated sons *(shown here)* in a $3 million Scarsdale mansion and a waterfront home in Sag Harbor, Long Island.

Each summer, beginning in 1921, the black elite sent their sons and daughters to Camp Atwater in North Brookfield, Massachusetts. These two photographs were taken in the 1940s, when the seventy-three-acre camp with its twenty-seven buildings was at its zenith. Its board includes Senator Edward Brooke, attorney Vernon Jordan, and former state department official Clifton Wharton.

In 1958, Rikki Stubbs is crowned at the Detroit Co-Ette cotillion. Stubbs's mother, Marion, was founder of Jack and Jill, and her stepfather, Dr. Alf Thomas, hosted Rikki's wedding on a family-owned island between Detroit and Canada. Holding the microphone *(center)* is Detroit philanthropist Mary-Agnes Miller Davis.

The cover of the brochure that New York brokers used to sell E. T. Williams's Upper East Side duplex apartment. He and his wife, Lyn, moved into the $2.5 million apartment in 1982 and were soon profiled in *New York* magazine.

With ties to Jack and Jill, the Boulé, and Comus, Auldlyn and E. T. Williams *(seen in both photographs)* stand in the yard of their four-acre Sag Harbor family compound, built in the 1840s. Williams's board memberships include the Museum of Modern Art, the Central Park Conservancy, Atlanta University, NAACP Legal Defense Fund, and the Schomburg Library.

The Rainbow Yacht Club includes black physicians and attorneys who own yachts and travel together along the East Coast. Harry Delany, head of surgery at New York City's Albert Einstein Hospital, is fleet commander. His wife, Barbara, integrated Cornell University's all-white sororities in the 1950s. The Delany family story was told in the Broadway play *Having Our Say.*

Dr. Asa Yancey of Atlanta *(back row, center)* helped hundreds of black doctors become surgeons in Georgia and Tuskegee, Alabama, when southern hospitals refused to train them. He is a member of the Atlanta Boulé, and he and his wife, Marge, have sat on the Atlanta School Board. Three of their children are physicians.

More than sixty years after its founding, Jack and Jill remains an organization that bonds children from black elite families. Members must perform volunteer community service and participate in leadership development projects. Here, Jack and Jill children from Chicago play in a backyard pool of a member's home in suburban Lake Forest.

U.S. congressman Harold Ford Sr. grew up in a prominent Memphis family where members served in the state senate, city council, and county legislature. His father founded a prominent Memphis funeral home. Ford is pictured here *(third from left)* in a meeting in 1977 with some other members of the Congressional Black Caucus.

My grandmother poses in 1920 for her engagement photograph in Memphis, the same year my grandfather started a trucking business.

Founded in 1946, the Links has included congresswomen, mayors, college presidents, wealthy socialites, and presidential cabinet members. My mother *(front, fourth from left)* and her Links chapter are photographed here during a White Rose Ball that she chaired with fellow member Dr. Betty Shabazz, widow of Malcolm X *(front, sixth from left)*.

Before she was appointed to serve as secretary of housing and urban development for President Johnson, Washington's Patricia Roberts Harris was Johnson's ambassador to Luxembourg. Here, in 1966, she joins the president on the campus of Howard University, where she was dean.

Harvard Law School graduate Reginald Lewis became one of the richest and most influential men on Wall Street when he purchased McCall Pattern Company in 1986 and Beatrice Foods in 1987 in a billion-dollar leveraged buyout.

Reginald Lewis and his wife, attorney Loida Lewis, stand on the lawn of their mansion in Amagansett, Long Island.

Before Reginald's death in 1992, he and Loida gave $3 million to Harvard Law School. Active as a Jack and Jill mother, Loida now runs her husband's business empire from New York and Paris.

Since 1903, Sigma Pi Phi (known as the Boulé) has been the most selective membership club for men in the black elite. Consisting mostly of doctors, lawyers, and wealthy businessmen, each chapter regularly hosts black-tie dinners and gatherings. I *(top row, center)* pose with my Boulé chapter members at a recent holiday party.

Harold and Helena Doley recently bought and restored the 1906 mansion of Madame C. J. Walker. Founder of the Wall Street firm Doley Securities, Harold was the first black individual to own a seat on the New York Stock Exchange. A native of New Orleans, he can trace his family back to eighteenth-century Louisiana.

bers his childhood in the 1960s and 1970s when he and his sister traveled to the white North Side, where they attended the private Latin School of Chicago.

"My sister and I went to the Latin School when there were very few blacks there," explains Chatman, who attended from the second through the eighth grade, "but there were many black families who chose private schools—like our friends the Dibbles. Ironically, I ended up in school with them when I went away to boarding school at Northfield Mount Hermon." Just as the Dibbles sent their kids to the school because they had attended, Eric attended because his father and uncle had gone before him.

Eric's experience in moving back and forth between the North and South Sides was one that did not happen with any regularity a generation before him.

For many years, the center of Chicago's South Side was Forty-seventh Street and South Parkway, a busy section that featured the Regal Theater, the Savoy, the South Center Department Store, and other establishments that were patronized exclusively by the black residents.

Sometimes referred to as "Bronzeville" by the newspaper *Chicago Defender* and others who lived there, the black South Side of Chicago had very strict borders before World War II because of restrictive covenants that kept blacks from living in areas that the city designated for whites only. The South Side was completely self-contained, with homes, businesses, schools, stores, and places of entertainment so that white Chicago could remain certain that blacks had no reason to cross its borders, regardless of how wealthy some of its black residents might be.

"When we were growing up on the South Side, we knew not to walk east of Cottage Grove," says Dr. James Jones, a physician who belongs to the Chicago Guardsmen. "Police would stop you, ask for your I.D., and then quiz you about what you were doing there. They knew blacks didn't live in those neighborhoods, and this was their way of intimidating us from even taking leisure walks in that direction."

"In fact," my mother recalls, "the only reason blacks had for going east of Cottage Grove—and the only explanation that would be acceptable to the police, would be our trips to swim in the all-black, segregated parts of Lake Michigan. We were always careful not to venture too far north or south along the Lake—beyond Jackson Park—onto the white beaches because we had heard stories about Gene Williams, the black kid who had been stoned and drowned a few years earlier after white swimmers saw him in the all-white section of Lake Michigan."

As black families were finally able to move beyond the black belt into areas that had been kept uniformly white with the help of restrictive covenants, they moved to such neighborhoods as Chatham for its houses and tree-lined streets, and to South Shore, which had many more apartments. They also moved a little west of South Shore to the Highlands neighborhood. They also moved south to Pill Hill, where many doctors settled. The most popular area for black professionals, however, was the older elite neighborhoods surrounding the University of Chicago: Hyde Park and Kenwood.

"Hyde Park-Kenwood has always been a sought-after area, and it finally opened up to us in the 1950s," says Jetta Norris Jones, a Chicago attorney who lives in the neighborhood with her husband, James, an obstetrician. Jetta, who sits on the board of the one-hundred-year-old Art Institute of Chicago, is a past president of the local chapter of the Girl Friends. She adds, "Because the neighborhood surrounds the University of Chicago, its residents included many progressive members of the faculty. This group was more likely to allow integration than many of the residents in the less affluent neighborhoods."

Although my Uncle Telfer's business as a bail bondsman was by no means a glamorous business—his office kept pace with the increasing arrests and prosecutions of black South Side "policy" (gambling) operators, as well as black residents who were fairly or unfairly arrested by an all-white and abusive police force—it did introduce him to many of the influential South Side attorneys and business owners. It also required him to have contacts among the white citizens who worked in the courts and police precincts. Because of those ties, certain members of my family were always privy to who was doing well, who was in trouble, and who in the community was being watched.

Most of the old elite supported their own black South Side establishments exclusively. One of the best-known old black businesses that maintained such loyalty was the Binga State Bank, which was founded by Jesse Binga. Originally from Detroit, the successful real estate broker opened the bank in 1908 and served South Side black businesses and families for many years. The funeral home most often selected by the elite families was Metropolitan Funeral Home. It was owned by the Cole family.

"I remember when my brother, Arthur, took the Cole daughter—Michele—to the high school prom. That family was such a big deal then," says Chicago native Ronne Rone Hartfield, who recalls when most of the formal parties and proms were held in another famous South Side establishment: the Parkway Ballroom.

Like the Coles, who owned the Metropolitan Funeral Home, and the families tied to Supreme Life, most of the elite became famous because of the connections to their high-profile businesses. One of those families was the Sengstackes, owners of several major black city newspapers. As a longtime TV commentator and newspaper columnist, Vernon Jarrett knew the publishing family quite well. "John Sengstacke just died, but he and his uncle, Robert Sengstacke Abbott, owned the *Chicago Defender*, the *Michigan Chronicle* in Detroit, and the *Tri-State Defender* in Memphis," says Jarrett, who has been a columnist with the *Chicago Tribune* and *Chicago Sun-Times* and continues to host a show on WLS-TV, the city's ABC affiliate.

"A lot of people credit Sengstacke for mobilizing the black migration to Chicago," explains Jarrett, "so the family name was known among blacks of all economic backgrounds. Blacks in the South were reading his newspapers and learning about job opportunities in the northern cities. Even though they didn't find everything they had hoped when they got here, they wanted to escape the sharecropping lifestyle and the harsher racial indignities that flourished in the South. And they credited the Sengstacke family for getting them here."

"The Sengstackes were dedicated people, but they weren't society people," says one of Jarrett's Boulé brothers. "The prior generation of the family—the Abbotts—they were the Republican society types. But this next generation was more liberal and more concerned about using its money and power to advance blacks."

Another famous family on the South Side was the Barnetts. Etta Moten Barnett, a longtime grande dame who was well known in the city, was a jazz and concert vocalist and a collector of African art. She was also married to Claude Barnett, who founded the Associated Negro Press, served as an adviser to President Franklin Roosevelt, and counseled many black and white political leaders on labor and agricultural issues. One of their daughters, Susan, married into the prominent Ish family, a well-respected Little Rock, Arkansas, family involved in medicine and insurance. Now widowed, Etta remains an important voice in Chicago old-guard life. "Some of the Links are giving Etta a ninety-sixth birthday party and inviting a lot of Chicago celebrities," explained Fern Jarrett, a longtime Link, when I asked her to update me on the local parties among the South Side crowd. "Etta and Claude knew everybody in this city. They lived a truly global life because of her musical and theatrical work and his travels to Africa. They were rare in their ability to bridge the black and white communities here. For the most part, you didn't see whites express an interest in blacks who lived on this side of town."

Other talked-about families include the Dibbles and the Cooks. And these are names that I had heard often at the all-black parties I was attending at Harvard, Wellesley, and Amherst. The Dibbles and the Cooks had offspring at all of these schools when I was in college and law school, and I recall many of my black friends adding names like Hillary Dibble, Mercer Cook, and Janice Cook to their party guest lists even when they didn't know these classmates. What we did know was that these were famous last names from black Chicago, and this was sufficient to warrant an invitation.

"Although we all think of them as Chicago people, Eugene and Ann Dibble actually grew up in Tuskegee, Alabama," says a Chicago Link, "where their father was an important doctor in charge of the V.A. hospital, and their grandfather was the architect who built most of Tuskegee Institute." Their uncle, Robert Taylor, who managed the Rosenwald and other real estate in the city, was memorialized when the Robert Taylor Homes development was erected on the city's South Side. At the time, the sixteen-story homes were the largest housing project in the United States.

"But Ann and Eugene both got married and raised their kids here," the woman continues. "Ann got married to Mercer Cook and later to Vernon Jordan. Mercer's family were Howard University professors, and everybody knows who Vernon Jordan is." It is hard to find a member of the old guard who does not know Eugene Dibble, the financial adviser and real estate investor, or any of his children. There are still high-profile Dibbles and Cooks living in the city and throughout the country.

As in other cities, the black elite in Chicago has had its favorite churches. When I talk to the established people today, two congregations are mentioned repeatedly. Not surprisingly, one is Congregational and the other is Episcopal. As an Episcopalian, I am well aware that America's black elite has a preference for these two denominations—in spite of, or perhaps *because of* the black mainstream's attachment to the Baptist and Methodist denominations.

"Although Linda Johnson got married at Fourth Presbyterian Church on the Near North Side," remarks Theresa Fambro Hooks, who covered many black society weddings for the *Chicago Defender*, "the two churches that were most closely identified with the city's old black elite on the South Side were Good Shepherd Congregational at Fifty-seventh and Prairie, and St. Edmund's Episcopal at Sixty-first and Michigan. The Dickersons and the Bousfields went to St. Edmund's."

While it is not as important as it was during my parents' generation, church affiliation in Chicago says a lot about an individual. It can confirm one's place in South Side society, and it can also confirm one's place outside of it. When people learn that my parents were married at the substantial but distinctly nonsociety Grant Memorial Church on Drexel Boulevard—a large stone Methodist church—they are not surprised to learn that they were new to the city and transplanted from the South. But occasionally, in the past, there have been old elite families that have shocked their old-guard neighbors with their church selections. Linda Johnson, the daughter of John Johnson, publisher of *Ebony* magazine and the richest black man in Chicago, disappointed many old-guard blacks who were hoping that she and her family would select their church for the ceremony.

"To be honest, a lot of us were kind of surprised that Linda got married at a church that was as white as Fourth Presbyterian," says a third-generation Chicagoan who belongs to Good Shepherd Congregational, "but then I have to remember that her parents had so much money living over there on Lake Shore Drive that of course they had a more integrated life. He probably didn't feel attached to our churches. He was the wealthiest black man in America, but he wasn't old Chicago."

Although some older people may not refer to her family as old-guard, it is only because her father was not born into a well-connected family. Despite the fact that he graduated from DuSable High with the children of many old-guard families and is a generous underwriter of many black and nonblack causes in the city, there is a segment of the black elite that considers him an outsider. The wedding at Fourth Presbyterian Church confirmed that for them.

"John is a brilliant, self-made man who has done more for blacks in this country than virtually any other person of his generation," says a member of Johnson's Boulé chapter, "and some of these old-guard types just can't stand seeing his success. If he wants to live on Lake Shore Drive and give a wedding on the North Side, he should be allowed to do it, but some of these people are unforgiving about that Fourth Presbyterian thing."

"My wife and I were married by the minister from Good Shepherd because my father belonged there," says Truman Gibson Jr., a member of one of those old-guard families that you'd expect to find at a South Side Congregational or Episcopal church. "Today, my wife and I go to Trinity Episcopal, where I've been a vestryman for many years."

Maudelle Bousfield Evans has fond memories of St. Edmund's. "Not only were my parents very active there but most of my friends like Katherine Dickerson and the Thatchers went there. I wouldn't have had my children go anywhere else."

Another St. Edmund's family is Fern and Vernon Jarrett, who sent their sons to the church's grammar school. "They've put the most beautiful stained glass in that building, and many of us bought windows to honor certain historic black people," says Fern Jarrett, who remembers going to the church before it moved to its present location.

Another church that the elite favors is St. Paul the Redeemer, which the Taylors and Dibbles have attended for several years.

The black elite in Chicago made some major changes during the early 1970s after the close of the civil rights movement. Even though many of the older families maintained their ties to their churches on the South Side, many of the elite and their adult children started to move to places they'd never lived before.

"Even though Lake Meadows had been very fashionable, what became even more popular for some people was living downtown," says Ronne Rone Hartfield. "Many families moved into buildings near Wacker Drive. And buildings like Harbor Point and Lake Point Tower were also desirable because of their proximity to the water."

This is when Maudelle and Leonard Evans moved downtown to an even more glamorous address: the sixty-fourth floor of the John Hancock Tower. It's also when *Ebony* magazine founders John and Eunice Johnson moved onto North Lake Shore Drive, into what is still considered the best address in the city. As the richest black man in America, he has been able to afford to set high standards in this crowd. A member of the Chicago Boulé, Johnson moves effortlessly between the black and white elites.

"I am optimistic for the future, but the Chicago metropolitan area is still a very segregated place," says attorney Jetta Norris Jones, who served as director of external affairs in Mayor Harold Washington's administration. "You still do not see large numbers of blacks in the north side of the city along the lake or in the northern suburbs like Winnetka, Kenilworth, Wilmette, or Lake Forest. Although brokers and residents will certainly sell their homes to blacks, the larger middle-class and professional-class black population has either stayed in Hyde Park-Kenwood or moved to the South Side suburbs like Flossmoor, Homewood, Country Club Hills, and Olympia Fields."

"And if they move to the north suburban area," says Jetta's husband, Dr. James Jones, an obstetrician-gynecologist on staff at the University of Chicago, "they will move to Evanston. That's been known as one of the first liberal suburbs here."

"And of course, you'll see a few Jack and Jill members in the northern suburbs," adds Ronne Rone Hartfield, as she sits with her husband, Robert, a math professor at the University of Chicago, and points out a recent *Chicago Tribune* article profiling a Jack and Jill garden party where children are playing around a pool in the backyard of a large colonial home owned by Ronald and Jacqueline Irvine, a young black family in the exclusive north shore suburb of Lake Forest.

Dedra Gourdine Davis, a Jack and Jill mother in the south suburban Chicago chapter, spends most of her time in the suburbs outside of the city. "We now have black families in every suburb around Chicago," says Davis, who is a pharmacist and just saw her youngest daughter graduate from their suburban high school in Homewood. "My husband and I used to live in Pill Hill, but we moved to the suburbs in the seventies," she explains, "and our children are a testament to the fact that you can raise black boys and girls with a positive black image in the Chicago suburbs despite what people on the South Side think. Our youngest daughter, Jaime, not only was in Jack and Jill here but was also an AKA debutante, as I was, and she's now at Fisk. If she lacked a black identity, she wouldn't have made such choices."

Jetta and her husband, who raised three kids in the Chicago Jack and Jill, nod in agreement. But what Jetta finds more interesting is how the politics of Chicago has also affected the wealth of black Chicagoans. "When Mayor Washington was in office, there were clearly black entrepreneurs and other professionals who were having their first chance at starting businesses because they had not had these chances with prior mayors." Jones also says she's optimistic. "We are seeing a new generation of blacks growing up in the city who still have a strong interest in its black political and economic development."

Many of the new black stars that people talk about in the city are, in fact, not so new at all. Many of them are the children of Chicago businesspeople or politicians who laid the groundwork before them. Among the group is forty-year-old Langdon Neal, an attorney who is chair of the board of elections. His father and grandfather were respected attorneys as well. "Langdon's father, Earl, was so connected with the old Mayor Daley," says a Neal family friend, "that he was able to help Langdon's grandfather get named as a judge." Langdon's father, Earl Neal, distinguished himself in Chicago as a lawyer who not only represented the city at times but was also the head of the First Federal Savings Bank and the first black to head the University of Illinois board of trustees.

Another family that produced a member of the new elite is the Davis family. Allison Davis Jr. is a lawyer and major real estate developer. A former head of the NAACP Legal Defense Fund in the city, he lives a life that straddles the elite black and white worlds. His father, Allison Davis Sr., who died in the early 1990s, was a tenured sociology professor at the University of Chicago and a vocal activist on the South Side. "When I first moved to Chicago, I thought Allison was white because of his complexion and his wife," says a young attorney who lives in Hyde Park, "but then I was seeing him at all the black events and hearing about his work with Reverend Arthur Brazier. He's everywhere." Allison's brother, Gordon, is a high-profile Wall Street attorney who served as New York City's commissioner of parks in the 1980s.

John Rogers is a third-generation member of the Chicago elite. Son of the recently deceased attorney Jewel Stradford Lafontant, he was a few years ahead of me at Princeton and now heads Ariel Capital, one of the nation's most successful money management firms. His grandfather, C. Francis Stradford, graduated from Oberlin and argued many major civil rights cases in the city. His recently deceased mother, Jewel Stradford Lafontant, was a prominent attorney who sat on the boards of Mobil Oil, TWA, the Equitable, and Revlon. "Jewel was a powerhouse and she made us all proud. She was deputy solicitor general and then an ambassador under President Bush," says Washington friend Savanna Clarke, who met Jewel during their volunteer work for the Urban League. "A lot of people know her because of her political and business activity, but she did a tremendous amount for blacks in Chicago and Africa. George Johnson, founder of Afro Sheen's Johnson Products, was a good friend of hers, and when we all reunited at her funeral at Fourth Presbyterian Church, we saw how much she meant to Chicago."

George Johnson, founder of Afro Sheen and a developer of other products, is a name that Chicagoans like Dedra Gourdine Davis know very well. "Everybody in Chicago knows about the company that George built with his wife, Joan. Their son, Eric, recently bought Baldwin Ice Cream, an old Chicago black business." Davis notes that Eric and his wife, Pamela Johnson, are in her Jack and Jill chapter.

There is also Valerie Bowman Jarrett, an attorney who heads the Planning Commission under Mayor Richard M. Daley, son of former mayor Richard J. Daley. Valerie is a graduate of the University of Michigan law school, her father is University of Chicago pathologist Dr. James Bowman, and her father-in-law is the journalist and commentator Vernon Jarrett.

And clearly, the most famous and respected member of the new elite is Linda Johnson Rice, CEO of Johnson Publications, a company that not only publishes *Ebony* and *Jet* but also produces a successful line of cosmetics. The daughter of founders John and Eunice Johnson, the Northwestern Business School graduate has run the $200 million company for several years.

And of course there are many in the new elite who are not children of old-guard families. They are entrepreneurs, attorneys, politicians, and physicians like Raphael Lee, who is a plastic surgeon with a Ph.D. from MIT; Bill Curtis, the architect; Marty Nesbitt, an entrepreneur; Teresa Wiltz of the *Chicago Tribune*; Peter Bynoe, an attorney who sits on the board of overseers at Harvard; and Lester McKeever, an attorney and CPA whose son is a record executive in Los Angeles. Many of the new black professionals have no ties to the South Side. Some live along the "gold coast" in Lake Shore Drive apartments, while others live in the mayor's neighborhood of Dearborn Park, or in town houses in the Printers Row section. Others have ventured south to suburbs like Olympia Fields or north to suburbs like Evanston and Glencoe.

Because they live more integrated lives than the generation that preceded them, they support cultural institutions in both the white and the black community. So in addition to aiding the patrician Art Institute of Chicago, the Field Museum, or the Chicago Zoological Park, they are also supporting the Black Ensemble Theatre, the DuSable Museum of African American History, and the Bud Billiken Day parade, an event that was created by the founder of the black newspaper *Chicago Defender* and has become the third-largest parade in the country. There is a particularly upscale black crowd that contributes to the activities of Abena Joan Brown's ETA Creative Arts Foundation, a South Side performing arts center that stages black cultural presentations. And since one of the

Links' best-known debutante cotillions in the country takes place each year in Chicago, the new elite are sure to be seen table-hopping at this black-tie affair every spring.

"Some say the young professionals are a modern version of Earl Dickerson, Truman Gibson, and Midian Bousfield," remarks a Boulé member in his seventies, "but these young people who are here from out of town are, for the most part, far removed from the life and history of Chicago's South Side. They may be upper-class, but they don't know anything about our old social clubs like the Snakes or Druids. They don't know what we endured when we were ghettoized into a couple of dozen square blocks. They never had a policy with Supreme Life. They never set foot in St. Edmund's. They came here after Harold Washington died, and they didn't have the chance to see a black mayor face down a city filled with racial animus. I'm glad to see them here, but they need to know that we have a history here."

As Chicago businessman and author Dempsey Travis points out, "Chicago's black history lives on. It didn't stop with DuSable."

Black Elite in Washington, D.C.

"What's with the broom?"

Trying to appear distracted, I tugged uncomfortably at the sleeve of my tuxedo and then looked up at the woman. "I beg your pardon?"

"The broom under your chair?" she asked as she lifted a corner of the tablecloth and peered underneath the table where I was seated.

The attractive woman's floor-length pink silk chiffon dress instantly made me realize I owed her some kind of answer: She was obviously one of the eight bridesmaids. "Oh, that's Lenny's. I hadn't even noticed it."

"But I saw you walk in with it," she added in an almost accusatory manner. She knew I was lying.

I looked at another guest seated next to me. Then I turned back to the woman. Not wanting to embarrass this white bridesmaid in front of a table filled with black guests, I stood up and quietly told her about "jumping the broom," a tradition that had existed for generations in African American culture. "It started during slavery when blacks were not permitted to have formal church wedding ceremonies," I explained politely, "so the man and woman would jump over a broom. And that would symbolize the—"

"Yeah, I know the history," she said, cutting me off while rolling her eyes, "but do your *brother* and his *wife* know you're doing this?"

I looked around the room as the eleven-piece swing orchestra began to launch into another number. It was a warm May afternoon in Washington, D.C., and my brother and his new wife had just been married in one of the most ornate hotels in the city, at one of the most formal weddings I had ever attended. Two hours earlier, the wedding party had stood in black tie and tails and pink silk listening to the young couple give their vows in front of an Episcopal priest surrounded by many members of Washington's elite.

With the exception of some New Yorkers from our side and some women who had gone to the same Los Angeles private girls' school as the bride, the room was wall-to-wall Howard and wall-to-wall Washington.

"How dare that white girl come over here and roll her eyes about our traditions," I said to a black woman attorney at the table, who had flown in from New York for the event.

"Tell me about it," the attorney responded in mock exasperation.

Almost two hours later, halfway through the reception, I had worked my way around the crowded hotel ballroom, finally getting to converse with the friends and colleagues of my new in-laws. Earlier, at first glance, I had estimated that more than half of the 250 guests—most of them Washingtonians—were white. But after meeting these new faces, it was now apparent that my initial assessment of the group was incorrect.

As I introduced myself to some of these "white" guests—including the accusatory bridesmaid who had approached me earlier—I discovered that they were, in fact, not white at all.

"Yes, I went to Howard with your brother's wife," said a young attorney who looked as if he were either Italian or Greek.

"My father is head of surgery at Howard Hospital," remarked a straight-haired woman with pale skin and hazel eyes.

"Our mothers pledged AKA together and our fathers are friends from Camp Atwater," said an older man with straight gray hair.

There was even a green-eyed guy with nearly blond locks who said to me, "I think we met at a Jack and Jill party a few years ago."

Table after table, people revealed their racial identity by the remarks they made and the stories they shared. Only one or two times before had I been surrounded by so large a group of people whose ethnic identities I had completely mistaken. I quickly got the feeling that such scenarios are not uncommon among the Washington black elite. Nevertheless, my new discovery emboldened me to step quickly to the microphone and coax my brother and new sister-in-law into performing the hundred-year-old broom-jumping ceremony in the center of the ballroom.

The ceremony lasted no more than five minutes, but it turned into a total disaster. For all of us. What had normally been considered one of the few lasting traditions from the pre-emancipation black southern culture was suddenly seen as a hostile and unwelcome gesture.

"Why would he bring a niggerish thing like that in here?" announced a woman as my new sister-in-law's dress was just clearing the elevated broom handle.

"These country-ass blacks always have to drag in this slave history crap," added an older gentleman sitting just off the dance floor. "Jesus Christ."

Welcome to Washington's black society.

Feeling defeated and humiliated, I quickly picked up the broom and walked off the dance floor and out of the hotel onto Sixteenth Street, where I waited for fifteen minutes. When I came back in—with the broom no longer in sight, I slunk back to my seat to face the wrath of several new in-laws and their friends. For the most part, the Washingtonians in attendance ignored me. Others patted me on the back. A couple of my brother's doctor friends thanked me for the dose of black culture in the otherwise lockjawed ceremony and reception.

What I remember most from that incident was a comment from a light-complexioned blonde woman in a bugle-beaded floor-length gown who spoke to me on her way out.

"I just want to tell you," the woman said, "that what you did was very courageous and I appreciate it." She squeezed my hand in a gesture that caused me to focus closely on her cloudy blue eyes. "Sometimes we have to be shaken up and reminded where we came from."

The woman, whom I now presumed to be black, smiled and then walked off in a mist of Chanel No. 5.

Two years later, I got married in my own formal wedding—this one on Fifth Avenue in New York City. After our 260 guests were transported in a phalanx of black limousines to our Upper East Side reception, and after dancing to Cole Porter and George Gershwin, my wife and I jumped the broom. Members of my family have been doing it for generations—regardless of our incomes or the makeup of our guest lists. We did it because we were proud to be black. We did it because it was one of the few lasting symbolic gestures that paid homage to our slave ancestors. The fact that I would be ostracized or even congratulated for "having the courage" to embrace this tradition still leaves me stunned. Though I have had many since then, it was a Washington moment that I shall never forget.

When one talks to Marjorie Holloman Parker in her living room at the exclusive Watergate complex, it becomes obvious that her family's life represents the collision of two high-society worlds: one black and one white.

She and her family have not only earned the top black Washington credentials but also maintained the ties that make her an important force among the white elite as well. She was the daughter of a prominent minister, and her husband was the son of a successful attorney.

A former member of the D.C. City Council, Marjorie met her husband, Barrington Parker, when they were both students at the city's elite, all-black Dunbar High School in the 1930s. After pledging AKA at Minor Teachers College, she went on to earn a Ph.D. at the University of Chicago. After pledging Omega at Lincoln University, her husband went on to graduate from law school. Around the time that Marjorie became the only black board member at the all-boys prep school St. Albans, her husband received a lifetime appointment to the federal bench. "My husband was appointed as a U.S. District Court judge shortly after President Nixon appointed me to the nine-member city council," explains Parker, who has long been active in Washington's political and civic affairs. Her credentials and demeanor are distinctly Republican and old guard.

In addition to serving four years as the national head—or Grand Basileus—of Alpha Kappa Alpha, the country's oldest black sorority, and spending the last eighteen years as a member of the black women's group the Links, Parker raised two sons who received degrees from Princeton, Johns Hopkins, and Yale. Jason, the elder, worked in the foreign service after receiving his Ph.D. in Near Eastern studies at Princeton and marrying Toni Trent, the daughter of the first head of the United Negro College Fund. Younger son, Barrington junior, is a federal judge, like his deceased father. He gained a great deal of renown as the judge assigned to the famous Texaco employment discrimination and obstruction of justice case involving white executives and black employees.

"I guess I've had a very interesting life," admits Dr. Parker, a modest woman who has grown accustomed to being surrounded by successful people of every age—whether they are contemporaries of her husband or of her granddaughter, who recently started Yale after graduating from the 143-year-old St. Paul's School in New Hampshire.

"My father became the minister at Second Baptist Church in 1917 and remained there for fifty-five years," she says, "and he always expected his family members to succeed at whatever we pursued. In a place like Washington, you are surrounded by accomplished black people, so I never thought of our family as being unusual."

Although Parker's world includes Watergate neighbors of the stature of former U.S. senator Bob Dole and American Red Cross president Elizabeth Dole, and although she and her family have contributed more to black history than most African Americans, there are more than a few members of the D.C. black elite who would exclude superstars like her.

"Oh, Marjorie is *new,*" says a fourth-generation Washington woman who has followed Parker's career. "She's not old Washington. Not old at all."

"And plus, they always lived in the wrong neighborhoods," sniffed another matron. "They're surrounded by white people."

Washington is a tough town. From my own experiences and interactions, I am convinced that the old black Washingtonians like it that way. I remember that during the first weeks that I was serving as a White House intern in the Carter administration, I was told by a group of White House staffers that "blacks here in D.C. are real snobbish, so don't take it personally if you don't get accepted." What made this advice shocking was that the White House staffers who offered it were white.

It was one thing for blacks to be aware of the cliques and elitism in our own community, but for whites to know it was surprising and unsettling. Unlike the whites in other cities, the whites in Washington are acutely aware of the black upper class around them. In this city of 580,000 people, the history of black success stories is obvious. And even whites can tell which blacks are "too new" to be accepted by the black old guard.

So, as I found out during my White House tenure in Washington, being left out of parties or conversations because one is lacking in Washington lineage is just one of the basic challenges of being black and accomplished in the District. In fact, there are many other reasons to be excluded. For example, many well-to-do blacks are chastised for living outside the black gold coast—a tony neighborhood of fifty- and sixty-year-old brick and stone houses in the upper northwest quadrant of the city—particularly if they take up residence in certain Washington,

Maryland, or Virginia neighborhoods that are deemed too white. Among these areas are very expensive ones like Parker's Watergate, or the million-dollar neighborhood where she and her husband had previously lived, on the west side of Rock Creek Park. Or Georgetown. Or Potomac. Or McLean.

In a relatively dense city that is 65 percent black, some would suggest that such criticism over choice of a neighborhood is a reflection of a unique and petty Washington-style parochialism. Others say that the neighborhood issue is important, but not nearly as important as the lineage conflict.

"In this town, you don't even *count* unless you're at least third or fourth generation," says a physician who grew up as a fourth-generation Washingtonian in the North Portal neighborhood referred to as the black gold coast.

"People are lying if they tell you anything different," he adds while walking slowly down tree-lined Kalmia Road and pointing out some of the famous names living in the redbrick colonials and Tudors in this section of northwest Washington. "A lot of the most important members of the old guard in Washington look at three basic defining characteristics: skin color, ties to Howard, and number of generations your family has been here. Of course you'd better be associated with the right social groups and events—Jack and Jill, Links, the Tuxedo Ball, etcetera—and you'd better have some doctors in your family. But even if you've got those things, you'll never be considered an 'insider' if you don't have those basic three D.C. qualifiers. And if you don't have them, you're out of the running—just like that," the physician says with a sudden snap of his fingers, "and there's nothing you can do to change it."

Benaree Pratt Wiley is a fourth-generation Washingtonian who was born in LeDroit Park, attended Howard University Nursery School, and graduated from Howard University before moving to Boston. "Because of Howard and because of the close ties that people have with each other and many of the institutions, Washington really feels like a small town," says Wiley, whose sister Sharon Pratt Kelly was the first woman mayor of the District.

"Of *course* I know Bennie Pratt," says ViCurtis Hinton, a Washingtonian who raised her kids in the Washington chapter of Jack and Jill during the 1950s and 1960s. "My daughter, Audrey, used to play with Bennie. In fact, I remember when Bennie and her sister both went to Howard, and when Bennie started at Harvard Business School." A few dozen other residents can rattle off the accomplishments of Bennie and Sharon, not because various members of the Pratt family had been in Washington since the late 1800s but because their kids were all playmates, and these people all went to schools like Howard, belonged to groups like the Links or the Boulé, took cotillion preparation lessons from the same teachers, and recognized each other as members of black Washington's extended family. Though accomplished in their own right, the Pratt sisters owned the basic three credentials that made them accepted members of the upper-class community. For some members of this city's old guard, those were the only credentials that mattered.

The history of blacks in Washington is an interesting one—very different from other southern cities and totally different from northern ones. When slaves were brought to Washington, a large concentration of them lived and worked in the Georgetown area. From there, they built roads and erected many of the government buildings and monuments. In the 1790s, Benjamin Banneker, a free black man, surveyed the city and designed the grid for the city's main avenues and streets. Banneker's status as a free black reflected only 20 percent of the blacks in Washington, since the other 80 percent remained slaves until approximately 1805.

By 1835, as white residents began noting the increasing number of black churches, black-owned businesses, and schools that were patronized by the free black population, white segregationists were becoming alarmed by the power and independence of these free-moving black neighbors. Their racist paranoia caused them to lead an antiblack riot during which they burned and destroyed homes and establishments owned by black Washingtonians. While the incident deprived the black community of their independence and some of their mobility, they did not abandon the District. In fact, by 1861, the year before Washington abolished slavery within its boundaries, there were more than ten thousand free blacks living in the city—representing around 75 percent of that city's black population and 25 percent of the city's overall population.

By 1876, blacks were more than one-third of the population, with some of them gaining the opportunity to work in government jobs and start small businesses. In fact, some of the old guard who live there today can trace their ancestral roots to this early community. It was during Reconstruction—the 1870s and 1880s—that Washington became home to two black U.S. senators and eight black members of the House of

Representatives from several southern states. In many respects, the presence of these accomplished black officials in Washington emboldened the well-to-do black community to raise expectations for itself and the people who wanted to belong to it.

But as whites in Washington and elsewhere feared the rise of this new black electoral power, wealth, and independence, they began implementing strict rules of segregation, removing blacks from many jobs as well as from elected and appointed offices. They established separate "colored schools" and pushed the black residents into separate neighborhoods—many of them in crowded and unsanitary alley housing.

Those groups of blacks who escaped the alley life and avoided generations of poverty were primarily those whose families had been free for at least two or three generations before emancipation or who might have been light enough to pass or who had amassed great wealth elsewhere or through unusual circumstances like inheritances from white relatives. This first group of families formed the beginning of Washington's black upper class.

It has long been said that in Washington, beginning sometime in the late 1800s, there was a "Black Four Hundred" or, in other terms, a defined group of four hundred elite families which made up the city's black aristocracy. Bebe Drew Price belongs to a modern version of that group. The oldest daughter of world-famous Dr. Charles Richard Drew, inventor of the blood bank and blood plasma storage, she grew up in a large four-story house on the Howard University campus knowing the children of other important families. Despite her famous father, she insists that nobody among her social group noticed the difference from one family to the next.

"Of course there were many prominent families on and around the Howard campus, but they were more like an extended family," says Price, who has seen several generations of her family pass through Washington as well as the prestigious university. "We grew up around some of the most accomplished black families in the world, but because there were so many, it simply seemed commonplace at the time." Price's sister, Charlene Drew Jarvis, sits on the city council today and continues to bring the old family name into a place of prominence. As candid as Price is, I venture to guess one reason she believes elitism is not as rampant as others might suggest: She's very comfortably on the inside and isn't privy to the aggressive back stabbing that goes on among those who are on the periphery trying to break into the D.C. elite. All her life, she has been a sought-after name among the old guard guest lists and has never been in the position of being an outsider.

According to Alice Randall, a graduate of Georgetown Day School, "the old elite families in this town appreciate the importance of being rooted—of having a cultural tradition that they value and that they know will be repeated." Although she married into an old family that is associated more with Nashville and with New York's Harlem Renaissance—the Bontemps—Randall's childhood in Washington gave her a real understanding of who the D.C. old guard includes as its own. "There are actually only 150 families that make up the core of Washington's black elite, and a number of them trace their ties to the city as far back as six or seven generations," says the Harvard graduate, who spent most of her years in the District before moving to Nashville, where she joined the Links and the Junior League.

Among the families that are included in Randall's 150 names would be such people as the Bruces, Bullocks, Cardozos, Francises, Grimkes, Hilyers, Hollands, Pinchbacks, Terrells, and Wormleys, and, of course, the Syphaxes.

I grew up hearing and reading stories about the Syphax family. I eventually came to know them when I got older. Although she now lives in Baltimore, Carolyn Syphax-Young and her family have long been associated with Washington, D.C., as the descendants of Martha Washington's white grandson. "I grew up in Arlington, Virginia, as did most of my family members, but you will find many Syphaxes living in the capital, or along the outskirts," says Syphax-Young, when considering her ancestors, who date back to an important eighteenth-century family. She has been a member of the Links for the last fifteen years, and she and her husband, Harold Young, are friends with others who travel in elite circles—people like their neighbors: Boulé and Links members Dr. David and Amy Dalton. Like many of the younger descendants of old Washington families, Syphax-Young exudes a modest reverence for her family's history and recognizes the importance of preserving the story. "My cousin, Evelyn Reid Syphax, in Arlington, is one of our family's best historians. She keeps some of the best records on our ancestors," says Syphax-Young, acknowledging that her family's history is, indeed, substantial enough to require a family historian to get it right.

As has been well documented, George Washington Parke Custis, the grandson of Martha Washington, owned slaves in the early part of the 1800s. One of his slaves, Charles, worked as a butler in the Custis home.

The father of Charles was William Syphax, who was also a slave but who had recently purchased his own freedom. Because of Charles's relatively high status as a Custis butler, he did not join his father's acts of buying his way out of slavery. A fellow slave working with Charles at the Custis home was a mulatto child named Maria whom Custis had fathered with Arianna Carter, a black slave who also worked in the Custis house.

Eventually, the two Custis slaves—Charles and Maria—married each other after Custis approved the union, arranged a wedding in his home with an Episcopal minister, and promised to release them from enslavement. Since he had only one legitimate white child—a daughter named Mary Custis, who later married Confederate officer Robert E. Lee—Custis gave his black daughter and her black husband a fifteen-acre plot of land that formed a part of his estate in Arlington, Virginia, as a wedding gift, and left his white child, Mary Custis Lee, the remainder of his estate when he died.

Charles and Maria Syphax, upon their marriage and release from slavery, owned the fifteen-acre plot that adjoined what was later to become Arlington National Cemetery. The couple had ten children, and during the 1850s, 1860s, and 1870s, several of the children gained important jobs in politics. The oldest son, William II, worked in the Interior department and later became superintendent of the special black segregated school district. This was at a time when a large black community resided in the Georgetown section of the city. A congressional bill was passed in 1866 acknowledging that the Syphax land was given to the family because of its relationship to George Washington Parke Custis.

In the 1940s, the federal government wanted to expand Arlington Cemetery and asked the Syphax family to exchange their land that abutted the cemetery for land in another part of the District. Also, in the exchange the Syphax family cemetery was moved to Lincoln Memorial Cemetery. In recent years, Syphax family members have attended Harvard, Howard, and many other schools, and have held such jobs as chief of surgery at Howard Medical School, foreign affairs officer in the State Department and legal posts in the U.S. government. As a child, my parents had told me about the Syphax family's roots when they first introduced me to Washington, D.C.'s history. It is a family history that everyone in the D.C. black elite knows.

"The Syphaxes aren't the only ones around here who keep track of their family history. People here will tell you who they are related to in a heartbeat," says a fourth-generation Howard graduate who grew up knowing the family names of old black Washington. "A lot of women will tell you their maiden names, and it's pretty common for them to incorporate a maiden name in the child's first or middle name—just so the historical ties can be made with little effort. It sounds pretentious, but knowing somebody's background is taken very seriously."

In Washington, a few dozen names carry so much history that it's rare to find a member of the old guard who can't give you at least a few details on the family's significance. For example, in addition to the Syphax history, most will know that the name Bruce relates to Senator Blanche K. Bruce and his wife Josephine Willson Bruce. Blanche was a black U.S. senator from Mississippi from 1875 to 1881 who also served as registrar of the treasury and recorder of deeds for Washington, D.C. They will also know that the Wormley family owned a hotel and other businesses. If someone claims ties to the Cook family, Washingtonians will know that during the mid-1800s that person's ancestors established the Fifteenth Street Presbyterian Church, graduated from Oberlin and the University of Michigan medical school, and ran Washington's colored school system, and that by the end of the nineteenth century, one of the ancestors, John Cook, had assets of $200,000, making him the richest black man in the city.

The merger of two prominent names—the Church family from Memphis and the Terrell family of Virginia—served to create an important dynasty in 1891. Mary Church, the daughter of millionaire Memphis businessman Robert Church; and Robert Terrell, an 1884 graduate of Harvard College and Howard Law School, became an important couple in black Washington society, as Robert served as an officer in the Capital Savings Bank, held government positions, and held a judgeship that he was given by President Theodore Roosevelt. The Terrells summered with other well-to-do black families at Highland Beach in Maryland and at times included affluent whites among their friends.

What most of these first elite families had in common was not just their wealth and social status but also their lineage, which included ancestors who were free people of color or who were able to gain their freedom at least one or two generations before slavery ended. While some were concerned about the status of poor blacks, they often did not associate themselves with the larger black community, which numbered around eighty thousand by the end of the 1800s.

As in many cities that had a black elite, one of the most devastating periods for these families was from the late 1890s through the first two decades of the 1900s, when new Jim Crow laws eroded their access to many jobs, public accommodations, and schools. Even though these people had money, education, and social status, the white community began to shut them out and limit their ability to move as freely as they had before.

But even with the setbacks that occurred during and after Reconstruction, there was a core group of Washingtonians who carved out a community that was almost untouched by the demeaning Jim Crow limitations. In fact, during the early decades of the twentieth century, segregation caused this group of elite blacks to become more and more insular—eventually forcing the creation of a subcommunity within the larger black middle-class community.

"There was a time in Washington when you really felt that everybody knew everybody," says Charlotte Schuster Price as she recalls life in the 1940s and 1950s as a Howard graduate. "The Howard community that lived in and around LeDroit Park, or that was in certain professions, all knew each other. You heard a maiden name or even an unusual first name and you just knew the family tie."

A short conversation with Price, a former library archivist at Howard, and the mother of National Urban League executive director Hugh Price reveals the intricate web that often connects many of the most prominent black elite Washington families. Within the web, the individuals each have dazzling Washington credentials, and they each have ties to some of the same top schools, social clubs, and friends. Her family, in particular, weaves together four prominent Washington families of doctors: Charlotte Schuster grew up in a family of ten children where many became dentists and physicians. "Long before I met my husband, Kline Price, he had grown up with his cousin, Frank Jones, in LeDroit Park. In 1927, he graduated from Dunbar High School," says Charlotte as she recalls the prominent high school that claimed many within the black elite who continued on to Howard. "The two schools had some of the most accomplished blacks in the country." Charlotte and Kline eventually met at Howard, where she pledged Delta and where he pledged Alpha before continuing on to Howard Medical School. Following his cousin, Frank, who eventually became head of the urology department at Howard, Kline entered medicine, also became a urologist, and served as clinical professor in the field at Howard.

Teaching at the medical school during those years were such luminaries as Dr. W. Montague Cobb, an expert in anatomy, and Dr. Charles

Richard Drew, who served as chief of surgery and worked on blood plasma. "Dr. Cobb and Dr. Drew were already important figures at the time," says Price.

Charlotte and Kline Price's two sons, Hugh and Kline junior, grew up in Jack and Jill at the same time as Dr. Drew's children, who lived on the Howard campus. After older brother Kline junior graduated from Tilton Academy, Howard, and Howard Medical School like his father, he ended up marrying one of the Drew daughters, Bebe. Younger brother Hugh went off to Amherst College—the same school as his father's colleagues Dr. Cobb and Dr. Drew—and then graduated from Yale Law School. Hugh tied together another prominent Washington family with his own when he married Marilyn Lloyd, the daughter of Dr. Sterling Lloyd, a prominent thoracic surgeon, who was also the son of a physician. Marilyn's mother, Ruth Smith Lloyd, earned a doctorate at Howard and became a well-known anatomy professor after her own sister, Hilda Smith, married Dr. Montague Cobb, the head of Howard's anatomy department. Dr. Cobb had been the one who had originally encouraged Ruth to enter the field of anatomy.

The way in which prominent doctors were linked together in the Price-Drew-Lloyd-Cobb families is not an unusual phenomenon in Washington's black elite circles.

Similar dynasties have been created in the areas of law, education, politics, and the clergy.

"I'm just not used to knowing people who have no interesting family ties," says a housewife who lives on upper Sixteenth Street. "In fact, I don't know if I can really trust people who don't come from some real background. How can you know what to expect from them? In the world I grew up in, everybody has some respectable person in the past that they can trace back to—somebody that gives you a clue of who they are. Sometimes it's a grandfather or grandmother who did something notable in business or education—or even a grande dame in society. There has to be something."

Background has long been important to many members of the D.C. old guard. For many years, one of the grande dames of black Washington was Benetta Bullock Washington, who was the daughter of George Bullock, the minister of Third Baptist Church, and the wife of Walter Washington, the city's first elected mayor. Although they lived in it even after Washington was named mayor in 1967, Walter and Benetta's LeDroit Park home was originally bought by Benetta's father during the heyday of that historic neighborhood. While her husband was not a native of the city, Benetta had ties to the District's most important institutions as a graduate of Dunbar High and Howard and a principal of Cardozo High.

Howard graduate Alberta Campbell Colbert agrees that there is a core group in Washington who have ties to the same institutions. In her world, those institutions seem to include Howard University, Dunbar High School, the Bachelor-Benedicts Club, and for her kids' generation, Georgetown Day School. "My husband, Frank, was born in Washington and graduated from Dunbar High School in its heyday. His classmates included Senator Edward Brooke and lots of other respected people in government." Colbert and her husband, Frank, were Washington pharmacists and began operating Colbert's Pharmacy in the 1940s. Their son, Craig, went to Georgetown Day; and their daughter, Doris, was a debutante who was presented by the Bachelor-Benedicts, a men's social club that included the city's most established black professionals. The family operated in the midst of the city's most important families. When Frank Colbert died in 1996, he had been a member of the exclusive Bachelor-Benedicts club for fifty years.

In talking to black Washingtonians, one gets the distinct impression that there have been so many prominent names over the past few decades that people don't even attempt to rank them in their minds. "Instead we just group them into the doctor crowd—people like the Leffalls, the Rayfords, the Spellmans, the Clarks, or the Freemans," says an elderly man who has belonged to the Bachelor-Benedicts club for more than twenty-five years. "And there was always the lawyer-government-policy crowd—the Brantons, the Brimmers, the Duncans, the Webbers, the Lynks—and of course, Vernon Jordan."

The man paused for a moment. "But actually, Jordan isn't one of us. He's new." The man laughed. "My wife would know it better because I never really paid much attention to lawyers—especially the new ones like Jordan or Brown."

I've learned well not to interrupt or ask for full names when members of the old guard talk about their neighbors and colleagues. To break the flow and reveal that I am unfamiliar with a name or position is counterproductive when I am in the company of members of this very insular group. Although I do know some of the individuals to whom this doctor refers, there are many that I don't; and to acknowledge this would suggest to him that I am "new," and should therefore not be privy to such conversation—and even worse, it would be insulting to suggest that I did not already recognize and appreciate the distinctions between old Washington names and new ones.

The fact that he dismisses a power broker who is as important and relevant as Vernon Jordan—former head of the National Urban League; partner at the law firm of Akin, Gump; and close confidant to President Clinton—as being "new" reveals what it requires to be taken seriously by some members of the old guard in this city. Equally outrageous was his lack of enthusiasm over one of my mentors, Ron Brown, who was living in Washington and serving as the U.S. secretary of commerce when he died in 1996.

"Upstarts," the man explains. "Sure they're in the Boulé and they know where to buy a house, but every four years—with every administration—they come and go. No roots, no history, no plans to stay. Why should I invest the time in knowing them all?"

Because Washington is a city of politicians and government officials, there are many blacks who have received national prominence from blacks and whites outside of the District yet little acclaim from the black elite who have lived in the city for multiple generations. The most clannish residents will admit that Jordan and Brown were clearly accepted into the group, but will note that the old guard is usually less likely to be enamored with new government appointees who come in from Little Rock, Atlanta, or New York after being appointed by the newest president who is sworn into office. Instead, this group prefers to adopt the permanent professionals—the doctors, the lawyers, the economists, the intellectuals and, to some extent, the entrepreneurs who come to the city to live and stay.

The lawyer crowd is probably the largest of all the society groups in Washington. "Because of the many government offices and the prominence of so many law schools like Howard, American, George Washington, and Georgetown," says Judge Henry H. Kennedy of the superior court in the District, "black Washington has always had more lawyers and judges than other communities." The graduate of Princeton and Harvard Law School grew up in a Jack and Jill family where all three children received the same Ivy League college degree. His brother, Randall, graduated from St. Albans School and Yale Law School and was a Rhodes Scholar. He was later one of my professors at Harvard Law School. His sister, Angela, two years behind me in college, graduated from the elite, all-girls National Cathedral School. I remember shamelessly circling her name and photo in the *Freshman Herald*—the Princeton face book that listed each student's name, address, birthdate, and high school—a few days after she arrived on campus.

Although he is modest and unassuming, Judge Kennedy grew up in a rarefied world where his childhood playmates included people like Dr. C. David and ViCurtis Hinton's daughters, Audrey and Diane, who were both celebrated Girl Friends debutantes in the 1960s, as well as Frank Spellman, who became a successful physician; Ernest "Chico" Williams, who was the son of a Howard dean and who later became a professor at the University of Maryland; and of course, a large contingent of attorneys and judges.

In addition to the Kennedy brothers, other Washingtonians who have played an important role in the successful lawyer crowd over the years are Karen Hastie Williams and her husband, Wesley Williams, a partner in Covington and Burling. Karen's father, Judge William Henry Hastie, was a class of 1925 graduate of Amherst and a class of 1930 graduate of Harvard Law, and in 1937 he became the first black appointed to a federal bench—a job he held before being appointed governor of the Virgin Islands. Karen's brother, Bill Hastie, a Boulé member, is also a successful attorney, practicing in San Francisco.

The old guard among the government and policy leaders has included such people as Patricia Roberts Harris, who was a Howard dean before she served as President Johnson's ambassador to Luxembourg, as President Carter's secretary of HUD, and then as secretary of health, education, and welfare. "Now that woman had class," says a woman as she took a break between sessions at a Delta Sigma Theta sorority gathering that I recently addressed. "Pat Harris was smart, stylish, and was going straight to the top. We just died when Marion Barry—with his coarse rhetoric and attacks—beat her in the mayoral race." An attorney and government official who achieved many "firsts," Harris died prematurely of cancer in 1985. A few years before Harris's death, the controversial Barry had run a vicious political campaign that labeled Harris as a representative of Washington's light-skinned elite. It was a simplistic and prejudicial label, but for the city's large working-class black population it was a characterization that stuck long enough to sink her mayoral ambitions.

During recent years, other government, policy, and civil rights people who've been accepted into the black elite ranks have been the first black cabinet member Robert C. Weaver, Secretary of the Army Clifford

Alexander and his wife Adele Logan Alexander, Children's Defense Fund founder Marian Wright Edelman, Dorothy Height, Sharon Pratt Kelly, Charlene Drew Jarvis, U.S. senator Edward Brooke, Nobel Prize–winner Ralph Bunche, Rockefeller appointee Ersa Poston, Secretary of the Army Togo West, journalist and former ambassador Carl Rowan, Eric Holder, Federal Reserve member Andrew Brimmer, air force officers Benjamin O. Davis Sr. and Jr., and Eleanor Holmes Norton.

Many of these individuals have broken into Washington's black society despite the fact that many of their early accomplishments have taken place outside of the District. For example, Edelman and Norton—both Yale Law School graduates from the early 1960s—travel back and forth within the oldest and most established black and white Washington circles even though some of the most important activities of their careers took place in cities such as Boston and New York. Edelman, who was the first black woman on Yale's board of trustees, worked for the NAACP Legal Defense Fund in New York and for the Harvard Center of Law and Education in Massachusetts before she eventually came to Washington and founded the Children's Defense Fund. Norton, although born in Washington and now the congresswoman representing the District, received her first high-profile appointments in New York, where she served under Mayor Lindsay and Mayor Beame as head of the Commission on Human Rights, before moving back to D.C. as head of the EEOC in 1977.

"Well, of course we've also adopted some of the people who are worthy of being in our crowd," says a somewhat haughty Washingtonian who claims to have dined at least twice at the home of Andrew Brimmer, a board member at the Federal Reserve. "Most people just think of Brimmer as being a real Washingtonian. But he's actually from Louisiana—and only from modest means. But frankly, it makes us look good to have him in our group. He's a Harvard Ph.D., he's on lots of *Fortune* 500 boards, he's got lots of honorary degrees, he was a Fulbright Scholar, he lives in a good neighborhood, and he's got lots of class."

Because I know Brimmer's daughter, Esther, I am somewhat aware of how little she or her prominent father have to do to elicit such praise and adoration from some of their Washington neighbors. I know that they generally operate above the fray, but I wonder if they know or care about how much their family résumé means to the people who dine in their home and claim to "adopt" them.

But just as interesting is the list of accomplished Washingtonians who are not embraced by the group. General Colin Powell is one of these people. Supreme Court justice Clarence Thomas is another. "Is it because they aren't originally from Washington?" I ask a D.C. native with whom I had worked when Ron Brown had invited us down to work on the 1992 Clinton campaign.

"Absolutely not," my friend explained. "Remember that Ron was accepted and he wasn't from Washington."

I ask if it's because of their political party affiliation, even though I know that a significant number of the old guard hold onto their Republican status.

My friend shakes his head. "No, they have nothing against Republicans."

"Then what is the distinction?" I ask. "Why would black Washington accept outsiders like Ron Brown and reject a Supreme Court justice and a serious presidential contender who enjoys national popularity?"

"Sometimes it's an issue of who demonstrates black pride," my friend explains, as he shares some of the assessments that his parents and their friends—all old-guard insiders—have made about people like Powell and Thomas. "They don't like people who are too black, like Jesse Jackson, but they also don't like people who seem to avoid embracing black people and black institutions."

Like others whom I speak to, my friend says that while Ron Brown was not from Washington, he long ago aligned himself with black causes such as the Urban League and belonged to black organizations, including the Guardsmen and the Boulé. His wife, a longtime member of the Links, has also been very visible in the black political and social scene. And their kids are too.

"But Powell and Thomas," says a Washington Boulé member, "have almost nothing to do with black people in this town. They don't join our groups and don't attend black events unless they are the featured speakers."

"And being married to some low-class white woman from God knows where," adds the Boulé member's wife, "is hardly going to help Clarence Thomas make a transition into our community. He doesn't like us and we don't like him."

Colin Powell, while not a member of the important black men's groups and not a graduate of a black college and not a participant in any of the old-line black civil rights organizations, is a more complex figure to dismiss. Although I admire him immensely, some feel differently.

"I keep hoping that Powell will surprise us and suddenly start connecting with the black community," says a former black congressional aide who grew up in Washington and says that he once admired Powell, "but he's really just a slickly packaged white guy who has just enough melanin and makes just enough references to his past as a black kid in New York to seem empathetic to black people. He's like an Ed Brooke, but without the white wife. Instead, Powell's son has the white wife."

So, it seems that in this city—a town that is 65 percent black, there remains a black elite community who bend the rules for some and hold the line for others. Whether it's because of an individual's marriage outside the race or his or her failure to associate with black institutions and causes, nowhere is the double standard more apparent than in the way that certain political and government figures are embraced or rejected by the city's black elite.

In the case of educators who are accepted by the elite, the rules seem to be more straightforward. For example, in general, those blacks who fall into the elite group of Washington educators had some tie to, or success at, Howard University or Paul Lawrence Dunbar High School, a school where the best-educated black teachers came because of their dedication to black children and because of the difficulty in getting jobs at northern or southern schools that hired only white teachers and administrators.

Among the Howard or college faculty group have been James and Samuel Nabrit, James Cheek, Mordecai Johnson, George W. Cook, Broadus and Lillian Butler, James and Lillian Lawson, and John and Paquita Attaway.

Dr. Alethia L. Spraggins belongs to a family of educators who have long been tied to the city's schools. A fifth-generation Washingtonian, she belongs to the Washington, D.C., Links along with her mother, Alice Spraggins. "My mother's father, Henry Lee Grant, was an instrumental teacher at Dunbar High School and taught Billy Taylor and Duke Ellington," says Dr. Spraggins, who recently retired as a high school principal in the D.C. public schools. Her mother's grandfather, Henry Fleet Grant, was the first black music teacher in the entire city school system. Spraggins's father, Dr. Tinsley Lee Spraggins, was also an educator and was one of the leading figures in black voter registration drives of the South and was recognized by President Lyndon Johnson for broadening his work on a national level.

Others in the Washington schools have included Francis Cardozo,

Dr. Winfield Montgomery, Dr. John Washington, Sadie Tanner Mossell Alexander, Eva Dykes, Charles and Thelma Baltimore, and Mazaline Baird.

"And then there are just names you always hear, but don't necessarily associate with any particular profession," says a man who belongs to the Bachelor-Benedicts club. "There are last names that when someone mentions them, you just ask 'Are you of *the* Pinkett family?' There are names like French, Epps, Diggs, Latimer, Gregory, Mazique, Moultrie, Wesley, Banks, Dunmore, Gilliam, and Bullock."

In recent years, one of the biggest rising stars on the Washington social and civic circuit has been Savanna Clark, wife of millionaire urologist Dr. C. Warfield Clark. Because of his long-standing ties and her vigor and personality, the Clarks have become one of those couples who live with a foot in two different worlds: one black and one white.

"That woman is everywhere," remarks a young Jack and Jill mother in a tone of admiration. "Every time I open the *Washington Post,* I see her organizing fund-raisers for scholarships or balls at the Kennedy Center. She's like no other black woman in this town." While they belong to the black society groups like the Boulé and the Links, Savanna Clark and her husband are actually even more prominent in the city's white society because of their status as multimillionaires who place an emphasis on philanthropy. I first met Mrs. Clark at the Harvard Club in New York after she and Altovise Davis, Sammy Davis Jr.'s widow, had come to New York for the city's annual Fashion Week.

A charter member of her Links chapter and of the Hillbillies, Clark has degrees from Prairie View A&M and the University of Oklahoma and has held teaching positions at several universities. Although she was not born in Washington, as was her husband, she is a die-hard D.C. supporter—both heading and serving on numerous committees at the Kennedy Center, the National Symphony, the Corcoran Gallery, and many other high-profile groups. Clark is often seen in couture gowns designed by Valentino, Bill Blass, and Chanel, and she gained a great deal of attention when she and her husband used their own resources to give special dinners on behalf of Ella Fitzgerald, Count Basie, and other blacks who had been honored by the Kennedy Center.

"My husband, Warfield, was born here and went to Dunbar, the University of Michigan, and Howard Medical School before he started practicing as a urologist, so he really knows this town," explains Clark, as she walks into her magnificent home in the Forest Hills section.

"Although I am not originally from Washington, I feel that it's important for me to contribute to this community."

Not only have the Clarks given to the cultural institutions, but they've invested in an eclectic group of black artists and have helped raise scholarship monies for black and other minority students at Georgetown Medical School.

Having just returned from the Chicago funeral of her friend, attorney Jewel Lafontant, Clark sits in her home in the Forest Hills section—a neighborhood on the west side of Rock Creek Park and an area of mostly white neighbors, except for Carl Rowan, Andrew Brimmer, and one or two other black millionaires—and she considers what she and her husband are trying to accomplish.

"It is extremely important for black people to contribute to the social, cultural, and educational institutions that are around us. If we want to have an impact on this city, and if we want to be taken seriously as a people, we have to participate in our own black organizations, but also in other groups as well." As the Clarks continue to participate in their Links and Boulé chapters, along with the many high-profile white charities, they are helping to redefine the new black elite.

Benaree Pratt Wiley remembers the quiet gentility of the LeDroit Park neighborhood that abuts Howard University. "We moved from LeDroit when I was quite young, but I remember that several of the kids in my neighborhood attended the Howard nursery school with me," explains Wiley, a Howard University graduate who was the fourth generation of her family to live in Washington. "There is a lot of history in those houses," she adds.

"When I was little, my mother used to write to her friends in Detroit and tell them we lived in LeDroit Park, but she was lying," says a third-generation Washingtonian in her seventies. "We actually lived about a block and a half outside the neighborhood, but Mother would say to my dad that she could see the boundary from our living-room window, so we should be able to say we lived there. This was in the early 1930s, LeDroit's heyday, and everybody important lived there. When we finally moved into the neighborhood, my mother had personal stationery engraved so that the neighborhood's name was emblazoned between her name and our address."

Developed by a white Howard University professor, Amzi Barber, who named the forty-two-acre neighborhood after his father-in-law,

LeDroit Langdon, the tree-lined residential neighborhood was located just south of Howard University. It is bound by Second Street on the east, Sixth Street on the west, W Street on the north, and Florida and Rhode Island Avenues on the South. This one-square-mile area of private homes was originally inhabited by whites when it was built in the 1870s. Worried by the presence of the all-black Howard University just a few blocks north, the white neighborhood maintained a fence between itself and the school.

Blacks first became residents of the neighborhood in the 1890s, with Mary Church Terrell and her husband being among the first to integrate the white area.

Alberta Campbell Colbert remembers when LeDroit Park was a sought-after neighborhood for blacks. "When I first got married, one of the most desirable neighborhoods for blacks was LeDroit Park. There were these beautiful Victorian structures, as well as handsome brick town houses," Colbert adds, noting that many of the residents had some sort of tie to Howard University.

Although several of the houses are historic landmarks, many members of the black elite began moving out of the area as desegregation opened up other neighborhoods and as LeDroit lost its cachet to newer sections that looked more affluent.

One of the first streets to be considered a "gold coast" for blacks outside of LeDroit Park was Blagden Avenue, a small street that runs on the east side of Sixteenth Street, just below Rock Creek Park. "We were the *original* 'gold coast' during the 1950s, and we had top-drawer people like William Thompson of the Urban League and Dr. Clarke, who used to have ambassadors when they entertained," says an eighty-year-old Washingtonian who recalls when she first moved near Blagden, a street that was lined with the homes of physicians and their wives and children. "Our houses weren't as big or palatial as the houses on the northern end of Sixteenth Street, but they wouldn't sell to blacks up there when we were looking."

Since the late 1960s, the term "gold coast" has been used to refer to yet another area on upper Sixteenth Street, Northwest, which sits above Walter Reed Army Medical Center and runs all the way to the Washington-Maryland border. This large neighborhood of early-twentieth-century colonials and Tudors includes the most affluent and best-connected blacks in Washington. Many of the streets are named after trees or flowers, including Hemlock, Primrose, Orchid, Juniper,

Redwood, Sycamore, and Spruce. "Ninety percent of the zip codes on my invitation lists for parties are 20012," says the octogenarian, who left Blagden and now lives in this newer gold coast neighborhood. "We actually call this the *platinum* coast."

The elite families who have lived there include doctors like William Matory, Charles Epps, and Howard Hospital's Linwood Rayford, as well as many judges and lawyers, such as Henry Kennedy; Vincent Cohen; Thomas Williams; and Covington and Burling partner Wesley Williams and his wife, Crowell and Moring partner Karen Hastie Williams, who both recently moved to the Watergate. While members of black society consider this to be "their neighborhood," it still consists of about 40 percent white residents on many of the streets.

"Everyone lumps us together with the North Portal neighborhood, but as you can see, I don't live in a 1950s split-level," says a physician who resents being grouped together with even this very fashionable neighborhood. "I live in a colonial that was built a long time before they even *thought* about North Portal Estates." Like a handful of the black elite who live on the upper end of Sixteenth Street, this doctor notes that there are actually three separate neighborhoods in the area.

There is Shepherd Park, which is the furthest south of the three neighborhoods. It runs along the northern side of Walter Reed Medical Center. Further north and over on the western side of Sixteenth Street is North Portal Estates, which features newer houses that were made available to blacks after housing desegregation in the early 1960s. Just to the north of North Portal is the neighborhood known as Colonial Village.

"As anyone can see, I've got older, more established houses in my neighborhood," adds the physician. "You may call us North Portal, but we're not the same thing."

When blacks started moving to the suburbs in the late 1960s and early 1970s, many moved to Silver Spring, Maryland. Later they moved to other well-to-do Maryland towns, such as Chevy Chase, Bethesda, and Potomac. "Although Virginia was just as close," says a woman who moved into a five-bedroom Silver Spring house after living in the District most of her life, "it was pretty obvious that we were more welcome in the Maryland suburbs." She continued, as she sat across from her bridge partner, who was dressed in a dark-blue silk chantung suit, "Until the mid-1970s, the only Virginia towns that really let us in were Alexandria and Arlington. Arlington has all those Syphaxes. That's their town, and there's no point in moving there unless you're related to them."

And what about the ritzy McLean, Virginia? "Absolutely not!" says the Silver Spring woman. "People say that's Bobby Kennedy territory, and that's good, but the truth of the matter is that it's Republican territory. So we never move to McLean. Never."

"I wouldn't say never," added the woman in the blue silk chantung jacket.

The woman paused and looked over at her bridge partner. "That's true. There's those Robinsons who moved out to McLean years ago. I don't know what they were thinking. They were light, but they sure weren't going to be passing for white."

The woman's bridge partner gave her friend a frown. "Now, don't be talking about Jackie like that. She's one of our own—and that girl faced enough tragedy."

The Silver Spring woman waved her friend away. "I loved Jackie too, but let's be honest. That son of hers married white because of that whole McLean mentality. Black people who move to places like that aren't going to raise children who appreciate their blackness. Now there, I said it. I know I shouldn't be talking about the dead, but they're all gone now."

"Jackie died two years ago," the other woman added with a pained expression. "And her son, Alvin junior—the one who married white—he was a doctor. Well, he was diagnosed with cancer many years ago. The family was just beside themselves when they found out—Jackie's husband was a doctor too. Well, it seems that young Alvin was determined to find a cure. And he and his wife were in a plane on their way to Mexico in order to have some special operation or get medicine, something like that. Well, the plane went down and they were both killed. They had a child back here. Left her an orphan. It was awful. Absolutely awful."

When the subject turns to Prince George's County, a popular new area that sits to the southeast of Washington, strong opinions are shared without reservation.

"Nobody important lives there," says the Silver Spring matron. "Believe me, I've heard about all these five-thousand- and six-thousand-square-feet houses—I even went out there with my daughter and her husband to look at one, but gawd—" the woman added, bursting into a laugh. "The whole damned place looks like Disneyland!"

"She's right," says the bridge partner. "Every house looks the same—or it's a mirror image of the one across the street. The trees are so small they look like toothpicks with garnish on top. The streets are wide like airplane landing strips. They call it suburban and spacious, but it's got no character. I think it's great that we are building our own respectable communities for the kids to be around other educated, well-to-do blacks, but go to Colonial Village—go to North Portal. Who ever heard of Fort Washington or Clinton, Maryland? Not long ago, the police out there were arresting black folks if we just looked at them wrong."

Many members of the Washington black elite are becoming more aware of the growing number of affluent blacks in nearby Prince George's County in Maryland. It was a trend that became obvious in the early 1980s, when development after development of large, expensive houses was attracting black professionals. In 1970, the mostly suburban county was 85 percent white and less than 14 percent black. Within twenty years, the county had become 43 percent white and 50 percent black. Although the original influx of blacks led to resistance from many white residents, it is one of the few counties in the country that have high income levels among both blacks and whites. The large new houses with their large yards have become a magnet for families. But, some say, not for the families of the old guard.

One of the reasons members of the old guard remain in the District is that their businesses and their livelihood and their important institutions are there. The Mitchell family, for example, has one of the city's oldest black-owned businesses. Cynthia Mitchell remembers how proud she was when her son, B. Doyle Mitchell Jr., was a sixteen-year-old student at Georgetown Day School and began working part-time at the bank her husband ran. "It was wonderful to see another generation of our family being introduced to a business that had served so many people in this city," says Mitchell, who attended Howard in the 1940s.

Industrial Bank of Washington was founded in 1934 by Cynthia's father-in-law, Jesse Mitchell, after he raised the original $250,000 by selling stock. Jesse's son, Doyle senior, took over the company as chairman in 1955 and became a major figure in several Washington-area organizations. "Like many of our friends, my husband graduated from Dunbar High School and Howard University," says Cynthia, who often joined him in his activities in the Boulé and the Bachelor-Benedicts club.

As one of the more prominent black business leaders of his time in the city, Mitchell served on the boards of many organizations, including Georgetown University and the D.C. American Red Cross, before his death in 1993. Today, the bank has assets of more than $270 million and nine offices in Washington and Maryland. Although it was originally established to serve a predominately black customer base, Industrial has long since moved into a mainstream customer base. Third-generation bank president B. Doyle Mitchell Jr. now runs the black-owned commercial bank—one of the nation's largest—from the main office on Georgia Avenue. "I went into this business because of its early mission to serve a population that was neglected," says Doyle, who serves on the board of the D.C. Chamber of Commerce. "My father and grandfather started a tradition that I want to continue."

Other older, prominent businesses that have held an important place in black Washington have been the newspaper *Afro-American*, the newspaper *Washington Bee*, Scurlock Photography Studio, the Dunbar Hotel, Lee's Flowers, the Howard Inn, the Lincoln Theater, and Independence Federal Savings, which is headed by William Fitzgerald and ranks as the country's second largest black bank.

Among the list of black businesses supported by this crowd are two select funeral homes—McGuire's and Stewart's. "Those are the only two places that the old guard will consider when a family member dies," says a college professor whose family has used McGuire for three generations. Another institution that the old guard has enjoyed is Lil Your Hairdresser, a salon that opened in the 1940s and was the popular spot for debutantes and socialites. Located on Ninth Street, it was just one stop within the debutante's schedule, which also included dance lessons at Therrell Smith Dance Studio and cotillion lessons from Jacqueline Robinson or Alicia Lanauze Webb, two prominent women who belonged to the Girl Friends social club.

Today, there are many larger Washington-based businesses, such as Robert Johnson's $170 million Black Entertainment Television cable network. In fact, Johnson has turned his BET franchise—television, publishing, and restaurants—into an important name for both blacks and whites in the city. The class of 1972 Princeton graduate sits on the board of Hilton Hotels and supports many charities in the city. Other notable Washington businesses include Earl Graves's $61 million Pepsi-Cola Bottlers of Washington; Digital Systems, a $90 million technology company; as well as numerous law, accounting, and consulting firms that are owned by blacks yet serve a clientele well beyond the city's black community.

Although many of my friends are now practicing lawyers and bankers in Washington, they tell me that despite their important professional positions, they still have a difficult time breaking into the elite. They say that beyond their position in the business world and the status of their neighborhoods there are even greater factors that determine one's place in Washington society. They say that one's affiliations are crucial. Included among those affiliations are one's church, school, and social clubs.

"Where someone goes to church or where they attended school," explains my friend Alice Randall, a graduate of Georgetown Day School, "can say a lot about that person and what his or her family happens to value."

One of my banker friends in the District happens to agree with what Alice says, particularly when it comes to churches.

"I remember meeting a girl at a Jack and Jill party when I was in the tenth grade, and my mother and I drove her home," explains the banker, now in his forties. "She didn't belong to Jack and Jill, but she seemed like a nice girl. So when I told Mom that I was going to ask her out, I lied and told Mom that the girl attended St. Luke's Episcopal Church. I knew that if Mom thought she belonged to a good church, then I'd be allowed to ask her out."

Over the years, the churches that have been popular among the black elite have included St. Luke's Episcopal Church at Fifteenth and P Streets, Asbury Methodist Church, Nineteenth Street Baptist Church, Metropolitan Baptist Church, and People's Congregational Church on Sixteenth Street, where the minister, Reverend Stanley, is considered to enjoy a particularly special status because he is married to former U.N. ambassador Andrew Young's daughter.

Certain schools have also played an important role in bestowing the right families with good credentials. Initially, it was the M Street School—subsequently named the Paul Lawrence Dunbar High School—that remained the black elite school of choice for Washingtonians for several generations.

"My son, Craig, went to Georgetown Day School in the 1960s, and, at that time, it was not uncommon to find a few black children there, but in the generation before him, the most ambitious black children were primarily directed toward Dunbar," says Alberta Campbell Colbert as she recalls her son's private school experience before he went to the University of Michigan. "When my husband, Frank, was going to Dunbar High School in the 1930s, he had the kind of classmates that caused Dunbar to rival the private schools." Colbert, a retired pharmacist who graduated from Howard's School of Pharmacy, remembers that one of her husband's Dunbar classmates was Edward Brooke, who later became the first black U.S. senator.

"By the time our kids were starting high school, we had to find alternatives to Dunbar because it wasn't what it used to be." Like Colbert, many Washingtonians remember the prominence of Dunbar High School, which opened at First Street between N and O Streets in 1916 and remained the premiere high school for blacks until the late 1940s. The school, which was originally known as the M Street School when the city established an entirely separate school district for blacks in the 1870s, had black teachers who had been educated in the best colleges and graduate schools in the Northeast. Many of the faculty had doctorates, and among the Dunbar graduates were students who went on to Harvard, Amherst, Yale, and other top universities.

"We had black teachers who pushed in the same ways that our parents did," says a class-of-1944 Dunbar graduate. "White teachers don't have the same high expectations."

During the late 1950s, the popular public high school for this group had become Roosevelt High School at Thirteenth Street NW, just a few blocks east of the Blagden Avenue gold coast neighborhood.

Beginning in the late 1940s, the black elite began attending Georgetown Day School. Like Colbert, they recognized that Georgetown Day was more liberal and progressive about welcoming black children than were the other private schools. Georgetown Day is located off of upper Wisconsin Avenue in a neighborhood famous for its private schools, and its tuition is set at fifteen thousand dollars a year. Among its alumni are the children of former mayor Walter Washington, former U.N. secretary Donald McHenry, and Secretary of the Army Clifford Alexander.

Although it lacks Georgetown Day's early history of integrating and having black members on its board, another relatively liberal private school is Sidwell Friends. Although many famous children had attended before her, Chelsea Clinton, daughter of President Bill Clinton, brought the school national attention. Clifford Alexander's daughter, Elizabeth, also graduated from the school.

"You actually see more diversity and a greater willingness for black and white kids to interact in private schools than at many public schools," says Paul Thornell, a 1990 graduate of Sidwell Friends. A member of an old society family—his grandmother, Frances Vashon, helped charter the Links; his grandfather, Dr. Nolan Atkinson, graduated from

Howard Medical School; his mother, Carolyn, graduated from Vassar and Harvard; and his father, Richard, has degrees from Fisk, Princeton, and Yale Law—Paul recalls how well he and his brothers interacted with other black and white students at Sidwell.

"Many of the public school districts in the city were uniformly black, and those in counties like Fairfax were uniformly white." Recently graduated from the University of Pennsylvania and now working on Capitol Hill, Paul feels that many of his black private school friends had a positive experience that allowed them to maintain a strong black self-image even though they made up a small minority at the elite day school. "In fact," he recalls, "almost half of my black classmates decided to attend historically black colleges."

Today, there are also many affluent black children at the most "blue-blooded" of the D.C. private schools: the all-girls National Cathedral School and the all-boys St. Albans School. Located on Wisconsin Avenue near the Washington Cathedral, National Cathedral School is considered the most expensive private school in the District, with an annual tuition priced at over fifteen thousand dollars. Judge Henry Kennedy—whose sister, Angela, is an alumna—remembers escorting a National Cathedral student, Virginia Brown, to the Washington Girl Friends cotillion when they were in high school. "Virginia's father was the archbishop of Liberia, and among her classmates was Luci Johnson, daughter of President Lyndon Johnson. That school has always attracted the daughters of influential families."

"My granddaughter, Tiffany, had a wonderful time at the National Cathedral School and it prepared her well for the University of Pennsylvania," says grande dame ViCurtis Hinton as she considers how much things have changed in the city since her own kids were graduating in the 1960s. "But what her parents made sure of while she was at that school was to keep her active in programs with black children. She was teen president of her Jack and Jill chapter."

The equally elite all-boys St. Albans School is as Waspy as National Cathedral and includes such alumni as Vice President Al Gore, *Washington Post* publisher Donald Graham, and former Indiana governor Evan Bayh. But at the same time that blacks such as Marjorie Holloman Parker were joining the school's board, black students were beginning to attend St. Albans, which now includes such black alumni as author and Harvard law professor Randall Kennedy from the class of 1974 and two young black congressmen, Jesse Jackson Jr., '84, who represents Illinois, and Harold Ford Jr., '88, who represents Tennessee. ViCurtis Hinton's grandson, Phillip, was also a graduate of the school. "All of my grandchildren have been able to balance these worlds while maintaining their identity as black people," says the proud matriarch.

Of all the D.C.-area private schools, the only one that appears not to have won popularity among the black elite is Madeira, the all-girls boarding school in nearby McLean, Virginia. Already respected in conservative WASP circles, it received national attention in the early 1980s when its headmistress, Jean Harris, was imprisoned for murdering her lover, Scarsdale diet doctor Herman Tarnower. "We took our daughter for a visit there several years ago and the students were ice-cold to us," says a Jack and Jill mother who settled on National Cathedral instead. "Madeira reminded me of being in the deep South. Lots of blonde, blue-eyed girls. Lots of tradition. And lots of talk about horses and getting married. They seemed to have a good academic program, but it didn't seem like they were really ready for black people yet."

But even beyond the schools and the institutions that keep Washington's old guard rooted to the city are the selective social groups that have linked elite black families for generations.

"We are the third oldest Links chapter in the country," explains Dr. Joan Payne McPhatter, president of the Washington, D.C., Links, as she recounts the distinguished history of the city's leading group of black women. A Howard University professor of communications, McPhatter is particularly proud of the causes that her chapter of fifty-two women has served and contributed to over the last fifty years. Because there happen to be six different Links chapters in the Washington metropolitan area, some women in this town are obsessed with the notion that the original group—McPhatter's chapter—holds the premier position in terms of history and prestige. The other chapters—while not as well known—have some impressive members in their ranks.

Although McPhatter grew up in Nashville, her Washington ties make her equally dedicated to the Washington community. She has four generations of ties to the city's Howard University: Her son graduated from the school; she graduated from Howard in the class of 1968; her mother, Dr. Gretchen Bradley Payne, was assistant dean of women at Howard from 1943 to 1947; and her grandfather, Roland Bradley, was also a Howard graduate. Among the members of the Washington Links are Alma Brown, the widow of Ron Brown; and Ann Dibble Jordan, wife of attorney Vernon Jordan.

In addition to having Boulé, Girl Friends, Guardsmen, Northeasterners, Drifters, Jack and Jill, and more chapters of the Links than virtually any other metropolitan area, Washington also has some very small elite organizations that have been in existence for generations. One of those groups is the Bachelor-Benedicts club, which includes prominent men from different backgrounds. Formed in 1910, the group was made up of black professional men who gave dinners and elaborate debutante cotillions. "The name of the group has an interesting history," says Alberta Campbell Colbert, whose husband, Frank, joined in 1943. "Five young bachelors founded the social organization, but after a few years, one of them got married, and they labeled him a 'benedict.'"

Colbert and her husband were successful pharmacists who frequently attended the group's dinners and cotillions with Alberta's uncle, William Lee, an attorney who had joined in the 1930s. "At that time," explains Colbert, "virtually everyone was a judge or a lawyer. They limited the group to one hundred members, and they usually had three formal affairs each year." Although the club hosts summer outings and group trips, the most popular event was always the annual cotillion. "Each year, the group would present around twenty girls," says Colbert. "My daughter, Doris, came out in 1962, and I still remember her marching in with her escort, Rodney Savoy. His father was also president of the group." The cotillion, which was held in various large hotels in the city, remained popular until the late 1980s.

A later cotillion that more than rivaled the Bachelor-Benedicts ball was the annual Girl Friends cotillion, which required white tie and tails and was held in major hotels, including the Sheraton Park and Washington Hilton throughout the 1950s, 1960s, and 1970s. It was an event that got covered not just by the black press but also by such publications as the *Washington Post* and the *New York Times*.

In recent years, the new "in" party to attend has been the Tuxedo Ball, which takes place every year in Washington and selects young black men and women from around the country as guests. One cannot simply buy tickets to attend the formal event; one has to be formally invited by the twenty-member ball committee. "My kids went one or two years to the Tuxedo Ball," says Bebe Drew Price, "and it really is a tremendous affair. My friend Rikki Hill has been very involved with it as well."

Rikki Hill is actually Fredrika Stubbs Hill, the daughter of Marion Stubbs Thomas, founder of Jack and Jill. "The Tuxedo Ball is an activity that brings together young people and families each year for a large formal party," explains Hill, who was a debutante in Detroit before she married physician Dr. Delany Hill, "but the event now incorporates seminars and workshops that help the young participants prepare for graduate school and look for summer jobs, and offers other activities that our young people need to be considering."

Sidwell Friends graduate Paul Thornell was both a member of Jack and Jill and a Tuxedo Ball committee member. "I've gone to the ball many times," he explains, "and it has served as a great vehicle for young black students to build a network that will last beyond high school and college. The founder of the group, Dr. Carlotta Miles, came up with the idea in 1986 because she recognized that we all need to be a part of networks—both social and professional."

As new organizations, new neighborhoods, and new schools become popular among Washington's black elite, the older representatives of this group begin to express mixed feelings. While they know of no other city that has such a concentration of old-guard families, they are concerned about how to preserve what makes their city and their group special.

"Sometimes I honestly believe that segregation was better for us," says a grandmother who thumbs through a photo album filled with graduation pictures of herself and other family members. "This is me when I was graduating from Dunbar, and here I am again when I finished Howard. Do you notice the difference between these pictures and the ones from my grandkids' graduations?"

When the woman lays out the sheets of photographs, some differences are immediately obvious. In the older pictures, the woman is surrounded by a sea of black classmates and older black teachers in elegant graduation robes.

"Now, look at these," she says with a trace of sadness in her voice. "These are my grandkids when they finished prep school and graduated from college."

The newer, color pictures show a black student surrounded by a sea of white faces. Each photograph shows the same black boy and girl surrounded by white prep school classmates and by white college classmates.

"Where are my grandkids' role models? Where is their support system? Where is their tie to our people?" The woman paused. "At Dunbar and at Howard, we had examples to follow and black classmates that we connected with. They were people with background—people we'd know for a lifetime. My grandkids don't know about fraternities or black roommates. They never even had a black teacher. They grew up in the best city in the country for rich blacks—and I feel like we somehow let them get away."

Young Washingtonian Paul Thornell has a different philosophy about where his generation is headed. "Even though my brothers and I attended private schools that might not have been available to prior generations of black people, our involvement in Jack and Jill and our ties to our black friends and important black institutions and causes allow us to be just as involved and connected to our culture as the ancestors that preceded us. We know who we are and where we came from."

Black Elite in New York City

"Girl, it feels like the end of an era."

They came from Harlem, from Brooklyn, from New Jersey, from Washington—from as far away as Los Angeles and from as nearby as Scarsdale and New Rochelle. On this damp fall evening, close to nine hundred people jammed the sanctuary as dozens more lined up outside to make it standing room only at the seventy-year-old white limestone Grace Baptist Church in suburban Mount Vernon, New York. It was a somber event that brought together disparate groups from New York's black society world. It was the funeral of Nellie Arzelia Thornton, national president of Jack and Jill of America.

"I can't believe it. She was just fifty—just fifty," said a woman who held a sprig of green ivy in her hand as she stood in one of the front pews with the other AKA members and looked over at the Links and Jack and Jill sections of the church.

As my father and I took an available seat in the back half of the sanctuary, we saw long lines of solemn black women in full-length mink coats, diamond earrings, and white dresses being led down front to view the body and catch a final glimpse of the woman who had united most of us. I could see my mother in the fourth row with her fellow Links members. She could have sat with the Jack and Jill women, since that was the group that had originally brought her and Nellie together, but they had actually become closer friends through the Links chapter they had chartered, along with their longtime friend Betty Shabazz. Although each of the three organizations of women would eventually make its own presentations that evening—the Links with candles, the AKAs with the sorority song, and the Jack and Jillers with flowers—everyone there wanted to say good-bye personally.

"Why isn't this in New York? This should be in the city," remarked a tall, light-skinned woman with a mink hat. She seemed to be in her late sixties. "If they really wanted people to show up, they should have had this at St. Philip's."

"Yes, but she's Westchester," remarked a younger woman standing nearby. "She's not from New York."

The tall woman turned back to the younger woman and shook her head. "Not from New York? That's the problem with you new people. You haven't been out here a hot minute, and you Negroes act like you forgot where Harlem is. They could have had this at St. Philip's—or even Abyssinian—and really had people show up. All this suburban foolishness. She belongs in the city—not mixed up with all these West Indians and Westchester wannabees"

"You know, I kind of agree," another older woman interrupted. "Not that she's really one of us, but she *is* representing Jack and Jill, and a New York church would have been able to really bring the right people in. I'm sure Wyatt would have let them do this at his church." The woman turned around and looked at the unfamiliar white people and the occasional row of dark faces she didn't recognize. "After all, the *Amsterdam News* isn't going to send Cathy Connors out here to cover a group like this."

The tall woman nodded in agreement. She liked Grace Church and its head minister, Franklyn Richardson. He and his wife, Inez, were Boulé people—Links people—*her* kind of people. And Grace had a strong and accomplished membership. But this was still not familiar territory to her. It wasn't the Harlem clique, and it wasn't St. Philip's. Since my mother and several of our cousins had been friends of Dr. Miriam Weston, wife of St. Philip's rector, Dr. Moran Weston, I'd long understood the historical significance of the 180-year-old St. Philip's Church on West 134th

Street and the position it once had within the world of old-guard New York blacks.

"Exactly," the woman with the mink hat continued. "At things like this, you want to *see* people. You want to know who shows up. Like the Girl Friends, for instance, are they even here? And the Boulé? How can you even tell? I know she wasn't from an old family or big money, but they really should have set this thing up right for the rest of us."

The younger woman, a suburbanite, shrank back, not quite knowing how to respond—and possibly not comprehending the points that seemed to be so important to the two older ladies, two hard-boiled New Yorkers who saw only one way to do things.

"And there ought to be a section for the Fisk people—she went to Fisk—and the Washington people," added the increasingly annoyed woman as she attempted to impose her own form of order on the occasion.

As I sat there with my father, thinking about Nellie Thornton and her credentials—Fisk graduate, school principal, AKA member, Links chapter founder, Jack and Jill national president, wife, mother of two, owner of a white turn-of-the-century mansion—I thought about how rigid these women might have seemed to an outsider who didn't understand their world.

"And what's *this* all about?" the woman removed the fur hat as the church suddenly fell silent. She sat back in her pew and grimaced as four young, dark-skinned black girls outfitted in black leotards walked to the front of the quiet congregation and stopped just to the right of the casket. After a few moments of silence, music began. The young girls, who seemed to be no more than twelve years old, looked out at the somber audience and began to sway in unison. They swung their heads sadly in circles and then shook their slender bodies to a rising rhythmic African beat. Moving their legs and arms with colorful flowing scarves, they introduced an almost whimsical counterpoint to the harsh, chilling music that played behind them.

Some of the people in the congregation seemed to watch approvingly, while others—many of the ones I knew—sat looking somewhat astonished, unsure of the course of the evening.

"What's with this jungle music?" asked a middle-aged man in front of us.

"I guess they are children from Nellie's school," a woman whispered back.

"I'm not sure I like this."

The woman paused tentatively and then nodded. "I don't think I do either."

The girls in the black leotards shook and shimmied, then bent down toward the floor, still writhing and shaking to the beat of the loud drum.

"Bringing that African stuff into a church. Did she go here?"

The woman shrugged. "I heard somebody say something about St. Philip's."

"Well, I doubt that."

As I looked around the sanctuary, it became obvious that here was a clash of New York cultures.

The rules of black society could be no more complex than they are in New York. In southern cities like Washington and Atlanta, there is a core group of old families whose names have dominated the social scene for four or five generations. In those places, family history rather than family wealth is what determines one's importance. In other cities, like Chicago, Memphis, and Detroit, the elite can often be traced back to old-line businesses like Chicago's Supreme Life Insurance or Memphis's Universal Life Insurance. In most of these communities, there is less of an influx of new families with new or different backgrounds.

But the rules are different in New York. Its sheer size, its black ethnic variety, and its complex levels of wealth have made it a hybrid city for the black elite. With a metropolitan area that is three times more populous than any of the other cities, New York has more than one historically black area of town: It has Manhattan's Harlem and Brooklyn's Bedford-Stuyvesant as well as sections of lower Manhattan and parts of the Bronx, all of which developed independently from the others. And while most of the other cities have black elite populations that rose almost exclusively from American black southern roots, New York has the older southern-born families as well as those old and new West Indians who contribute their Jamaican, Haitian, or Trinidadian traditions to an already complex community of cultures. And because New York is also a financial center, with so many opportunities for wealth that go beyond the traditional black career paths of medicine, law, or family-run insurance and funeral home businesses, there are many examples of individuals who have quickly and recently catapulted themselves into the elite through the financial success of just one generation's activity on Wall Street or in the fast-moving media businesses.

The very earliest representatives of black society in New York City had established themselves rather firmly by the late 1800s in the boroughs of both Manhattan and Brooklyn. While they were only a fraction of the sixty-two thousand blacks living in all of New York City, they found their financial success through businesses tied to catering and small restaurants. One example of this was Samuel Fraunces, who ran a tavern and catering business during the late 1700s in the Wall Street area. A black man born in the British West Indies, Fraunces served prominent British and American political figures such as George Washington. Decorated with eighteenth- and nineteenth-century American memorabilia, Fraunces Tavern still operates today in its original location at Pearl and Broad Streets and serves a clientele made up mostly of affluent white bankers and Wall Street executives.

Since New York City had no black universities with which to attract or foster a black intellectual community in the way that Atlanta and Nashville did in the late 1800s, New York saw its elite, during this period, grow mostly from successful caterers who served a wealthy white clientele, or from doctors, dentists, and other professionals who served an exclusively black customer base. Among those first wealthy black New York families were names such as White, Bishop, Mars, Lansing, and Van Dyke. Most of them belonged either to Manhattan's St. Philip's Protestant Episcopal Church, which was eventually to move to Harlem, or to Brooklyn's St. Augustine's Church. Eventally, there would also be a St. Philip's in Brooklyn that would have prestige similar to the one led by Manhattan's famous rector Hutchens Bishop.

As a child growing up in suburban Westchester County—just thirty minutes north of New York City—I discovered how firmly people drew distinctions between the disparate groups and cliques that make up New York's black upper class. Over the years, relatives, family friends, and new acquaintances have made it clear through their questions and remarks about various other black New Yorkers that there was an "us against them" dynamic at work in this town. It did not matter that we all looked alike. What mattered most was who came first and where we came from.

Rather than come together as a cohesive group, black New Yorkers have divided themselves into separate cliques—cliques that hailed from the American South, from Jamaica, from Haiti, from Barbados, from Grenada, from other parts of the West Indies, and from Africa. Additional divisions drawn between people are based upon their current geographical residences: Harlem, Brooklyn, Queens, the Bronx, Long Island, Westchester, and northern New Jersey.

With an overall black population of 2.5 million people—the largest in the United States—in the New York metropolitan area, some believe it would be unwieldy and illogical for all the black professionals in these many communities to come together in the same social and civic clubs. Others say that because of the differences in culture between blacks from the South and blacks from the separate West Indian locales, there are obvious cultural conflicts that make it difficult for these people to interact with ease. What I have noticed over the years, however, is an explanation that is much more distressing. I have concluded that the primary reason for the separate groups is that many members of the various cliques within the black professional class truly feel superior to the others and couldn't imagine diluting their group with "those Jamaicans from Brooklyn's Bedford-Stuyvesant" or "those southern blacks from Harlem" or "those Haitians from Queens" or "those wannabees in Westchester."

"The best way to understand New York's black upper class is to study its origins," says Dr. Chester Redhead, a New York dentist who was born in Harlem and divides his life between the city, his suburban home in Scarsdale, and a summer house in Sag Harbor. "The original black groups in New York are really the families from Harlem and from Brooklyn."

From his dental office in the Riverton apartment complex at 135th Street, Redhead points out the many generations of blacks who have fanned out from Manhattan and Brooklyn's Bedford-Stuyvesant neighborhood.

"But don't ever confuse the Brooklyn crowd with the Harlem crowd," says William Pickens III, whose family has lived in both places. He agrees that the groups were quite distinct two generations ago and remain so today. "There are different churches, social clubs, cotillions, events, and families associated with each." Pickens knew the cliques on both sides while growing up in the Brooklyn chapter of Jack and Jill. His mother, Emilie, was a well-known Brooklyn socialite as well as the national president of Jack and Jill in the late 1940s. His father was a New York attorney and the college roommate of poet Langston Hughes. His grandfather was a class-of-1907 Yale graduate who helped found the NAACP. With such credentials, he was part of the groups that were held in high esteem in old New York.

Although my own parents didn't arrive in New York until the 1950s, by the time I was born, in 1961, they were fully aware that the important and authentic black experiences were to be found in Harlem and Brooklyn, where most of their friends lived. They, like many black New Yorkers of the 1950s and 1960s, had chosen to raise their kids in the less-congested Westchester suburbs, but while doing so they found themselves faced with yet another black New York clique to satisfy.

Like many black families who had moved to the suburban areas of southern Westchester, northern New Jersey, and parts of Queens after the late 1950s, we encountered blacks who looked down on inner-city blacks with no ties to white America. These upwardly mobile blacks—often called "bourgie" because of their bourgeois attitudes—conducted themselves and ruled their group's "members" with complicated and schizophrenic standards. The rule here was to embrace your black identity but to be sure to balance it with the white American culture as well. *Be black, but don't be too black.* Those black suburbanites who were able to master this balance were also able to find a place within the upper-class clique of suburban blacks.

I remember how schizophrenic our existence was for all of us as we participated in these activities: Friday night parties with the white neighborhood kids; Saturday expeditions to black plays and museums in Harlem with the black Jack and Jill kids; tennis outings at segregated Westchester country clubs with a white classmate's family; a first date with a black girl from the local NAACP teen chapter; weekend sleep-aways with the kids from the white Boy Scouts troop; summertime sleepovers with cousins and black friends from church. It was all a cultural seesaw.

Keeping the groups separate and getting along with both, while also maintaining our distinct black identity, was a skill that the black suburban cliques valued most of all. Mastering this skill was a requirement I had to fulfill if I was going to be accepted by the black New York clique that we referred to as "our friends."

Although Brooklyn's black community offers a regal history, most New Yorkers will agree that the city's black society became renowned only when it finally settled in Harlem. In fact, many will also agree that the original roots of Harlem's black aristocracy can be found in the old families affiliated with St. Philip's Protestant Episcopal Church. Once known as the richest black church in the country, St. Philip's was founded in 1818 and included some of the best-educated, most powerful, and most affluent blacks in New York.

"My grandmother, Mae Walker Perry, was married at St. Philip's in 1923 at a very lavish ceremony that was reported on widely because of its cost," says Radcliffe graduate A'Lelia Perry Bundles, who works as a television producer in Washington. The fifty-thousand-dollar wedding ceremony and reception were hosted by Perry's mother, Harlem socialite A'Lelia Walker Robinson. Although it was Perry's mother who had brought social status to the Harlem family through marriages to two physicians, and through an active social calendar, it was actually Perry's grandmother, the famous entrepreneur Madam C. J. Walker, who had made the family rich.

"Madam Walker, originally from Louisiana, had become the nation's first self-made woman millionaire after turning her cosmetics and hair care firm, the Walker Company, into a national success in the 1890s," explains great-great-granddaughter A'Lelia Bundles. So staggering was Walker's wealth that she was able to build a twenty-thousand-square-foot stone mansion on an estate overlooking the Hudson River in the affluent Westchester County community of Irvington. It was located in the same neighborhood as mansions that belonged to such white millionaires as Gould and Rockefeller. Bundles, an alumnus of Jack and Jill and a member of the Links, notes that her great-great-grandmother also maintained two Manhattan homes as well—one on West 136th Street and one on Edgecombe Avenue.

As the granddaughter of the country's richest black woman and the daughter of a socialite who supported many writers of the Harlem Renaissance, the attractive Mae Walker Perry had Spelman College credentials and the hand of a young Chicago physician—thus becoming the right kind of black woman to be married in a high-profile wedding at the socially obsessed St. Philip's Church. She defined Harlem society.

Not only had St. Philip's been an important institution that attracted the wealthy and socially elite black Episcopalians for its Sunday services, but it had also been a catalyst in getting many blacks to relocate to Harlem permanently during the early 1900s.

"My parents moved to Harlem the same year that St. Philip's did," explains Dr. James Jones, as he relaxes in the living room of his elegant Harlem town house on 138th Street—one of the two blocks that make up the prestigious Strivers' Row neighborhood. A retired oral surgeon who was among those who interviewed and initiated me when I joined the Boulé, Jones explains that Harlem was actually a middle-class white community before the important black churches arrived. "Black New

Yorkers didn't start off in Harlem," says Jones. "In the early 1900s, the black community and particularly the black professionals—were found on the West Side of the city—around Fifty-second and Fifty-third Streets."

The son of a doctor and a schoolteacher, Jones was born in 1914 and notes that his parents lived in midtown prior to that time. "My father and mother had been living on West Fifty-third Street until 1911, when a black migration started moving north to Harlem." He and his parents first lived on West 139th Street—not far from where he lives today with his wife, Ada Fisher Jones. "My parents were hard-core New Yorkers, and until they moved to Harlem, most of their life had been spent downtown, where my father's practice was located and where my mother had gone to Hunter College."

Although white Harlem residents initially put up strong resistance to the influx of blacks in 1911 and 1912, there were outspoken and well-connected black ministers like Hutchens C. Bishop, who had been rector of St. Philip's, and Adam Clayton Powell Sr., minister of Abyssinian Baptist Church, who publicly argued for equal access to housing in this northern community of Manhattan. "St. Philip's was run by the Bishop family for seventy years, and together they made Harlem a welcoming place for black people," says Dr. Moran Weston, a 1930 graduate of Columbia University who was named rector of St. Philip's in 1957. "It had been obvious that blacks in Harlem needed the backing and the organization of a strong black church to establish themselves, their housing, and their voice in this new community. So they were fortunate to have the clout of St. Philip's." Fighting against those efforts were several white churches, which voted against the establishment of many of the Harlem churches that blacks were building in the early 1900s.

Weston, who has held board positions at Columbia University, the NAACP Legal Defense Fund, and Mount Sinai Medical Center, points out that the black church was a major force in developing political and economic power for Harlem blacks of every income group. Between 1910 and 1920, the period in which St. Philip's relocated to Harlem, the city's black population grew dramatically—by 66 percent—as blacks from southern states such as Virginia, South Carolina, and North Carolina moved north for better jobs. Weston himself was responsible for building over twelve hundred housing units for Harlemites. Other high-profile Harlem ministers, including Wyatt Tee Walker, head of Canaan Baptist Church on West 116th Street, continue to be outspoken in their Harlem activities. Walker, who served as Dr. Martin Luther King's chief of staff and resides in nearby Westchester County, has also developed a record number of residential buildings for the Harlem community. Although I was a suburbanite, I grew up knowing this Harlemite, since his kids were in my Jack and Jill chapter and his wife, Anne, was in my mother's Links chapter. "The old Harlem community has a rich history, and the churches played a major role in building it," says Reverend Walker. "Those first churches and residents opened up the community for the rest of us."

Because today's Harlem is seen as the consummate black community, a fourteen-square-mile area that includes blacks of every economic group, it is hard to believe that many parts of it were once off-limits to blacks—even into the 1930s. For even those black elite families, all-white apartment buildings and all-white theaters and restaurants along 125th Street made the historic neighborhood a checkerboard of all-black and all-white fiefdoms.

"I get mighty annoyed by these West Indians and Hispanics who showed up here in the 1960s and turned their noses up at us native Harlemites," says a retired attorney who was born in Harlem in the 1920s. "They think this community is just open to anybody who wants it. They have no idea how hard we had to fight white people to earn a place here."

This attorney's remarks reflect some of the resentment that members of the old Harlem elite harbor against the bigoted white residents who came before them and the West Indians and Hispanic groups who succeeded them. In fact, he and others argue that if it weren't for the old black southerners, there would have been no old black Harlem. Like many blacks who were born in the Harlem of the 1920s and 1930s, he believes that it was those residents who came from the American south who arrived in Harlem and fought to desegregate the neighborhoods that are now open to minorities of every background. While this view gives short shrift to the West Indians who had also arrived in the early years (by 1935, 22 percent of black Harlem was West Indian), this is the basis for some of the resentment that exists between certain cliques among the city's black elite: the debate over who got there first and who fought for the right to live there.

"I remember when many parts of Harlem were still segregated," says my cousin Dr. Robert Morton, who began his career as a surgeon in New York after graduating from Meharry. "Until the 1935 race riot here, it didn't matter how wealthy you were or how well you dressed. Blacks were simply not allowed to rent in certain buildings, eat in certain restaurants, or go to certain movie theaters along 125th Street."

Morton recalls when even Harlem Hospital would not allow black surgeons to operate on patients. "Even though there were many black physicians in the community," he explains, "the only institution that would allow us to perform surgery was the Edgecombe Sanitarium—and that was a building that didn't even have elevators. Patients had to be carried up and down the stairs on stretchers."

Although the March 1935 race riot in Harlem involved working-class residents who were incensed by the treatment of a black teenager accused of shoplifting in a white department store on 125th Street, the results of the unrest had a great impact on well-to-do blacks. In response to the unrest, affluent whites and brokers relinquished their hold on previously segregated well-to-do apartment buildings and town houses, thus making it finally possible for the black elite of Harlem to separate themselves from their working-class neighbors and relocate to more expensive streets and buildings.

"One of the first important addresses for well-to-do blacks in Harlem was Strivers' Row, a two-block area on West 138th and 139th Streets between Seventh and Eighth Avenues," says Cathy Connors, a longtime Harlem society columnist who covered the black elite for the *Amsterdam News*. As Connors leans over her desk pointing at a Harlem map, her pearl necklace drops gracefully around the collar of her blue suit. "Now that's where you'd find the wealthiest and best-connected black people over the last fifty years—people like Louis T. Wright and Booker T. Washington's grandson, and I believe Dr. Powell too," she says with a wry smile.

A member of the Links and the Northeasterners, Connors recalls attending many town-house parties along the tree-lined Strivers' Row with her now-deceased husband, John Connors, the former publisher of the American Express magazine group. "Thirteen-foot ceilings, marble staircases, towering brass and copper chandeliers," Connors says while thumbing through photographs of social events in various homes over the last thirty years.

Dr. James Jones lives in one of those million-dollar town houses that Connors points out. "These homes were designed by top architects like Stanford White and built in the 1890s when whites were the primary residents," says Jones, who bought a house on the street in the early 1950s. "Blacks were still not permitted to buy on these streets even after the wealthiest whites moved out. The realtors decided to keep them empty rather than upset the remaining white residents. I think Dr. Wright was one of the first blacks allowed to buy on this street."

I am tempted to interrupt Jones and ask if "Dr. Wright" is the same person as the "Louis T. Wright" that Cathy Connors referred to earlier, but I don't want to reveal myself as another "bourgie" Westchester black who doesn't know my black New York history. So I simply say, "Oh, yeah, Dr. Wright."

Jones looks at me skeptically.

"You *do* know who Dr. Wright is—don't you?"

I take a guess. "That would be Louis T. Wright, if I recall correctly."

Jones rolls his eyes slightly and breathes out impatiently. "Then I'll just tell you who he is," Jones adds without missing a beat.

In our short conversation, I learn that the Louis T. Wright, who lived in one of the town houses on Strivers' Row, was the first black physician to join the staff of a New York hospital. A 1915 graduate of Harvard Medical School, he eventually founded a cancer research center at Harlem Hospital. "He was the head of surgery at Harlem," recalls my cousin Robert Morton, "and he was a role model for a lot of us up-and-coming surgeons who had few mentors and even fewer opportunities."

I go back to Connors, and she gives me a piece of advice: "If you want to act like you know what you're talking about when you talk to Harlemites, there are some key people you'd better know about," she says. "It's Dr. Louis T. Wright, C. B. Powell, Arthur Logan—and, of course, the Delanys."

"Hmmph, the light-skinned doctor crowd," says a retired schoolteacher when I repeat the four names. The teacher says she knew Wright's daughters in Harlem during the 1930s. "So superior—so high-yellow."

The schoolteacher's female companion frowns. "They weren't all that yellow," she remarks. "Actually, I always thought Dr. Powell was rather dark. Yes, as a matter of fact, *quite* dark."

"Well, not those Wright girls," the schoolteacher snaps. "So superior—going off to their private schools."

The women recall that the Wright girls grew up in a well-to-do, high-profile family and that both ended up doing what few women—black or white—were doing in their generation. Educated at the exclusive private school Fieldston—one of the favorite private schools then and now for elite blacks—Barbara Wright went off to Mount Holyoke College and Columbia Medical School.

"And she married that really dark lawyer who was in President Reagan's cabinet," adds the schoolteacher. "What was his name?"

"Samuel Pierce?" I ask, recalling that President Reagan had appointed the New York attorney Sam Pierce as his secretary of housing and urban development.

"That's the one," the teacher says.

"And her sister, Jane," adds the friend, "went off to Smith College in the thirties—and when she was done with medical school, she came back and was an internist and then a surgery professor and a dean. They were really an amazing family."

"I still say they acted superior," the teacher adds. "As if Howard and Spelman weren't good enough for them. What kind of Harlem black sent his kids to Fieldston and Smith back then? Maybe now, but *then*?"

As I work my way around Harlem and other parts of the city, visiting with and talking to friends like Dr. Redhead, socialites George and Mary Lopez, Girl Friends founder Anna Small Murphy, and former Manhattan borough president Percy Sutton, I return to the list of important Harlem people that Cathy Connors gave me as soon as I hear one of the names mentioned.

Arthur Logan?

"Now, *that* was one sad story," someone says at a dinner I attend over on Sugar Hill's Edgecombe Avenue.

"Isn't he the one they named that hospital after?" asks the host. "I think he lived at 409 or 555. He was definitely at a good address—with at least two of his wives."

"He was a big-time doctor here in the forties and the fifties," explains a guest who belongs to the Boulé, "good-looking, rich, and really political. I think he raised a lot of civil rights money too."

When I ask what happened to him, the room falls silent. The host of the party runs briskly into the kitchen. The Boulé member turns his back to me.

Later that evening someone whispers to me that the good-looking, rich, and really political Dr. Arthur Logan jumped off a bridge in Harlem in the 1970s, but I decide to do my own digging because I realize how the jealous New York cliques can revise history.

Like Louis Wright, Arthur Logan was a high-profile physician who had been educated at white schools and was identified as a member of the light-skinned Harlem elite. Born in 1909, he went to the private Ethical Culture School, a division of the exclusive Fieldston School, then graduated from Williams College in 1930 and Columbia Medical School in 1934. His sister Myra was a New York physician, and his other sister, Ruth Logan Roberts, along with her husband, Dr. Eugene Roberts, was a popular host for black intellectuals and socialites during the Harlem Renaissance.

"My father was particularly known in Harlem because he was Duke Ellington's physician," recalls Arthur Logan's daughter, Adele Logan Alexander, who was born in Harlem in 1938. "His practice on 152nd Street, the Upper Manhattan Medical Group, was opened in 1953, and the city eventually named a hospital after him."

Although she now lives in Washington and teaches history at George Washington University, Alexander grew up surrounded by Harlem's elite. Like Louis Wright's privileged daughters, she graduated from the exclusive Fieldston School. She grew up with other well-to-do black kids in Jack and Jill and then went off to Radcliffe before later earning her Ph.D. at Howard.

When I later talk to the Sugar Hill host who had avoided deeper discussion of Dr. Arthur Logan, I tell her I had the fortune of speaking with his daughter, Adele. The woman's face suddenly lights up.

"Well, you *do* know who she is—don't you?"

I nod slowly, suddenly realizing that yet another box is being opened. Even if she didn't have her facts straight on Dr. Logan, she was on to a story about his daughter.

"Adele's husband is Clifford Alexander, and he was secretary of the army under President Carter," the woman practically shouts. "I bet she didn't tell you that, did she?"

Native Harlemites keep up with their neighbors—current and former.

"Well, my sister used to have such a crush on him. He was this big escort at the cotillions," the woman adds, as she goes on to explain that Clifford and Adele were both students at Fieldston and were in Jack and Jill together, and both went off to Harvard in the 1950s. She even recalls that Adele's mother had gotten a master's degree from Columbia in the 1930s.

I nod as I recall that Adele, despite her modesty, had revealed some of those personal facts.

"But we weren't Jack and Jill material and didn't have a good address back then," the woman explains, while still avoiding any discussion of the circumstances surrounding Dr. Logan's death. "The Logans lived at 505 then, and of course the Alexanders, while not well-to-do, lived in the Riverton."

"If they didn't live on Strivers' Row, the black elite in Harlem lived at 555 Edgecombe, 409 Edgecombe, Lenox Terrace, or the Riverton," says Connors, who lives in Lenox Terrace, a set of redbrick high-rise buildings with doormen and awnings, which welcomes such Harlem residents as attorney Percy Sutton, Congressman Charles Rangel, and former New York secretary of state Basil Patterson. Located at Lenox Avenue and 135th Street, Connors's development has a high concentration of second- and third-generation black professional families.

"When I was a teenager in Harlem, there were two important addresses that everyone wanted to have," says Dr. Beny Primm, a New York anesthesiologist who is also a renowned researcher on drug dependency. "It was either Strivers' Row or 409 Edgecombe Avenue," adds Primm, who grew up attending St. Mark's Methodist Church at 138th Street and Edgecombe.

"At parties, people didn't even bother saying the street name," recalls Primm. "They just said 409—and they expected you to know. The same was true for 555 Edgecombe."

Standing thirteen stories high on the top of the highly elevated Sugar Hill neighborhood in Harlem, 409 Edgecombe is a large luxury building that had originally been inhabited by Jewish and Irish professional families. With its long green awning that runs to the curb, its large lobby, and its ornate chandeliers, the building is reminiscent of apartment buildings found on Park or Fifth Avenue. Although it is showing wear from the years, the building has historic landmark status, and its wide view of upper Manhattan and ornate cornices still make it an attractive place today.

Edgecombe Avenue's 409 and 555 "opened up to blacks several years after Strivers' Row," says Jane Wright, a longtime Edgecombe Avenue resident. "Even after blacks moved into central Harlem, there were still many whites who stayed up here on Sugar Hill. From the 1930s through the 1950s, there were many famous blacks like Thurgood Marshall, W. E. B. Du Bois, and NAACP leaders Walter White and Roy Wilkins living at 409 and 555."

Other important residential landmarks in Harlem included the

Riverton and Lenox Terrace. I grew up visiting many relatives and family friends in those buildings. "The Riverton and Lenox Terrace were built specifically for blacks," says longtime Harlem resident Percy Sutton. "Because of that, race finally wasn't a factor in getting an apartment."

Built in 1947 along Fifth Avenue and 135th Street, the Riverton's simple, dark-red brick buildings would not turn heads today, but when they were built, this was a choice address for the upwardly mobile. In the early 1960s, the more upscale Lenox Terrace co-op apartments opened across the street. With uniformed doormen and driveways, the development quickly gained popularity.

I first met Sutton, the owner of the historic Apollo Theater on 125th Street and several radio stations, when I was eight years old and spending weekends with my father and uncle as volunteers in his successful campaign for Manhattan borough president. The former New York state assemblyman recalls some of the landmarks that made the community special when he first arrived in 1943. "I first came to New York in order to propose marriage to my wife, and at that time in the forties," he explains, "the Theresa Hotel on Seventh Avenue was the only nice New York hotel that blacks were allowed to visit or hold functions in."

The Theresa was referred to as the "black Waldorf," and many members of black society held their important social functions there.

Other landmarks that have been popular among the group have included such churches as Abyssinian Baptist—headed by two generations of Adam Clayton Powells, now by the more conservative but equally respected Calvin Butts; St. James Presbyterian Church on St. Nicholas Avenue; and at the northernmost—and possibly whitest—edge of Harlem, Riverside Church, which is headed by Rev. James Forbes.

"And Harlem credits the Powells for some important New York institutions too," says oral surgeon Dr. James Jones.

Before I can speak, he continues.

"And no, I'm not talking about the Adam Clayton Powells. Everybody knows *them*."

I nod.

"I'm talking about C.B."

"C.B.?"

"Yes, Dr. C. B. Powell. Dr. Clilan B. Powell, the city's first black radiologist."

A 1917 graduate of Howard Medical School, Powell was a physician who became a millionaire businessman as the president of Victory

Mutual Life Insurance Company; the founder of Community Finance, an investment firm; and owner in 1936 of the black paper the *Amsterdam News*. In addition to receiving appointments from Governor Rockefeller and New York mayor Wagner, he established a charitable foundation and left three million dollars to Howard University when he died in 1978.

"He set his wife, Lena, up quite well with a twenty-five-acre estate, in Westchester," remarks Cathy Connors.

In addition to the Logans, the Wrights, the Powells, and the Bishops, who governed St. Philip's Church for several generations under Hutchens and his son, Shelton, there have been many other Harlem families who have established or led institutions. They include the Hudgins family, who played an important role in building both Carver Federal Savings Bank and Freedom National Bank. Working alongside St. Philip's Moran Weston, family patriarch William Hudgins helped found Carver Federal. He served as president of the bank in the late 1940s. "My father and I felt that black communities needed to establish their own banks so that there is a sympathetic ear for those individuals who are trying to start businesses and families," explains Alvin Hudgins Sr. when I talk to him during a Sag Harbor summer holiday party. "If we aren't willing to invest in our own enterprises, no one else will either."

Several years later, in 1965, Alvin helped to found another bank—Freedom National—which his father, William, headed as its first president. Although Carver continues to thrive as a bank with over $400 million in assets, the demise of Freedom National in the late 1980s is a sore spot for many of the wealthy black investors who supported the institutions. "One of the downsides to being a wealthy resident of Harlem," says a Harlem native, "is that we end up giving out funding—or in Freedom National's case, giving out loans—to poor people who can't afford to repay. Sometimes we have to be as detached and profit-focused as white investors. But that's hard to do when you want to help your own."

Today, a third generation of the Hudgins family continues to invest in real estate and work in the securities industry while William, Alvin senior, and their families divide their time between homes in Manhattan, Southampton, and Sarasota, Florida.

The Delany family also had an important name in Harlem beginning in the early 1940s and has been particularly renowned through books and a Broadway production that was based on the family's experiences in New York. *Having Our Say,* a best-selling book and a successful Broadway play, told the story of the Delany sisters, two black professional women in their nineties who look back on their experiences as blacks in the early 1900s in New York. One of their brothers, Hubert, was a prominent attorney and judge in the 1930s. Serving as New York City tax commissioner, Delany was a big name among the political elite.

Today, the family name still looms large among New York's elite. Dr. Harry Delany, chief of surgery at Albert Einstein Medical Center in New York and the son of Hubert, grew up in the glare of black New York's society parties and columns where his parents were often a major feature. While a student at Columbia University, he met his wife, Barbara Collier, at the 1954 Girl Friends Cotillion—the very night of her debut and of her being selected queen of the debutantes. "Harry just happened to be escorting his mother to the ball, so he had no date to dance with," recalls his wife, Barbara, who was starting Cornell at the time. They are just one of the couples who met for the first time at the Girl Friends Ball of Roses cotillions that brought together the daughters of the city's best families.

Today, Harry is a member of the Boulé—as was Hubert Delany—in addition to being the fleet commander of the Rainbow Yacht Club, a group of New York-area black professional men who own yachts and travel together along the East Coast. His wife, Barbara, is now a member of the same Girl Friends organization that introduced them.

"Like Harry, a lot of us met our future wives or girlfriends at the Girl Friends Cotillions," says Brooklyn realtor Earl Arrington, who, like Delany, was one of the young men selected to be an escort at the Ball of Roses in the 1950s. He, too, is among the group that socializes on the Delanys' sixty-foot cabin cruiser, the "Barbara D."

"What I get annoyed about is the lack of credit we West Indians get, when we were the hardest-working blacks in Harlem," says a retired accountant as we sit in Sylvia's, the Harlem soul food restaurant on Lenox Avenue. "They used to refer to us as the 'black Jews' because of how smart we were and the many businesses we opened." It is indeed true that people of West Indian extraction have had great success among New York's elite. They have included such people as Hulan Jack, the first black Manhattan borough president; Ford Foundation president Franklin Thomas; and federal judge Constance Baker Motley. A longtime member of St. Ambrose Episcopal Church on 130th Street, the popular upscale church for Harlem West Indians, the accountant adds, "For years, these southern blacks in

Harlem have been ignoring all the prestige we've brought. They treat us like we're from Brooklyn or something—like we're outsiders."

The history of Brooklyn's black elite does not often get the attention it deserves—least of all from black Harlemites, who consider it a second-tier clique. There are dozens of families and individuals whose careers and accomplishments in Brooklyn can challenge this view, and no two people better represent the high position that black Brooklynites have held in New York's black elite than the husband-and-wife team of John Procope and Ernesta Forster Procope.

As I sit in the conference room of the Procopes' Wall Street insurance brokerage firm, E. G. Bowman, I can see photos and tributes to what John and Ernesta have accomplished in American business and in the groups that they hold dear in Manhattan and Brooklyn. Before Procope helped Ernesta turn the company she founded into one of the first successful black-owned firms on Wall Street, John rose through the ranks of the newspaper business to become publisher of the *Amsterdam News*. "I had been head of advertising for the paper under Dr. Powell," says Procope, "and before he died, he sold to a group of Harlem people which included Percy Sutton. After I became publisher, Percy ran for mayor."

"He easily could have been pretentious—but like Percy, he continued to give back," says Ernesta, a native Brooklynite who has sat on boards of numerous nonprofits, colleges, and *Fortune* 500 companies such as Avon and Chubb Insurance.

The rarefied world that the Procopes travel in today is a more integrated one than would have been possible for Powell and his wife. Their résumé demonstrates a combination of old-guard black and old-guard white credentials: He's an Alpha and a member of the Comus Club, the prestigious black men's group founded in Brooklyn. She's an AKA and a mentor to many young black professionals on Wall Street. At the same time, they have memberships at the Union League and the Cosmopolitan Club on Park Avenue.

They start each morning with a swim in their pool before taking a private car from their home in Queens to their Wall Street office. This morning, as we sit in their offices with Ernesta's niece, Jacqueline Forster—a graduate of Miss Porter's School, Cornell, and the University of Pennsylvania law school—we talk about the role of black business in developing New York's black community.

"We were the first minority-owned company to advertise in *Fortune* magazine," says Ernesta as her husband holds up a full-page ad from the 1970s. "And we realized that for blacks to truly succeed in business in this town or elsewhere, we needed to reach out to more than just black customers."

Even though their insurance brokerage firm, E. G. Bowman, was successful in capturing many *Fortune* 500 clients, its groundbreaking advertisement also elicited a great deal of hate mail from some of *Fortune*'s white readers.

"This is what we were up against then," says John as he shows me some of the offensive comments that had been scrawled across copies of the ad and mailed to their office twenty years ago.

"Although Harlem is the most famous black community in New York," says Dr. Jones, "there has always been another black professional community developing alongside Harlem—and that was taking place in Brooklyn. And believe me, Brooklynites do not like being lumped together with Harlem. They have their own history, organizations, churches, and schools that make them unique."

Doris Cumberbatch Guinier couldn't agree more with that statement. "Brooklyn is a distinctive community, and while we all know the various groups and families of Harlem, there are specific ones that are unique to us," says Guinier, a retired guidance counselor and schoolteacher who was born and raised in Brooklyn.

Guinier is one of those Brooklyn names that are heard at gatherings of socialites, businesspeople, educators, and old families. I always remember hearing it during our summer visits to Sag Harbor. "Doris has all the right credentials," explains a Brooklyn matron as she points out Guiniers' summer home in the Sag Harbor Hills section of Sag Harbor, "but her credentials are Brooklyn ones—not Harlem ones."

"You know Doris's father, Joshua, was one of the first black licensed undertakers in the city," said a Brooklynite during the funeral of Dr. Vernal Cave, another well-known neighbor. "And her husband went to Harvard back in the thirties and later ended up teaching at Columbia. That's Lani Guinier's father."

"Doris pulled strings and helped us get our daughter into the cotillion at the Waldorf-Astoria," says another woman.

"When Doris and Hewart sent Clotilde to that fancy private Ethical Culture School, we thought, 'Who do they think they are?' Next thing we knew, Doris had gotten Clotilde into Wellesley—and she was just sixteen! She knew what she was doing."

"If she wanted, Doris really could have been one of those nose-in-the-air President-Street types. She was on the board of directors of the Y, she went to all the right schools, belonged to the right church, married the right husband, summered in the right place. She even had the right relatives. I remember her godparents on Strivers' Row. But she's as regular as can be."

Guinier could hardly be called regular, because she does have all the "right" Brooklyn society credentials. In addition to being a former president of the Brooklyn Girl Friends and the Brooklyn Links, she was a charter member of both groups—the Girl Friends in 1927, along with her longtime friend Hazel Thomas Gray, whom she still sees in Sag Harbor. "Hazel and I went to Girls High together with Dorothea Mason, and our friend Hazel Bunn asked us to join her in starting the Girl Friends chapter in 1927," says Guinier, as she unpacks after a two-week stay at her Sag Harbor home on New York's Long Island.

"And we already knew Anna—because she actually installed us."

I don't interrupt because I assume she means Anna Murphy—one of the group's national founders, who has a summer home on the same street as Guinier.

"So years after I became a Girl Friend, I founded the Links chapter in Brooklyn," she says, "which was probably the same year—in the early fifties—that my daughter had her coming out, with a group of lovely girls who were going off to the best schools. There was Clotilde, Millicent, Joan, Lynne, Clotilde's good friend Barbara—"

Once again I don't interrupt because I assume she means Barbara Collier Delany, who was Clotilde Guinier's Jack and Jill friend, and who was going off to Cornell.

"—and I think Lorraine and Alma were in the group too."

I assume Alma is the widow of Ron Brown. I remembered that Alma Arrington had been a Brooklyn child and debutante long before her and her husband's political life began in Washington.

"There's always a few links between Brooklyn and Manhattan in any family," says Ada Fisher Jones to her husband.

"That's right—Ron Brown and Alma Arrington," agrees Dr. Jones as he thinks of his former Harlem neighbors. "Ron's dad was manager of the Hotel Theresa, the most famous hotel in Harlem. Now that was a good-looking family."

"Believe me, they wouldn't let you work there if you didn't look good—and didn't come from some good family," adds an attorney whose father and grandfather had attempted to host a party for a group of lawyers in the late 1920s at the Theresa—a few years before the hotel became more liberal with its policies—but had been turned away. "They didn't like anything darker than white to stay at that hotel—and it was smack in the middle of Harlem. Our St. George's Hotel wasn't nearly as bad as those Harlem people were."

Just as Harlem has its important addresses and institutions, so has Brooklyn established its own distinctive locales. In addition to the Hotel St. George, where most black society dinners and cotillions took place, Brooklynites favored the elite President Street with its large homes, as well as Stuyvesant Avenue. The churches that have been favored by the Brooklyn elite over the years have been Concord Baptist, Cornerstone Baptist, St. Philip's (remember that Brooklyn has a St. Philip's too), St. Augustine's, Siloam Presbyterian, and St. Peter Claver's. And not surprisingly, Brooklyn's Bedford-Stuyvesant neighborhood lays claim to having established the borough's most exclusive men's clubs: the Comus Club and the Guardsmen.

"One example when we all pulled together was when we got David Dinkins elected as mayor," says Dr. George Lopez of Queens when asked if the black cliques in New York ever become unified. "When David was running for mayor in 1989, you saw blacks from all five boroughs, as well as from Nassau County and Westchester, working together on fundraisers and voter drives."

A lifelong resident of Queens, one of the less populous boroughs of New York, George Lopez, along with his wife, Mary Gardner Lopez, has been a pillar in numerous black society organizations and activities. A graduate of Howard and Meharry, Lopez is firmly entrenched in such elite groups as the Guardsmen and the Boulé. "Just because we live outside of Harlem and Brooklyn doesn't mean that we are cut off from the black social and political world," says Lopez as he sits in the living room of their large white stucco home.

His wife, Mary, agrees. She's been a society columnist with the newspaper the *New York Voice* for thirty years, and as a member of AKA and a founder of the Doll League, she has enjoyed their many years in Queens.

"I'm actually from Nashville, but I've never felt like an outsider here," Mary says. I suddenly recall that her maiden name, Gardner, is the same as that of Nashville's black society funeral home.

It is evident that people like the Lopezes don't see the boundaries between the different boroughs and their cliques. In fact, the very next time I run into them is at a Sag Harbor party attended by Harlem's Chester Redhead and Brooklyn's E. T. Williams. There are clearly people who refuse to allow the old divisions to come between them.

At a recent Boulé Christmas party at the Waldorf-Astoria, it occurred to me that more than ever the social organizations are trying to bring together the different New York chapters. If for no other reason, the members like it because it helps in their business networks. At that event I saw Brooklyn's Don and Saundra Williams Cornwell (Don is a former Goldman Sachs investment banker who now owns a string of TV stations; his wife is a corporate attorney) and Manhattan's Bruce Llewellyn, who went to my high school and belongs to my Boulé chapter. After owning the Fedco Supermarket store chain in the city, he bought the Philadelphia Bottling Company. Also present was Bronxville's Hector Hyacinthe, who invests in real estate, office furniture, and media businesses. He is a major Republican fund-raiser. He and his wife Phyllis used to host our Jack and Jill parties and meetings around the pool of their Tudor home. After growing up with their kids in Westchester, I was at Princeton with their daughter Sylvere.

Another face I saw was Harold Doley, head of Doley Securities, the first black owner of an individual seat on the New York Stock Exchange. He and his wife, Helena, host many events at their Irvington, New York, mansion, which was originally built by Madam C. J. Walker.

There are also some members of New York's black elite who do not operate within the most obvious social groups and neighborhoods. One of these people is Upper East Side realtor and socialite Alice Mason, who was born into the prominent Christmas family in Philadelphia. Her father, Dr. Lawrence Christmas, was a successful dentist and one of the founders of the Philadelphia chapter of Alpha Phi Alpha, the oldest and most prestigious black college fraternity in the country. She and her sister, Marie Christmas Rhone, became New Yorkers many years ago when Marie moved to the West Side of Manhattan and her sister became a successful Upper East Side real estate broker to the WASP elite. Alice's friends include such top celebrities as Mary Tyler Moore, Barbara Walters, and pianist Bobby Short.

I first met Ms. Mason, a consummate party giver, at one of the famous dinner parties she gives in her East Seventy-second Street apartment. Surrounded by a million-dollar art collection, the elegant Mason and her daughter, Dominique Richard, regularly entertain the most powerful people in business, media, and Democratic politics. They are also a constant fixture in the New York society columns.

While Harlem and Brooklyn were going through their various transitions, there were several communities of black middle- and upper-class families developing in the less densely populated borough of Queens as well as in suburban Westchester and New Jersey.

"Black people had been in the Jamaica section of Queens for many years," recalls Ersa Poston, an appointee of Governor Nelson Rockefeller, who now lives in Washington but who had lived in both Queens and Brooklyn during the 1940s, 1950s, and 1960s. "There were many successful blacks living in East Elmhurst and St. Albans. Later, there were also some of us moving into Hollis and Flushing."

Like many others, Poston recalls the way in which the more suburban and more white neighborhoods of Queens responded to the influx of blacks into the white areas. "I got to see what blockbusting was all about as brokers kept a tally of what streets had a black family."

Barbara Collier Delany remembers how some white neighbors responded when she and her parents moved into St. Albans. "I was in elementary school and there were only a half-dozen black families in the town," explains Delany, who now lives in Westchester with her husband, Harry, a New York surgeon. "One afternoon in broad daylight, someone threw a rock in our window and they had actually tied a note to it telling us to move out. This was in the late forties, but my parents decided to stay."

Delany, who often found herself blazing trails, was one of the few black students attending the selective all-girls Hunter High School in Manhattan and later made national headlines when, as a student at Cornell University, she became the first black to be offered membership in a white sorority.

Like many others who were raising children in Manhattan, Barbara Collier Delany decided to move to the suburbs of southern Westchester where nice neighborhoods, open space, and public schools were plentiful. One of the richest counties in the United States, Westchester is situated just north of New York City. Its two southernmost cities—Mount Vernon and Yonkers—are today its two most congested cities.

In the 1950s, the two favorite communities for the black professional crowd in Westchester immediately became Mount Vernon and New Rochelle. Both cities had substantial Jewish populations and were considered liberal. Mount Vernon, though not quite as affluent as New Rochelle, had large homes and good schools.

New Rochelle still continues to be a magnet for well-to-do blacks. Among its past and present black residents are American Express president Kenneth Chenault, National Urban League president and former United Negro College Fund head Christopher Edley, former Urban League president Whitney Young, columnist Carl Rowan, Ruby Dee, and Ossie Davis, as well as many attorneys and physicians.

Other communities in Westchester popular with members of the black elite are Scarsdale and White Plains. A few families are venturing north to the small town of Chappaqua, where actress Vanessa Williams lives and where high-profile surgeons like the Boulé's John Hutchinson and the Reveille Club's Bill Curry live. In northern New Jersey, the popular suburbs for black professionals have been Teaneck, South Orange, and Montclair, which happens to be particularly popular among a new wave of black bankers with ties to Harvard Business School.

In all of these New York communities, whether inner-city or suburban, what also remains a constant is the social clubs and their events that connect the many elite residents. Whether it's the Links, Jack and Jill, the Guardsmen, the Girl Friends, the Fellas, or the Boulé, there is a thread that ties the people from these communities together. Barbara Collier Delany agrees. "Three generations of my family have been in Jack and Jill—and we have lived in three very different parts of New York—Queens, Westchester, and the Riverdale section of the Bronx." With such organizations, it is almost possible to forget the geographical boundaries between people.

There is a new generation of young, black New Yorkers who refuse to be distracted by Harlem-Brooklyn or southern–West Indian types of distinctions. And they refuse to "dilute" their racial identity in the way that some suburbanites had done as they ran back and forth between white New York suburbs and the black inner city—always hoping not to seem too black or too white.

Like Jacqueline Forster, young black New Yorkers are now taking their experiences in stride without regard to the old black New York cliques. Still in her twenties, Forster lives in Manhattan, even though her family name is best known in Brooklyn. The titles and the boundaries are meaningless to her. She looks at her past experiences in Jack and Jill, in AKA, at Miss Porter's, and at Cornell not as black experiences or white experiences, not as Brooklyn ones or suburban ones. Instead she focuses on looking for mentors and building her career.

"I knew that I wanted to go into the insurance business since I was seven years old," says the University of Pennsylvania Law School graduate as she sits across from her aunt, Ernesta Procope, in their Wall Street office. "And I could never find a better mentor than my aunt. We live in a city where there are hundreds of black living legends. No city has more than New York. I'm just happy that I'm getting to know them."

Black Elite in Memphis

"There were only about thirty students in each graduating class, and most of us went to the same church, lived in the same neighborhood, and were friends with the same families," says Alma Roulhac Booth as she recalls the other children in her class at the private LeMoyne School in Memphis during the 1920s. "Because our circle was so tight-knit, Memphis felt like a very small town."

Alma Roulhac Booth grew up surrounded by other black children of the Memphis elite. Her father, Dr. Christopher Roulhac, had graduated from Howard University Medical School and arrived in Memphis in 1910—a time when the city was populated by a small group of black lawyers, physicians, and business owners on the South Side who all knew each other.

Among Booth's classmates and closest friends at LeMoyne was Roberta Church, whose family had been the wealthiest and most influential black family in the South during the late 1800s and early 1900s. "Roberta's grandfather, Robert senior, was the first black millionaire in the South when he opened the Solvent Savings Bank and Trust Company and invested in real estate throughout the city," says Booth as she sits in the dining room of her large gray-and-white colonial home, "and her father, Robert junior, was so prominent a contributor and activist in the Republican Party in Memphis that he was named a delegate to the national convention as early as 1912 and was offered membership at the all-white Congressional Country Club in Washington, D.C."

Although she does not consider herself as such, Booth is representative of the well-educated, well-connected black Memphis elite. She is the daughter of a physician and schoolteacher and the wife of a successful restaurateur, and her father and brother were both members of the Boulé. She is a charter member and former president of the Memphis Links, a Jack and Jill mother, and a former school principal. A quick peek into her life and activities reveals ties to the most important names and institutions in black Memphis: the Walkers, the Byases, the Churches, the Hayeses, the Hooks, the Riverses, the Ishes, the Pinkstons, the Speights, the Phyllis Wheatley Social and Literary Club, and Emmanuel Episcopal Church.

Because my own parents and many relatives—including my maternal grandmother, who was four years ahead of Booth at the LeMoyne School—were born in Memphis, I spent many childhood visits to the city hearing the same names that Booth knows so well. During those childhood conversations, I learned how whites who were led by Boss Crump had run the Church family out of town in the 1930s because the Churches had become too rich and powerful. I learned how the black elite families were allowed to buy plots in the otherwise all-white Elmwood Cemetery, but were restricted to a particular corner—far away from the white Confederate war heroes. And when one of my Memphis cousins was sent north to attend Phillips Exeter in the 1970s, I learned that this provincial city's racial atmosphere had not changed dramatically for even the next generation of privileged blacks. Even today, I agree with several of my relatives when they say that this city still has some qualities that make it similar to a small segregated town.

"If we go, it'll be for five minutes. No lingering. No pictures. And no talking to people out front."

My brother and I shook our heads in disbelief as our uncle lectured us in the backseat of his car.

Our aunt then leaned into the open window. "And don't let them sign anything either."

"That's right," he added in agreement. "And don't write your name down on anything. There's always some nut out there with a petition or a sign-up sheet—trying to get us to name some landmark after that character."

My uncle and aunt had lived their entire lives in Memphis, and they strongly resented the fact that their young nephews, visiting from New York, wanted to go see Graceland, the home of Elvis Presley. For two weeks, they had challenged our request—never quite explaining, but strongly hinting, that we, as blacks, shouldn't be seen standing in front of this famous landmark.

"And if I were you, Earlene—" began a neighbor who stood in my uncle and aunt's driveway as we began to perspire in the backseat of their Buick Electra 225. It was August—sometime in the mid-1970s, before the singer's death. "I'd get out of there before noon, before those tourists get there. You just may run into one of the teachers from your school. You know how these rednecks love that Elvis."

The contempt that middle- and upper-class black Memphians have for Elvis Presley—a man they have never met personally—is greater than one might imagine. As an adolescent visiting Memphis during summers away from home, I saw the resentment that my relatives and their black friends had for this Memphis icon. "Uneducated, crude, racist, trash" were the words that were used when we brought up his name. Over time, I learned that the revulsion was directed less at the man and more at the community that had overlooked the black talent that the city had produced for so long.

Today, when I return to Memphis—passing through the streets, looking at Graceland, or more often, at the homes, churches, and storefronts that I had visited as a child with relatives—I am disappointed. I am disappointed because of the lack of vision demonstrated by the white elite who controlled the city throughout much of the twentieth century. There was good reason for my relatives and their friends to resent Elvis and resent so much of white Memphis.

Since the beginning of the 1900s, blacks had been ghettoized—without regard to their income levels—in South Memphis, with whites living everywhere else, including midtown and the eastern side, where upper-income whites resided. My aunt recently showed me a letter that she had been sent by a real estate developer in the mid-1960s advertising an "all-Negro single-family neighborhood" that was being built for middle-class blacks in South Memphis. The language of the letter, written well after the courts had outlawed segregation, defined the almost plantationlike mentality of many white Memphians who, until recently, had no sense of how great their city might have been had they accorded the least bit of meaningful integration or respect to blacks who had the desire and the talent to contribute.

Instead, this small town has a national reputation within the popular culture for two reasons: the life of Elvis Presley and the death of Dr. Martin Luther King Jr. Among the well-educated black community, there is no irony lost in the fact that the most celebrated figure in Memphis history is a barely schooled white rock-and-roll star who lifted his musical and performance style from lesser-known black musicians. The feeling that their own people's talent had been "ripped off" and repackaged in a more racially acceptable form (i.e., white) has left them so resentful that the very idea of visiting his estate and posing in front of his front gates along a bustling, newly renamed boulevard does nothing but add insult to injury.

The irony continues for them when they note that the world's most respected civil rights leader was a black Nobel Prize winner—a Ph.D. who came out of Morehouse—whose lasting attachment to the town was that he was slain there. He is memorialized at the old Lorraine Motel (now the National Civil Rights Museum), which sits in a desolate section of town amidst old, crumbling warehouses.

"We could have been a Nashville, maybe even an Atlanta," my grandfather used to say as he drove us down what is now Elvis Presley Boulevard, a busy street that runs north to south through the ironically named White Haven section, an area that was all-white when I was visiting Memphis in the 1960s and early 1970s and has since become predominantly black. Although he loved the town, he used to say, "All we have here are a lot of churches and segregated schools."

Even today, one has the feeling that this town has dragged its feet in its physical, economic, and intellectual growth, given where it was at the turn of the century as a major commercial center for cotton distribution. But I am well aware that it is not the fault of the people in power today, or even those who have been in power during the last fifteen years. The greatest stagnation was the result of resistance to progress—racial and otherwise—that took place in the 1950s, 1960s, and 1970s, when other small cities were growing physically and culturally. For example, in 1960, the population of Memphis was almost 500,000. Twenty years later it had grown only to 650,000, as compared with Nashville, which more than doubled in size during those years. By 1990, Memphis had lost 5 percent of its population. And today, with fewer than 600,000 people, Memphis has little city life and little city development to reverse the decline. What are plentiful are the churches and segregated schools that my relatives were talking about thirty years ago. The churches are either black or white, and the schools are either black public schools or white Christian academies that were created from the 1950s through the 1970s in order to avoid integration. By hiding behind religion, these Christian academies managed to keep another generation of blacks and whites apart.

I am still a devoted fan of Memphis because of my childhood memories and because of the progressive people—both black and white—who I know are working together today; but like its black elite, who were educated elsewhere, I feel it is a town trying to overcome great odds. The thriving downtown that it once had along Main Street—between Beale and Jefferson—was killed in the 1970s when the whites abandoned the increasingly black city, which now is only 44 percent white. There is not a department store within ten miles of City Hall. Big stores like Gerber's, Lowenstein's, and Bry's are all gone now. The area surrounding the municipal buildings, courthouses, and county offices is littered with pawnshops, bail bondsmen, and vacant storefronts. What would have long ago been a well-developed Mississippi River waterfront in any other town is just now seeing walking paths, green grass, and trees. With the exception of a few tall buildings built by the city's superior hospitals—Baptist and Methodist—and by First Tennessee Bank and Union Planters Bank, one gets the sense that no major company or industry calls Memphis its home. Federal Express is there—several miles out of downtown, near the airport, but the headquarters for Holiday Inn and Cook Industries left years ago.

Even the city's premier hotel, the Peabody—as plush as it is, by Memphis standards—seems a bit corny and anachronistic. Founded in 1869 and rebuilt in the 1920s, the imposing brick structure attracts tourists to its main lobby each morning for a ritual that began in the 1930s and continues today, seven days a week. At 11:00 A.M. sharp, an elevator door opens on the main floor, and marching in line across the carpeted floor are five trained ducks. Marching in unison to taped music that plays over the lobby speakers, the small ducks waddle toward a small, ornate fountain and pool in the middle of the floor. One by one, they hop up to the fountain and then dive into the pool. The routine is repeated in reverse at 5:00 each afternoon. Since the hotel had a policy of segregation thoughout my older relatives' lives, it was not until we were teenagers that they permitted us to visit the building and view this amusing event.

"Memphis used to have the largest and most developed metropolitan area in Tennessee," explained a black former city councilman who acknowledges that a fear of integration is what kept Memphis small and rather underdeveloped. "It can't be blamed on the people who are in power today," he says, "but those who were making decisions in the 1950s and 1960s created a problem between the races and within the corporate community that was hard to correct."

The city's black elite seem to be painfully aware of how much better their black counterparts are doing in Nashville—a city whose metropolitan area had once been less affluent, less respected, and less populated than that of Memphis. In fact, most of the Memphis black elite who had grown up in the city prior to the 1960s had to leave town and go to Nashville in order to get their education. Although the town had the small, all-black LeMoyne College since 1870, it lacked the truly elite black institutions that Nashville had: Fisk University and Meharry Medical College. The black Memphians also lacked Nashville's Tennessee State University, a black public college that ran itself like an elite private school.

"Although I grew up in Memphis—a city that looked down on Nashville at the time," explains a sixty-year-old physician who attended Fisk, "I always had the feeling that Nashville was going to catch up and then leave us behind—intellectually and racially. Memphis had no premier schools for whites or blacks, and Nashville had Vanderbilt for whites and these other top schools for us. White Memphians—and even some black Memphians—seemed to get more backward and more provincial as other cities outgrew us. So few blacks here were able to break out of the box and really gain national exposure the way that blacks in Nashville did."

One of the black Memphis families that did manage to gain national exposure—particularly in politics—was the Ford family. In 1974, Harold Ford Sr. made national news when, at the age of twenty-nine, he became the first black U.S. congressman from Tennessee. Twenty-two years later, in 1996, his son, Harold junior, at age twenty-six, took his place in Congress and made national news again as the youngest member of the House of Representatives. Although he is a resident of Memphis, it is a stretch to say that this sophisticated, Washington-raised graduate of St.

Albans School, the University of Michigan, and the University of Pennsylvania Law School is a typical example of what Memphis's integrated schools were producing in the 1970s.

"Our family first entered politics when my father ran for the Tennessee State House in 1964," explains Dr. James Ford, an ophthalmologist who joined the Memphis City Council in 1980 and now serves as a county commissioner in Shelby County, the area that includes Memphis. "My brother, Harold, was the first member of our family elected—when he joined the Tennessee legislature in 1971."

"We were able to gain a foothold in a city that had once been run by Boss Crump because our family had owned property and businesses in Memphis for so long," says Barbara Ford Branch, the sister of James and Harold senior. "Our father, Newton, founded N. J. Ford Funeral Home in the 1930s, and our great-grandfather Newton had been a major landholder who donated the funds and land for Ford's Chapel AME Zion Church in the 1890s."

Like other members of the old Memphis black elite, Barbara and her siblings were sent to Nashville to get their college education. Among the generation of siblings, five brothers have held elected political positions. One brother, John, was on the Memphis City Council from 1972 to 1980, and then he was elected to the state senate. James served on the council from 1980 until 1994, when he was elected county commissioner. A third brother, Emmett, was elected to the statehouse in 1974. Joe was elected to the city council in 1994.

As a child visiting my Memphis relatives during the summers, it was a required activity that my brother and I go door-to-door with uncles, aunts, and grandparents to pass out literature for any one of the Ford politicians who happened to be running at the time. When Harold senior ran for Congress in 1973, he was taking on a longtime white incumbent, Dan H. Kirkendal, who my relatives were certain had no interest in serving the black constituents who populated his district in overwhelming numbers. "It's like living on a plantation," my grandfather used to say when we would sit in one of the restaurants that he and my grandmother owned along Florida Street. "These white people don't want educated blacks here. This place is home, but your parents did right by getting out of here and raising you in New York." Many times my relatives reminded me that divisions between local blacks and whites were based on race and not on class. Even the wealthiest of them had little or no interaction with whites.

"When I was growing up here, it was very obvious that the city was segregated enough to ensure that the old-guard black families did not socialize with anyone but blacks," says Ronald Walter, executive vice president of WREG-TV, the CBS affiliate in Memphis. "When I got married in 1987, we had black and white guests, but not very long ago, influential whites were ostracized for attending a reception or a wedding hosted by a black person."

Walter, who grew up in the Memphis chapter of Jack and Jill and whose wife, Marianne, is a member of the Links and the Junior League, is one of the best sources on the history of the black Memphis elite. He is a fourth-generation Memphian and was the first black to serve on the Tennessee Historical Commission. An unabashed supporter of the city and its residents, Walter acknowledges that Memphis is more like a small town than it is like Chicago or Atlanta. He recalls that one of the first times that prominent Memphis whites attended a black elite wedding was when Fisk graduate Lily Patricia Walker, the daughter of Universal Life Insurance chairman A. Maceo Walker, got married in 1961. "Even though the Walkers were wealthy, powerful, and respected by both races," explains Walter, "many white citizens in the city were upset that a white mayoral candidate had attended the wedding, and to send him a strong message, they refused to vote for him."

During the 1950s and 1960s, the Walkers probably had the only black name even acknowledged within the white community. And they were preceded only by the Church family.

"The most affluent and accomplished blacks in this town were the Church family, who were headed by Robert Reed Church," explains Walter, who worked with Church's granddaughter, Roberta, on the book *Nineteenth Century Memphis Families of Color*. Considered the first black millionaire in the South, Church was born in 1839, was raised as a free black, and then became an influential member of the Republican party. As a major property owner who purchased buildings, open lots, a hotel, a saloon, and a restaurant, he eventually opened the Solvent Savings Bank and Trust Company in 1906.

Although the family mansion was later burned down by white residents who resented the Church family's influence, Robert Church's ties were great enough by 1900 to make him a delegate to the Republican National Convention and to enable him to build an auditorium and park on downtown's Beale Street that would host President Theodore Roosevelt when he visited the city.

Church's son, Robert junior, was a graduate of Oberlin College in Ohio and eventually joined his father's bank. In addition to founding the Memphis NAACP, he established groups to encourage blacks to vote even though he received many death threats from white residents. A prominent Republican like his father, he also attended several national conventions.

"I remember when Boss Crump was so threatened by the Church family he literally had them chased out of town in the 1930s," says Alma Roulhac Booth, who attended LeMoyne School and Emmanuel Episcopal Church with Robert junior's daughter, Roberta. "They moved to Chicago and that's why Roberta went to Northwestern University for college. After that, they later moved the family to Washington. It's a shame after all that family did for Memphis. I'm so glad that the parkland Mr. Church gave to Memphis still carries his name." In prior years, some white residents attempted to remove the Church name from the downtown park, located near Beale Street.

Robert Church Sr.'s son was not his only accomplished offspring. His daughter, Mary Church Terrell, was probably the most famous member of the family. An 1884 graduate of Oberlin, she became a political activist who fought against segregation and became one of the charter members of the NAACP and the Delta Sigma Theta sorority. In 1891, she married fellow educator Robert Terrell, who had previously graduated from the Groton School, from Harvard University (class of 1884), and from Howard Law School (class of 1889). She became the first black on the Washington, D.C., board of education in 1895, founded the National Association of Colored Women, protested the Jim Crow laws through her speeches and articles, and worked with Susan B. Anthony and Ida B. Wells in their attempts to improve women's rights in politics and elsewhere. She continued to work on behalf of women and blacks until her death in 1954.

Many members of the old guard quietly comment on the fact that there is another branch of the Church family. "But they live as white people," says a physician who remembers hearing his grandparents tell stories about the white Churches and the black Churches, and how both families lived in Memphis. "I felt sorry for Roberta Church because she was light enough to pass and therefore was not always immediately embraced by blacks," explains the physician. "Although she always went as black, she often heard white people say the nastiest things about us. Her color put her in an awkward position in both groups."

But old stories about mixed races are also passed around about the Fords, the political family that owned acres and acres of property throughout Memphis beginning as early as the turn of the century. "There is no family more dedicated to helping blacks move ahead in this town than the Fords," remarked a former librarian who says that she has always voted for Fords—whether they were running for city council, statehouse, state senate, or U.S. Congress, "but one has to wonder how a black family was able to start off with so much wealth in the late 1800s, when they had been slaves—not free blacks." According to this longtime Memphian and others, the Fords had family ties, going back to the early 1800s, to a wealthy white judge—who was also a slaveowner. "Of course this is going back a long time, but old man Newton Ford was born a slave, and people just assumed that the judge was his father and left him property," says the librarian, before covering her mouth as if she had revealed something she shouldn't have.

Such ties to white families were a not at all unusual explanation when recently freed slaves had suddenly come into modest inheritances. While some living Ford descendants grudgingly confirm ties to this judge, they freely acknowledge that it was highly extraordinary that their great-grandfather Newton had owned such a great deal of property and had received a minor political appointment at the turn of the century in Memphis.

Although less well known by whites, the Hayes family is another old name in black Memphis history. It's a name that is first on my relatives' lips whenever a family member dies. Their name has appeared in virtually all of my deceased relatives' obituaries since the 1930s. "Everybody knew the Hayes family because they owned the society funeral home," says Dr. William H. Sweet, a native Memphian who belongs to the Boulé and grew up with my parents and their siblings. "The T. H. Hayes Funeral Home was in this palatial redbrick Victorian building on South Lauderdale Street with a goldfish pool in the front. And all the prominent blacks were funeralized by Hayes."

"Everybody who's anybody was buried by T. H. Hayes," agrees the retired wife of a physician who notes that the prestige of the Hayes family approaches that of the Churches. "A lot of people talk about the fact that the Hayes family burial plots over at the Elmwood Cemetery are right next to the Church family's mausoleum. Two important families—right next to each other," says the woman with reverence in her voice. "Being buried by Hayes was like getting an extra social credential. The next best thing you could get from that family was getting to join Frances Hayes's dinner club. And that's something I never got invited to do."

The modesty of Mrs. Frances Hayes, the current owner of the business, belies the family's stature. "My father-in-law, Thomas H. Hayes, started this business in 1902 because his friend Robert Church Sr. encouraged him. High society was never important to him—he simply worked hard to maintain a good reputation."

"Mr. Hayes had two sons, Thomas junior and Taylor," says Frances while standing in one of the large rooms inside the redbrick mansion on South Lauderdale. I had first met her when I was a child and my cousin, Addie Owen, had joined Frances's famous dinner club—an elite group of black women who shared recipes and dinner menus with many Memphis socialites, in addition to the *Memphis Commercial Appeal,* the city's primary newspaper. Since that time, Frances has talked me through the illnesses and deaths of many close Memphis relatives. "My husband, Taylor, became funeral director in 1940 and I took over when he died in 1968," she says, explaining how she got into the business. "My husband's brother, Thomas junior, was involved in other activities, and he made a great deal of money by owning the Birmingham Black Barons. It was that Negro League baseball team which had Willie Mays, back when black folks couldn't play in the major league."

"Not only did Hayes bury most of the Church family," adds a widow whose family always used the funeral parlor, "but they also buried the rest of the A-list—people like Dr. James Byas, Dr. Speight, Dr. Fred Rivers, Charles Hooks, Fred Hutchins, G. P. Hamilton, Blair Hunt, Dr. Hollis Price, virtually everyone who ever worked for Universal Life, and of course the Walkers." The woman paused while flipping through an old personal phone diary. "You know, the only big people they missed out on were Lily Pat Walker and Dr. Martin Luther King when he got killed here in 1968. I don't know what Lily Pat's husband was thinking, but he was probably so devastated when she died so young from cancer. And, as for Dr. King, he wasn't from Memphis—so his people just didn't know any better."

Frances Hayes acknowledges having buried many of the old families, but she believes they came to her out of friendship. "We belonged to Second Congregational Church with a lot of these families and we lived on the same streets and belonged to the same clubs. For example, my brother-in-law Thomas lived on South Parkway near the Walkers, the Riverses, the Byases, and the Speights."

While Frances and Taylor had no children to pass the business to, Thomas and his wife, Helen, had two daughters, Helen Ann and Tommie Kay, who both grew up in Jack and Jill and attended Fisk. "At one point," Frances adds with some sadness in her voice, "I thought Tommie Kay would want to come back to run the business." Today, Helen Hayes Groves is a well-to-do socialite in Los Angeles, where she has been active in Jack and Jill and the Links. Her husband, Wesley, is a physician from a prominent family in Idaho. Tommie Kay Hayes Armstrong, also a member of the Links, lives in northern California with her husband.

Prominent among the Memphis doctor crowd was the Byas family. Four brothers—Andrew, Arthur, John, and Thomas—all became physicians after graduating from Meharry Medical College in the late 1890s and early 1900s. Although Andrew was a well-known physician in Memphis, he invested much of his time in politics and buying real estate, ultimately owning a drugstore in the city. His brother John had a son named James, who also practiced medicine in Memphis after graduating from Meharry. Dr. James Byas was a trustee of LeMoyne-Owen College, and he and his wife, Orphelia, were active in the Memphis Boulé and Memphis Links. Their two children, James junior and Maye, later became the third generation of Byas doctors to graduate from Meharry.

But there is no Memphis family that elicits more mixed emotions than the family of Dr. J. E. Walker, founder of Universal Life Insurance Company, a business that has survived four generations. "That family has every story imaginable," says a former neighbor who lived near the Walkers on South Parkway East, "Money, glamour, murder, social status, drugs, tragic deaths—you name it, they had it."

As a child, I remember my grandfather pointing out that some members of our family lived next door to the Walker family on Mississippi Boulevard and then again on South Parkway, one of the most desirable streets for blacks beginning in the early 1940s. "Parkway," as it is known, snakes through the city going west to east, and you can trace the desegregation of Memphis by noting the years when upper-class blacks were allowed to move progressively east along the fashionable street. Although Dr. J. E. Walker, a physician who graduated from Meharry in 1906, founded the company, it was his son, A. Maceo Walker, who turned Universal Life and the family's Tri-State Bank into financial successes. In fact, by hiring members of the Olive, Willis, and Gilliam families, he gave them all additional status in Memphis society.

Born in 1909, Maceo Walker graduated from Fisk and received an MBA from New York University in 1932. His own status was later enhanced when he married into the prominent and well-to-do Ish family of Little Rock, Arkansas.

"I'm the one who introduced Maceo Walker to Harriett Ish," says Alma Roulhac Booth. "My father was a doctor who belonged to the Memphis Boulé and Harriett's father was a doctor who belonged to the Little Rock Boulé, so we knew each other fairly well. I introduced the two of them and they eventually married and lived next door to me on McLemore."

A graduate of Talladega College, Harriett Ish was far more than just a doctor's daughter, though. Her uncle was a graduate of Yale's class of 1909 and her family had ties to Supreme Life Insurance Company in Chicago. The stylish couple quickly climbed the social ladder, moving from one more prestigious street to the next—from Mississippi to McLemore to South Parkway East. They had three children, Lily Patricia, Lucille "Candy," and Antonio. And with the goal of keeping the business in the family, Maceo's sister's husband, Julian Kelso, was employed as Universal Life's medical director.

"But that family was cursed," says a former neighbor of the Walker's. "Not only did the mother, Harriett, die at a young age, but so did Lily Pat and Candy. And of course Maceo's father was shot one day in his office at Universal when a former partner had a dispute with him." The man paused slightly. "And don't even ask about Antonio, the son—he's the last Walker child living and he's hit all kinds of hard times. But Maceo's second wife is still living over on Parkway."

Adding to the Walker family mystique are the rumors that besides Maceo and his sister Johnetta Kelso, Dr. Walker had another child, but out of wedlock, while he was still married to his wife. Today, there is still speculation over who that child was and whether the child's birth had any connection with the murder of Dr. Walker years later.

"That story is just tearing this town apart—just tearing us apart."

The others sitting around the table nodded in agreement as Erma Clanton pushed her shiny black hair from her left eye and gestured over to a large book that sat on my Aunt Earlene's kitchen counter.

"Dr. Hunt was a fine man, and it's just a shame that he should be vilified by his own flesh and blood," added another woman who sat in the family room that opened up into the breakfast room where a dozen of us had collected the night before my uncle's wake at the T. H. Hayes Funeral Home.

Most of the fifteen or sixteen men and women who sat in the kitchen—including my parents—had grown up in Memphis. And they were fiercely protective of the highly esteemed black families who had played a role in the town's history. The person they were talking about was Blair T. Hunt, leader of two of black Memphis's most important black institutions: Booker T. Washington High School, the city's first black high school, and Mississippi Boulevard Church.

"Those stories are just tearing us apart," repeated Clanton, a recently retired professor who had taught drama for twenty years at Memphis State University. Like many on the South Side of Memphis, she was disturbed by a book that had recently been published by the granddaughter of Blair Hunt, the prominent educator and minister.

"I don't know how much we should believe anyway," one of the men snapped. "Isn't she married to one of these white rock stars anyway?"

My relatives shrugged as they listened to the well-substantiated rumor about how the beloved Hunt had allegedly placed his light-skinned wife in an insane asylum in the 1940s against her will. What was most shocking was that many of the old guard presumed that she had died in the 1950s, though when in fact she was still alive in the 1990s when her granddaughter wrote about her in the controversial book *Repossessing Ernestine.*

Much of what was said—and not said—in my aunt's kitchen really captured the essence of how black Memphians protect their local heroes, even when the evidence is stacked against them.

Because Memphis lacked the important black educational institutions that Atlanta, Nashville, and Washington had and also lacked the larger middle-class black populations of New York and Chicago, the black elite in Memphis was always rather small and parochial. In fact, its reliance on the church rather than on business and intellectual institutions made it seem second string to the more cynical and more sophisticated black elite in other cities. In fact, in other cities, during the 1930s and 1940s, the best-known members of the elite circles were attorneys, college presidents, professors, and rising politicians. Not in Memphis. Of course there were some business owners, attorneys, and physicians. But the biggest names in town at that time were ministers and teachers.

"People had great respect for Dr. Hunt," says native Memphian Dr. William H. Sweet. "As principal of Booker T. Washington High School and head minister of Mississippi Boulevard Church, Dr. Hunt was simultaneously running two of the most important black institutions in the city." In many ways, Sweet explains, Hunt was the bridge between the old and the new elite because unless someone was sent away to a boarding school, virtually every black person growing up in Memphis between 1925 and the early 1950s attended Booker T. Washington. "The Hunt family even have an official section of the Elmwood Cemetery named for them," says Frances Hayes, owner of the Hayes Funeral Home and a graduate of the famous high school.

My aunt, Earlene Graham, also a graduate of Booker T. Washington, recalls that Hunt was not a celebrity just in Memphis. "We all knew Dr. Hunt from high school, but we also saw him again when we became students at Tennessee State University," she explains while recalling that Hunt was a friend of the college's president. "He would come to campus in Nashville and give an incredible sermon entitled, 'Good-Bye, God, I'm Going to College.' He would talk about why we should not be dismissing religion as we advanced through school and our professions. Since his church attracted many of the successful families like the Walkers, this was an important issue for him."

The power of the church—in both the black and the white communities of Memphis—cannot be underestimated. Today, there are nearly one thousand churches in a metropolitan area of under one million people. This can be compared with a city such as Detroit, which has only around six hundred churches for its one million citizens. Many of Memphis's black elite, who are not considered as active in their church as the black working class, blame the city's lack of commercial growth on the black and white churches that dominate the street corners, the storefronts, and the radio stations in the city. "If these Baptist churches had their way," says a black attorney in his forties, "all our stores would be shut tight on Sundays, just like they used to be." This attorney does not identify with prior generations of black society who found their lives and livelihood in the church community.

As a part of the Bible Belt, church life in Memphis has played a major role in the city's history. Mississippi Boulevard Church's Blair Hunt was just one of the many early elite black Memphians who recognized the ties between the church and career success. In fact, although his church did include the top executives at Universal Life and Tri-State Bank, it was actually Second Congregational that attracted the most prominent black residents.

"You could call a quorum of the Links and Jack and Jill at a Second Congregational service," says Ronald Walter, a fourth-generation Memphian who grew up in Jack and Jill.

"At one time, that church was criticized for being solely for light-skinned blacks, but what it clearly had was the largest number of well-educated, affluent, and accomplished people." Among its members have been Taylor and Frances Hayes, who owned the funeral home; the Price family, who ran LeMoyne-Owen College; Addison Branch, a LeMoyne professor; and many who belonged to the Memphis Boulé. "They even had an indoor swimming pool in that church," says Dr. William Sweet, who learned to swim there as a child in the 1930s.

Tied with Second Congregational for historical prestige would be Emmanuel Episcopal Church, which was founded by the wealthy Church family. Alma Roulhac Booth grew up in the church and still belongs to it today.

Now that Memphis is a city with dozens of churches throughout the black neighborhoods, and with some well-to-do blacks attending white churches in the eastern, mostly white section, there are fewer distinctively black upper-class churches. The only others that have been placed in this top group are Metropolitan Baptist, Centenary United Methodist, and, for the few Catholics, St. Augustine's. St. Augustine's was unique because, while all the other churches had completely black clergy and black memberships, it was the only one of the group that had white clergy. "Our priest, Dr. Bertrand Koch, was white, as were the nuns at the church's school," recalls Dr. Sweet, who has attended St. Augustine's for five decades.

"Today, you don't find churches dividing themselves by skin color and by affluence with the same degree as they used to," says Walter who found himself forgoing a church and renting the historic Orpheum Theater when he and his wife, Marianne, got married in 1987 at one of the largest weddings in the city's history. Covered in the *New York Times* and the *Washington Post*, the wedding included over one thousand guests and required three florists and 250 candles to transform the stage into an ornate altar. Among the eighteen groomsmen and eighteen bridesmaids in the wedding party were former congressman Harold Ford and Tri-State Bank executive Jesse Turner, both members of old-guard families.

Over the years, many prominent families have been members of the clergy. Among the group are the Fullers and the Whalums, who also had ties to the Union Protective Life Insurance Company. In addition to churches, the other locales that have drawn the elite crowds were certain neighborhoods and streets.

Historically, blacks have lived on the South Side of the city, and the upper-class blacks of the 1930s and 1940s sought out homes on such streets as Mississippi, Lauderdale, and McLemore. "When I was growing up in the thirties," says my cousin, Anna Griffin Morton, a graduate of Booker T. Washington, Fisk, and Columbia University's graduate school, "we lived next door to Dr. Walker and his kids, Maceo and Johnetta. He had already started Universal Life Insurance, and at that time, Mississippi was a very desirable street." Now a New York resident, Morton recalls that her sister, Addie Griffin Owen, later joined the exclusive Memphis Dinner Club with Maceo's wife, Harriett.

"Beginning in the late 1940s, the street to live on was South Parkway," says Dr. William H. Sweet, "or at least any street that was just off South Parkway." That street—particularly the end of South Parkway East, heading toward Heiskell Farms, is still desirable today because of its deep front lawns and large old trees.

"Whites lived on the eastern end of this street, as well as the other parts of Parkway that wrap around the other parts of the city," explains Ron Walter as the electric gate opens up to the driveway running up past his redbrick home on the famous boulevard. Among the people who have lived on the street are the Walkers, the Hayes family, Mutual Federal Savings & Loan founder Chew Sawyer and his wife Helen, Dr. W. O. and Jewel Speight, and many others.

In more recent times, the street has included such residents as Judge and Mrs. Odell Horton, William and Addie Owen, Dr. James and Orphelia Byas, Dr. Fred Rivers, and Dr. Leland Atkins.

During the early 1970s, the entire street became available to blacks, and the western end became less desirable for single-family homes. Soon after that, blacks also moved east to the section known, ironically, as White Haven; and when the restrictive covenants were removed in Chickasaw Gardens, some integration took place there as well. During the 1980s, there was a large group of affluent blacks moving farther east to Hickory Hill, and then to suburban areas like Germantown and Cordova.

Perhaps the richest and most historic neighborhood in Memphis is known as Central Gardens. "I moved here because it feels like a suburb in the middle of the city," says Memphis attorney Earl Douglas, as he drives his taupe-colored Jaguar down a street of million-dollar mansions and pulls into his three-car garage. "This neighborhood was begun in 1903. But what is so special about these houses is that they have a lot of land and a lot of character and you cannot find that in East Memphis or the suburbs." A graduate of MIT and Columbia Law School, Douglas represents a wave of young professionals who have moved to Memphis in recent years. While Central Gardens remains more than 90 percent white, blacks have started to consider it, as well as neighborhoods close to the Mississippi River. Dr. Anita L. Jackson, an otolaryngologist with degrees from Princeton and Harvard, moved into Harbor Town, an island area that lies along the Mississippi River. The expensive development is one of the largest and newest sections of Memphis. "A lot of the young physicians who are working here need to be just a few minutes from the hospitals," says Jackson, who is originally from Augusta, Georgia. She is surrounded by other physicians and professionals who like being near the downtown activities. "The revitalization of downtown Memphis and the waterfront has brought a lot of black and white professionals into a more integrated environment," says Alex Coleman, anchorman at WREG-TV, who lives not far from the upscale neighborhood of actress Cybill Shepherd. Other young elites like Ella and Odell Horton Jr. and Amber Dancy Northcross and her husband Dr. Reginald Northcross have chosen the suburbs just east of Memphis.

In addition to favorite neighborhoods, the black elite in Memphis have always had their favorite institutions. As owners of the city's oldest and most prestigious funeral home, the Hayes family had no problem attracting the old-guard families. But besides T. H. Hayes and Sons, other prestigious funeral parlors have been S. W. Qualls Funeral Home, which was owned by Sam Qualls; and R. S. Lewis and Sons Funeral Home, which was run by Robert Lewis, who also owned the Memphis Red Sox baseball team and Lewis Stadium before selling the team to the Martin family. The Lewis Funeral Home handled the bodies of Dr. Martin Luther King Jr. and bankers Chew and Helen Sawyer.

"One reason why the Lewis name has some extra star power," says a retired attorney, "is that Robert Lewis Jr.'s wife, Ruth, was the daughter of Bishop Mason, who started the Church of God in Christ in the late 1800s."

Today, a major force in the funeral business and other aspects of Memphis life is the Ford Family, which includes several siblings who hold political office in the city and in the state of Tennessee. In fact, when Harold Ford Sr. decided, in 1995, not to run again for his congressional seat, he was able to just about pass the job to his son, Harold junior, who had just graduated from law school. "People respect the Fords so much in this town," says a retired dentist, "it didn't matter that Harold junior had spent most of his life growing up in Washington, D.C. They have fought for us and served us well."

For many years, black society relied on one particular photographer to cover their events. The sons of a popular music teacher, brothers Robert and Henry Hooks started the Hooks Brothers photography business and passed it on to their own sons. Though he did not enter the family business, Benjamin Hooks further distinguished the family when he became the first black member of the Federal Communications Commission and was named head of the NAACP. His wife, Frances Dancy, also comes from an old Memphis family.

The early years of Memphis did not feature the large number of black businesses that one saw in Atlanta or Chicago during the 1920s and 1930s. While there had been black businesses on Beale Street and certain other small businesses around the South Side—like Eggleston's Tailors, Buffington Tailors, Service Drug Store, and Coyne Shoes—the only large businesses were funeral homes and insurance firms. My grandparents owned three different restaurants from the 1930s until the late 1960s, along with farms, as did many other families, but blacks were locked out of most other high-paying professions except for medicine and law until the early 1960s. Some of the small neighborhood stores like Prescott Pharmacy on Bellevue near Parkway were segregated: for whites only.

Equally segregated are the social activities of many elite black Memphians. There are many groups in addition to the small, exclusive Memphis Dinner Club, which included fewer than twenty women who got together each month, and included such names as Bland, Hayes, Owen, Rivers, Walker, Reed, Hooks, Speight, Young, Byas, Trigg, Daniels, and Beauchamp. Former club president Mrs. T. J. Beauchamp had also been president of the Negro Girl Scouts before the white council allowed black girls to join the Girl Scouts.

Several local clubs for the black elite predate the Dinner Club. They include the Iroquois Club, the Whist Club, and the Primrose Club. Among the members of Primrose were Mordecai Johnson, the first black president of Howard.

In the city, there are also chapters of national organizations such as the Links, which includes well-known family names like Price, Dancy, Prater, Hulbert, Latham, Horton, and Pinkston. The Girl Friends has members like Maria and Toya Pinkston, Frances Dancy Hooks, Erma Lee Laws, and Dr. Chrystine Shack. Laws, a former society columnist for the Memphis newspaper *Tri-State Defender*, says that Memphis society has always been a much smaller and more cohesive group than affluent black groups in larger cities. "I had many friends in cities like Chicago and Detroit who belonged to these groups," says Laws, "but the groups lacked the small-town intimacy that you find in a city our size."

Given what Laws says, it is not a surprise that the Boulé has only approximately fifty members. "But they are the people you'd expect to find," says Margaret Rivers, who used to throw parties and Boulé outings in the home she shared with her deceased husband, Dr. Frederick Rivers. Included in that group is Mayor Willie Herenton, the city's first black mayor, who was also the city's school superintendent. There are also chapters of the Smart Set and the Drifters. And so popular has the Links become in Memphis that two other chapters—River City and Shelby County—have been established. "But the real old guard is in the original Memphis chapter," says a woman as she pulls up to the curb near Second Congregational Church and waves to a fellow Links member.

Many of these organizations hold annual black-tie dinner dances at places like the Peabody Hotel, the Memphis Yacht Club, or other upscale locations in the city.

There has been a Jack and Jill chapter there since 1946. My cousin, Dr. Angela Owen Terry, remembers growing up with Tommie Kay Hayes in Jack and Jill. "Even though I went off to Spelman and Tommie Kay left for Fisk, we still remained close friends because of our early years in Jack and Jill together." Angela, who now lives in Connecticut, often calls Tommie Kay in California. They both remember that Alma Roulhac Booth's son, Christopher, was teen vice president of their chapter.

Today, the black elite still use Jack and Jill for their children, but they have access to other organizations as well. Ronald Walter, who had been teen president of the chapter in the 1960s, now has his three children in the group. His wife, Marianne, belongs to the Memphis Links, but she also belongs to the blue-blooded Junior League organization. "I joined in the late 1980s and we have raised a lot of money for charity," says Marianne Savare Walter, who grew up in Memphis and was the second black woman to join the group.

Many members of the Memphis black elite were around when the old Malco Theater used to require blacks to sit in the balcony or when some of the major department stores like Goldsmith's didn't allow blacks to try on certain kinds of apparel, and they now see a town that could have been much more important than it is today. The old Goldsmith's that used to be downtown has moved a half hour outside of downtown on a section of suburban Poplar Avenue where the number of blacks is very small. Beale Street, the historic downtown avenue that had been popular among blacks who socialized among their own when black jazz musicians were developing their craft in the 1920s, has now become a mostly white tourist destination where Elvis Presley T-shirts are sold. The street, which was once the locale for the city's first black bank and the first park to be named after a black man, now has a faux "New Orleans meets Disneyland" look to it.

Unfortunately, when it comes to integrating certain all-white institutions, the community has also been slow to change—even for the wealthiest blacks. For example, many of the white society clubs in Memphis—the Chickasaw Country Club, Colonial Country Club, Memphis Country Club, and Memphis Hunt and Polo Club—had no black members as of 1996. The University Club had one black member, and that was Ronald Walter. For all of its small-town qualities, Memphis still has managed to draw a hard line between its black and white residents.

Successful Memphis entrepreneur H. Arthur Gilliam represents a link to the next generation of elite Memphians. The son of a Universal Life executive, Art grew up in the black South Side of Memphis and belonged to Jack and Jill with the rest of the local elite. But unlike the prior generation, he was educated in northern white schools, attending the Westminster School in Connecticut and graduating from Yale (class of 1963), and from the University of Michigan with a graduate degree in actuarial science.

"Memphis is changing, but more of us need to be in business if we want to have a voice during this transition," says Gilliam, owner of radio station WLOK and the first black to own and operate a radio station in the city.

Memphis attorney Rita Stotts agrees. "We cannot become complacent after our parents have done so much to get us where we are," adds the Vanderbilt Law School graduate. "Successful blacks like us need to remember to give back and speak out so that we inspire the next generation. Just because I have enough money to move to a suburb east of Memphis doesn't mean that bigotry no longer exists or that black kids no longer face major hurdles in this town."

Although Memphis is still considered to be an important hub in the American South, its black professional community has not kept pace with the elite black communities in other southern cities that started after it. As many of my older relatives in Memphis often comment, they are surprised that successful young blacks would want to return to or relocate in a community that has taken so long to embrace a segment of the population who have wanted to contribute.

I love Memphis because it feels like a small town when I return there and walk the neighborhoods and streets that my parents and grandparents knew as children. But I also find myself resenting Memphis because its small-town nature prevented it from developing into the kind of community that Nashville, Charlotte, and Atlanta have become. It is a town that has the old organizations and infrastructure that make it possible for the old guard to feel rooted, but it still needs the kinds of businesses, neighborhoods, and black cultural and educational institutions that will provide a vibrant future. It is a place that is clearly only recently finding itself, and I am hoping that it does so completely before the blacks of my generation turn their backs on it.

Without a trace of a southern accent, Gilliam recalls his father's years as the first black on the board of Memphis Light, Gas, and Water utility. "My father loved this town, and I do too. But we have to stay connected and involved," the media executive explains, "because if you are black in the South—even today—it can be a humbling experience. We cannot get intimidated or complacent. Our parents gave us these advantages and we have to pick up where they left off. We have to do things that are important and productive."

Black Elite in Detroit

Whenever my wife and I return to visit her relatives in Detroit, I find myself driving through streets and neighborhoods that look nothing at all like the image that has so often been painted of this city. During one of my earliest visits with her—while we were still both students at Harvard Law School—her family took me around a city that featured a wide range of neighborhoods. Some were urban and depressed, but some looked suburban and upper-middle-class. One area at the end of the tour included a neighborhood of quiet, winding, tree-lined streets with well-preserved colonial and Tudor homes.

"This is *still* Detroit?" I asked as we drove through three different neighborhoods in the northwest corner of the city. I gaped out of the window in disbelief.

"Yes, it's *still* Detroit," my wife answered, mocking my tone of incredulity.

"And black people live here?" I was starting to sound like some of the condescending white children I had grown up with: always stunned to learn that *some* black people weren't destitute, perpetually caught off guard by the notion that you can't believe what you see on the six-o'clock news.

"Of course black people live here," she said. "Black people *and* white people live here."

Even though my family and I knew professional black families in most of the country's largest cities, we had few connections to Detroit. Because of this, I not only had relied on the conventional wisdom about this town, but had also been completely convinced by the media's mostly negative portrayals of the city as a place of burned-out buildings, vacant boulevards, abandoned factories, and depressing housing projects. To the dismay of my wife and her relatives, I had already embraced the notion that when people left the city for outlying suburbs like Southfield, Bloomfield, and the various Grosse Pointe communities after the 1967 riot, the things they left were General Motors, poor blacks, and decrepit homes.

As we turned up a quiet street and passed the Detroit Golf Club and Palmer Park, we were suddenly enveloped by a neighborhood of large gracious homes with automatic sprinklers spraying over manicured lawns. The street was Pontchartrain Drive. Thick trees shielded parts of the road from the sun.

"This is Palmer Woods," my wife explained as my eyes fell on a towering oak tree shading a deep lawn. "These neighborhoods—Palmer Woods and Sherwood Forest—have had black people living in them since the 1960s."

Since that first tour of the city, I have gotten a better sense of the blacks and whites who lived in this town before the tumultuous 1960s. After talking to blacks and whites who have lived there since that time, I also have a better understanding of how class played a major role in the city's past racial strife as well as in the healing process that has taken place under current mayor Dennis Archer—a well-respected representative of the city's black elite.

Throughout the late 1960s and 1970s, when my wife was growing up in Detroit, there was a general belief that whites and blacks could not coexist in the city under any circumstances. Even in her small private school, it was rare for blacks and whites to cross the color line for even the most superficial of social interactions. They shopped at different malls, joined different activities, took separate buses, and in certain years had separate school proms. The white children from the suburbs were afraid of their black classmates, and the black children from the city were suspicious of their white counterparts.

When I was in college, I recall that whenever I met white classmates from the Detroit area, they always clarified their geographical identification. Unlike students who hailed from Chicago, Washington, Los Angeles, New York, or Philadelphia, these Detroit-area students immediately made it clear to new acquaintances that they did *not* actually live inside the city limits: they lived in the suburbs. Even if those suburbs were working-class communities like Taylor, Southgate, or other poorer white areas that the old-guard blacks derisively refer to as "hillbilly heaven," these students were quick to point out that they didn't call the city of Detroit their home.

But even though whites seem to be quite forthcoming with their opinions on the city and the blacks who live there, black residents are not nearly so free with such commentary. In fact, I have not found a black community more reticent than the one in Detroit. More than those in any other city—including Chicago, Atlanta, New York, and Washington—black professionals here are unwilling to freely voice their opinions on the issue of class among the black population.

"The black upper-class community is very small in this city," explained an attorney in his forties who grew up in Detroit and now works for a downtown firm. "We're not like Atlanta or Chicago, where there are lots of old families or new transplants. Not many ambitious, wealthy blacks decide to relocate to this town. And even though the city is three-fourths black, only a handful of us are well-to-do, and nobody wants to offend anybody. Where would we go if we did? It's too small a community for you to get lost in."

Even with the introductions I received from my wife's family and personal friends like federal judge Damon Keith, I found dozens of Detroiters who eschewed any discussion of the city's black elite. Many would speak with me only on background without being quoted. Others insisted that their names not be used, even though they invited me to dinner at their homes. And still others simply refused to talk to me at all after I reached them and explained my project. Remarkably, my letters and calls to a Detroit-based national officer of Jack and Jill—a group I admire and grew up in—would not respond even after I left messages telling her that I had interviewed Jack and Jill presidents going all the way back to the 1950s. All of this formed a stark contrast with the elite families I grew up with and knew in other parts of the country. In those settings, people not only were unapologetic about their views on class, but were also proud to acknowledge their roles in their communities.

"Possibly because of the contempt that Coleman Young exhibited toward affluent blacks during his twenty-year term as mayor," says a former reporter from the *Michigan Chronicle*, "people have become conditioned to keeping their mouths shut. Coleman didn't like whites, suburbanites, or upper-class blacks. He was an angry fellow and he had no use for them. He put up with them while he was in the state senate, but not later."

In this city of one million people, where 75 percent of the population is black, Young's twenty-year tenure as mayor had a long-term impact on people's views regarding race and class. And although he died in 1997, his influence continues to be felt by many black residents who came of age in the early and mid-1970s, when Young was most outspoken. Menacing and resentful toward many in the black professional community in a way that exceeded the tone set by Washington's Marion Barry, he sent a clear message to the post-1970s black elite that they should keep quiet and stay out of his way.

Coleman's tenure reflected a fiery zenith in the city's racial history: he was the first elected official who refused to allow blacks to be walked on or ignored. "Coleman Young was quick to point out that the treatment of blacks under former mayors like Louis Mariani had been a disgrace," says Michael Goodin, senior editor of the *Michigan Chronicle*. "Mayor Young often declared that neither the city nor the major companies operating in Detroit made any efforts to invest in or support black businesses or black entrepreneurs here. In 1973, the year before he took over, the city and its major companies had spent not much more than $30,000 with black businesses. But by 1993, when he stepped down, more than $200 million was being spent with black businesses each year. Mayor Young was not afraid of letting bigoted white citizens or businesspeople know what they owed black people. And he was also not afraid of telling certain black people what they should be doing."

Judge Damon Keith of the U.S. Circuit Court of Appeals is a Detroit native and agrees that Coleman Young served an important role during Detroit's difficult times. "Many white residents left the city after the 1967 riot," explains Keith, who practiced law in the city before being appointed to federal judicial positions by Presidents Johnson and Carter, "but Mayor Young solidified the black community and convinced many of them with businesses to stay in Detroit rather than leave for the suburbs."

Although many in the black elite community benefited from

Coleman's message, some were offended by his coarse style and by his rumored resentment of old-guard blacks. Many of these people ignored him or avoided him. In fact, there is a whole old-guard community that was in place long before the Coleman Young regime, and it members seem to have lived their lives and conducted their activities very far above his radar. That community includes people like Mary-Agnes Miller Davis.

"Mary-Agnes created a lot of bright moments for us girls when we were growing up in Detroit," says Fredrika ("Rikki") Stubbs Hill as she recalls the years that she and her sister, Patsy, spent in Mary-Agnes Miller Davis's Co-Ette Club, an exclusive social organization that included girls from all the well-to-do black families of Detroit. Rikki says that she can barely recall the circumstances around her being selected "Miss Co-Ette of 1954," even though it was covered in all the papers. But what she does remember is the positive influence that Mary-Agnes had in getting young black girls to join this group—a black version of the Junior League—and volunteer their time for public service in the city's black community.

"The Junior League was all white back in the 1950s and 1960s," says a mother whose daughter became a Co-Ette in the late 1960s. "But Mrs. Davis showed the community that young black women could have the same impact. She whipped them into shape and brought out our best." Whether she is known as Mary-Agnes or Mrs. Ed Davis, few people are better known among Detroit's old-guard black society than she and her husband. Since the day they moved onto West Chicago Boulevard in the historic Boston-Edison neighborhood, they have remained important figures in the tightly knit circle of the city's west side black elite.

Ed became famous in 1939 when he became the first black in the United States to own an automobile dealership. Mary-Agnes, a longtime member of the Girl Friends, has remained prominent in social circles since 1941, when she founded her well-respected social club for young women.

Briefly interrupting our conversation at their stately brick home in order to admit a television producer and camera crew, she remarks modestly, "CNN is here to interview Ed because he was just selected for the Automobile Hall of Fame." A few moments of conversation about Studebakers ensues.

The Davises have been profiled and toasted at major affairs throughout the Motor City, so such situations are not out of the ordinary for Ed and Mary-Agnes. They are used to standing out and being in the center of things. Moving into a neighborhood that had included auto entrepreneur Henry Ford, inventor Thomas Edison, the Kresge retail family, and the auto-building Fisher brothers, they surprised a lot of white west-side residents in 1949.

"When we moved here, we were the first blacks in the neighborhood association, and the second black family to move onto the street," explains Mrs. Davis as her husband leads the camera crew to a downstairs playroom. "In fact, our street has the oldest residential neighborhood association in the country, and when we got here, the all-white membership met to vote on whether they should let us in."

Davis had no problem getting the house, even though the area had been a white upper-class stronghold in the center of a city that had a fast-growing black population. It wasn't long before she was being feted in the pages of the *Detroit Free Press* and *Michigan Chronicle* for the contributions she was making, and the crowd of corporate benefactors that she was attracting through the Co-Ettes.

Over the last fifty-seven years, more than twenty-five hundred young women have belonged to the selective club. Although the Co-Ette members—all in high school when they join—are responsible for raising money for charities like the United Negro College Fund, maintaining high grade point averages, and volunteering their after-school hours to community groups, they are best known for the annual Co-Ette Charity Ball, a cotillion that brings together the city's old guard.

Flipping through the club's fiftieth-anniversary album, one sees the expansion of Detroit's black society from one generation to the next: "I remember all the old Co-Ette members, like the Stubbs girls—Patsy and Rikki," says Davis as she looks at some old photos from the 1950s. "And there's Gail Burton. Her father, Dr. DeWitt Burton—we called him D.T.—gave her a coming-out party that got her on the cover of *Ebony* magazine. She's a doctor now. And there's Leslie Brown and Karen Heidelberg. Both of them are doctors too. And there's Trudy DunCombe, who was a Co-Ette in the sixties. She's a judge now and married to our mayor. Trudy's sister, Beth, is a partner at one of our biggest law firms. And Beth's husband Joe was law partner with Damon Keith."

I listen, impressed that no detail regarding the women she has feted is missed by Mary-Agnes.

"And there's Camara Jones-Singleton from the seventies. She went to Wellesley and graduated from Stanford Medical School. She was a Henry Luce Scholar and is a doctor at Johns Hopkins Hospital now. And there's Nancy Farmer from the sixties—she's a judge. . . ."

But of course there are those who are not enamored of Davis's group. "A bunch of Plymouth Congregational snobs," remarks an elderly mother who says her daughter was shut out of the group thirty years ago because she was too dark and didn't go to the right church.

"I could understand that they didn't want Baptists, or girls from any of those working-lass churches," says the woman, "but people told me we didn't get in because we weren't in the right Episcopal church. I was just devastated. I mean, Episcopal is Episcopal. What was I going to do then, just jump up and switch churches? Weren't money, breeding, and good looks enough for a fifteen-year-old?"

Mary-Agnes admits that Co-Ette meetings were often held at Plymouth Congregational—a church that she and many well-to-do blacks attended—but it was by no means a factor in the selection process. Indeed, many of the young women in the club over the last six decades have belonged to a wide range of churches in the city.

Despite occasional snipes from onlookers who miss the point of black elite groups like the Co-Ettes, Mary-Agnes, along with her sister, Dora Miller Reid, continues to galvanize resources from the black and white community while continuing to teach girls the importance of public service.

Another experience that contrasts dramatically with the antiwhite, anti-black-elite mood that was later sometimes advanced by Mayor Coleman Young was that of Marge Dunbar Yancey, who grew up in the city when blacks were still a small percentage of the overall population. Yancey remembers that her experiences were much more racially integrated than those of her children thirty years later. "Although there were parts of Detroit that were exclusively white in the 1930s and 1940s—like the northwest part of the city," explains Yancey, who has fond memories of the years she spent as a Detroit adolescent, as a Wayne State University student, and as a teacher in the Detroit Public Schools, "I always had both blacks and whites in my east-side schools and neighborhoods. There were certainly a few restaurants along Woodward Avenue that would not seat blacks—and I remember when one of my white female friends in college asked me to join her for dinner and we got turned away at the door—but for the most part, there was interaction between the races."

Although her life is now in Atlanta with her husband, Asa Yancey, a prominent Atlanta surgeon, and includes her activities as a Spelman College trustee, Marge Dunbar grew up surrounded by Detroit's black middle and upper classes. Her family attended the black society church, Plymouth Congregational. Her father, Henry Dunbar, graduated from the Detroit College of Law, and in addition to being an attorney, held such positions as director of the Detroit YMCA and manager of the Brewster Public Housing Development during its very early and successful years as a big-city public housing project.

"From the time I was eight years old, when we moved to Grand Boulevard and Woodward Avenue near the Fisher Building," says Yancey, "I remember having a completely integrated life experience—nothing like the way people describe the city today. Breitmeyer Elementary and Northern High School were both integrated with blacks and whites from different economic groups. Back then, you saw whites without going to the suburbs—after all, blacks never ventured into the suburbs then. Of course this was dramatically different from today's Detroit and from the segregated Atlanta school experience my own children saw when they were growing up."

Nancy Tappes Glover, a member of the Detroit Girl Friends, represents a newer generation of Detroit residents who arrived at a time when blacks were allowed to live in the surrounding suburbs. She and her husband, Dr. Frank Glover, move comfortably between their home in suburban Dearborn, which lies on the east side of the mostly black Motor City, and the Girl Friends' events held inside Detroit. "Like most social clubs here, we hold many of our activities in the city as well as the suburbs," she explains. "Whether it's a picnic in a member's large suburban yard or a formal gathering at the Rooster Tail, downtown, overlooking the river, we try to take advantage of the best of this city."

For years, it was said that there were forty influential families who made up Detroit's early black history. Arriving mostly from Virginia in the 1880s when there were fewer than three thousand blacks living here, these families earned a living as attorneys, physicians, business owners, and investors in real estate. The names best associated with this group were Hackley, Beard, Ferguson, DeBaptiste, Cook, Purvis, and Pelham.

Best remembered of this group was the Pelham family, which today has its name on a school in the downtown area because of its ties to newspaper publishing, government, and education.

"My husband, Alfred, graduated from Harvard Business School in 1922—the second black to attend—and then was later the director of budget and finance for Wayne County and for the city of Detroit under Mayor Cavanaugh," recalls Doris Pelham. "His father was born in downtown Detroit in 1862 and got involved in local government at a time when there were very few blacks here." Probably the best known of the Pelhams, Doris's father-in-law, Benjamin, was appointed to the Wayne County clerk's office in 1895 and became county accountant in 1906.

A light-complexioned black man who practically controlled Wayne County—which included Detroit—and its finances by the time he retired in 1937, Ben Pelham was often referred to by black and white Detroiters as the czar of Wayne. Some of his power also derived from the *Detroit Plain Dealer*, the newspaper he published. Like many of the "top forty" black elite families, the Pelhams arrived in Detroit from Virginia in the 1850s. Today Pelham descendants still reside and work in the Detroit and Grosse Pointe area.

"To understand the experiences of those first blacks in Detroit," says *Michigan Chronicle* editor Michael Goodin, who grew up in the city's Russell Woods neighborhood, "you have to look at the old distinctions between the city's east side and west side. When I was born in the early fifties—and during the time that my parents and grandparents were here in the 1930s and 1940s—most of the homes and businesses owned by blacks were located on the east side. As blacks became more affluent and as restrictive covenants were challenged, we were able to move farther and farther west."

The borderline between east and west in Detroit is the major thoroughfare Woodward Avenue. "You didn't find blacks living west of Woodward in the thirties and forties," explains Joseph Brown, a Detroit attorney who grew up in the city. "Blacks lived, worked, and socialized in the neighborhoods around Hastings Avenue."

That neighborhood of thriving black-owned establishments was alternately referred to as Elmwood Park, Paradise Valley, or the Black Bottom. Before the 1950s, all blacks—whether working-class or upper-class—conducted their business and social activities in this east side area.

"The Black Bottom-Paradise Valley neighborhood was an entrepreneurial oasis for blacks," explains Michael Goodin. "Black doctors and lawyers, like John Roxborough—Joe Louis's attorney—had offices there. Ed Davis's car dealership, the Paradise Theater, Sonny Wilson's Bar, Sidney Barthwell's drugstores, insurance offices—all the black-owned businesses were there. It's ironic, but segregation helped us to build our own businesses."

When large numbers of blacks were arriving in Detroit from the South in the 1920s, they established neighborhoods on the east side of the city close to downtown. They chose an east-side street, Hastings, as a main thoroughfare for their stores and businesses. Diggs Funeral Home, an upscale parlor, was just off Hastings. John R. Street was the area for many of the nightclubs and bars.

Great Lakes Insurance Company, one of the city's largest black-owned firms, had its own building on Woodward Avenue before it was folded into a large insurance parent.

Owned by the Sengstacke family, the *Michigan Chronicle* newspaper has been an important link among members of the black community.

Another institution was the Gotham Hotel, considered one of the most elegant black hotels in the country. "Harlemites like to say that the Hotel Theresa was elegant," remarks Judge Damon Keith, "but it couldn't light a match to the Gotham and its lobby with oil portraits of W. E. B. Du Bois, Mary McLeod Bethune, and other black historic figures." Owned by John White, the Gotham hosted black fraternities and celebrities like Lena Horne and Joe Louis, since white hotels in the downtown area didn't admit black people or their organizations until the late 1950s.

An influential family that ran many of their businesses in the black east-side neighborhood were the Diggses. Charles Diggs Sr. headed Diggs Enterprises and owned a string of businesses including an insurance firm, a limousine company, the House of Diggs Funeral Home, and the Diggs House of Flowers on Mack Avenue. His son, Charles junior, graduated from Wayne State University and the Detroit College of Law before being elected to the Michigan State Legislature. Elected to the U.S. House of Representatives in 1954, he remained a congressman for more than two decades.

Although people like the Pelhams and the Diggses became a part of the elite because of their activity in government and public service, most of Detroit's black elite developed through the legal and medical professions, which served their own community exclusively, since many of the other high-paying employment opportunities were closed to blacks. The city's largest industry—automobile manufacturing—did not offer well-educated blacks office or management positions because until the early 1940s, many of the white-run auto unions were resistant to giving blacks opportunities. It wasn't until blacks were hired as strike-breakers that they got a serious foothold in the industry. Unlike Chicago, Washington, and Atlanta, Detroit in its early years—the period preceding the 1930s—lacked the sizable black population that could support the early growth of major black-owned banks and insurance companies. Another reason the medical and legal professions dominated the elite was that the city lacked an important black educational institution that could attract and establish an intellectual elite such as the ones flourishing in Atlanta, Washington, and Nashville.

Among the prominent physicians were Dr. DeWitt Burton, who headed Mercy Hospital; Dr. Remus Robinson, who was one of the first blacks on the board of education; Dr. Alf E. Thomas Jr., who was a major NAACP fund-raiser and the husband of Jack and Jill founder Marion Stubbs Thomas; Dr. Darnell Mitchell, who was married to Helen Cox Mitchell, a prominent socialite and schoolteacher; and Dr. Marjorie Peebles Meyers, who was in private practice before joining Ford Motor Company. Although favored in the east-side black community, these "establishment" black doctors also had great access to powerful whites.

Deborah Fitzgerald Copeland grew up in one of the establishment doctor families in the Boston-Edison neighborhood, a west-side neighborhood where many of the old-guard blacks began moving in the 1950s. "A lot of the kids I grew up with in Jack and Jill and the Co-Ettes in the fifties and sixties were living in integrated settings and attending integrated schools. Most of these black friends, however, were going to the same black churches," explains Copeland, who grew up attending the establishment Plymouth Congregational Church.

Although she now lives in the Palmer Woods neighborhood, a far northwest section of the city that would have been off-limits to even prominent blacks like her parents forty years ago, Copeland and her husband carry on the traditions of old-guard Detroiters: Her kids are in Jack and Jill, her husband is a member of the Boulé, and she is now in her mother's Girl Friends chapter. "My mother was president of the chapter in 1959," she says, "and I recently became president too because it's important for the next generation to keep supporting groups that have aided our community." In her role, she has helped raise money for Meharry Medical College, as well as many charities in the city.

Like those prominent physicians, certain lawyers in earlier years were able to bridge the east-side black and west-side white communities, including Hobart Taylor, a prosecutor and White House staffer in the

Johnson administration; George Crockett, who served as a judge after holding high labor law posts in the United Auto Workers union; Judge Elvin Davenport, who served on the court of common pleas; and John Roxborough, who represented the prizefighter Joe Louis.

The older of these professional men were in the Detroit chapter of the Boulé, which was founded in 1917 by nine doctors and one engineer. During the 1950s, many of their wives and daughters established local chapters of the Links and the Girl Friends. The men who came of age in the 1950s and 1960s joined social groups like the Rogues, a local group, and the Guardsmen. Old-guard names like Dickson, Arrington, Mitchell, Milton, Scruggs, and Loomis appeared in these organizations. Surprisingly, there was one name that almost never appeared in these rosters. And although that name is almost synonymous with Detroit, the Gordy family, which founded Motown Records, was never embraced by the city's black professionals.

"When Berry opened Motown in 1959 on West Grand Boulevard," explains a retired auto executive, "he lost any hope of moving into the upscale black crowd. Since his family already owned businesses—cleaners, bakery, drugstores—a lot of us already knew them. But Motown appealed to this lower-class element that made it impossible for us to ever really let him into our circle."

Although Motown eventually grew into America's largest black-owned company, neither Gordy nor his recording stars ever became a part of Detroit's black society. *Michigan Chronicle* editor Michael Goodin agrees on this assessment of how Detroit's black society viewed the Motown celebrities. "Their national stature did not translate into status among the elite. They were perceived as a jitterbug crowd that wealthy blacks didn't appreciate."

"Music people would never have fit into my parents' crowd," says a third-generation resident of Detroit who remembers her family's circle of friends, who belonged to the Links, the Deltas, and the Omegas. "The Motown people were uneducated entertainers, so they were somewhat coarse. We certainly asked them—as celebrities—to sit on the dais at our events—you know, for publicity purposes—but that's as far as things went. The best of them was Diana Ross because she wanted to improve herself, and just look how well she's done." The woman pauses for a moment and stares out her living room window at a mostly white tour group walking through her neighborhood. Her mind drifts back to Diana Ross, and she continues, "But then again, she was a bit coarse too, because you know she also grew up in the Brewster Housing Projects. My parents were professionals, so they certainly weren't talking to people in the projects."

Because my visits to Detroit often bring me together with professional blacks who belong to the national groups that my wife and my family participate in, I find myself socializing almost exclusively in bucolic neighborhoods that don't look anything like the descriptions that most people hear when this city is discussed.

In fact, much of Detroit's current reputation comes from the devastation suffered from the 1967 riots, and the way in which the city moved from being a prosperous city with black and white residents to a stagnant metropolis struggling to keep uneasy white residents and the auto industry within its boundaries. Fighting hard to hold onto a population of a million people, black professionals in Detroit say they are starting to see a renaissance that makes the loyal old and new members of the black elite grateful.

"We finally have one of our own running this city," explains a retired Detroit physician as he stands in the foyer of a spacious Tudor home in the exclusive northwest neighborhood of Palmer Woods. "I don't blame the whites and corporations for leaving during the 1970s and 1980s. Before Dennis Archer got here, people saw this city as an anti-white town that was being run by an angry, unsophisticated black mayor whose head was still in the 1960s."

Like many well-to-do blacks who have remained in Detroit or who live just outside in Southfield, Bloomfield Hills, or the other suburban areas, this physician says that former mayor Coleman Young gave blacks pride and direction during the 1960s and 1970s when he was a state legislator and mayor because he stood up to the racist whites who ran many parts of the city establishment, "but he overstayed his welcome. He was loud and difficult from the moment he got elected in 1973. I'm sure that he had to be tough in that job during the early years, but he went too far. He despised successful blacks just as much as he disliked the whites who had left for the suburbs. But plenty of us stayed because we believe in this place. My wife and I could have run to the Pointes, but this is home."

It is obvious why the black elite quietly call the city's new savior, Mayor Dennis Archer, one of their own. Although he did not grow up in one of the privileged families, he has clearly moved into their circles. A former Michigan Supreme Court judge, he is a member of the Alphas and the Detroit chapter of the Boulé and is an articulate and capable liaison with the corporate and mostly white suburban communities. His wife, Trudy DunCombe Archer, also an attorney, has long been a part of the elite and is a member of the Detroit Girl Friends and the Detroit Links.

More typical of a local member of the elite who gained national stature was Wade McCree, who became U.S. solicitor general under President Jimmy Carter. A graduate of Fisk University and Harvard Law School's class of 1944, McCree served as judge of the Michigan Circuit Court, judge of the U.S. District Court and then a judge on the U.S. Court of Appeals in the sixth circuit. His daughter, Kathleen McCree Lewis, is now a partner at Dykema Gossett, one of Detroit's best-known law firms.

Like Kathleen and her father, other current elite in the legal profession include recent mayoral candidate Sharon McPhail and federal judge Damon Keith, who sits on the U.S. Court of Appeals. I first met Keith, one of my mentors since graduating from high school, through his support of a special NAACP competition that brings together black high-school students and black professional role models from around the country to compete in the areas of essay writing, engineering, architecture, and chemistry.

"What I've always liked about the black professional community in Detroit was that it was always small and intimate enough that others in our group took a personal interest in helping each other in their careers," says Keith, who was mentored by such Detroiters as 1918 University of Michigan graduate Alfred Pelham. A former Wayne County official before joining the federal bench, Keith has been honored by virtually every major black or white Detroit organization in addition to receiving honorary degrees from schools like the University of Michigan, Morehouse, Georgetown, and Yale. "When I began practicing law here in 1949, all the black attorneys were in the same building at 1308 Broadway—and almost all of us were criminal lawyers. We belonged to the same groups and mentored each other because we're all we had."

Today, the medical and legal communities in Detroit are not what they used to be. They are not as tight-knit as they were when segregation kept all the doctors at DeWitt Burton's all-black Mercy Hospital or all the attorneys in the Tobin Building on Broadway. Now, some are in white firms and some are in black firms. Some are downtown and some are in the suburbs. In addition, they no longer form the wealthiest black cliques that once ruled the community.

"Black-doctor money hardly reaches my radar screen," says a black resident who owns a firm that supplies products to the auto industry. "People like me who own businesses have turned this town upside down because we make real money and can afford to not only to move into the top suburban neighborhoods, but also to outspend the white neighbors."

Like this self-admitted millionaire manufacturer, there is a growing group of wealthy black entrepreneurs who have somewhat eclipsed other professionals because they own businesses tied to the city's primary industry. In fact, although the auto business had been closed to blacks in its earliest years—with occasional jobs being handed out through an odd patronage system that involved Henry Ford and William Peck, a high-profile black minister—the most successful black businesses today serve the auto industry as vendors and suppliers. "Henry Ford was never enamored of black people, but he was friends with Reverend Peck," explains Michael Goodin, "and when Ford visited Peck's church, the reverend would sometimes point out some lone person in the pews who had lost his job, and he'd ask Ford to give him a job in the assembly plant. It was a condescending and insulting approach, but that's how those first jobs were received by blacks."

Today's black Detroit business leaders are now taken quite seriously because of their ties to the industry and their important role in the auto fields. In fact, among the twelve largest black-owned business in the Detroit area, nine of them produce components or services for the auto manufacturing industries. Former Detroit Pistons basketball player Dave Bing owns and runs a $190 million empire that includes a steel processing company, a firm that manufactures and assembles automobile bumpers, and another that manufactures foam for cars and trucks. Founded in 1980, the Bing Group employs over six hundred people and includes Ford, General Motors, and Chrysler among its clients.

Also in the city is Mel Farr, who owns a $600 million auto dealership empire that sells Ford, Toyota, General Motors, Volkswagen, Hyundai, and Mazda cars. Founded in 1975, the company employs over eight hundred workers and forms the largest black-owned dealership in the United States.

Although he now spends most of his time in New York, former Ben and Jerry's chairman Bob Holland has owned Detroit-area auto manufacturing and beverage distribution businesses for many years. Before serving as chairman of Spelman College's board of trustees, he was a partner at the international consulting firm McKinsey and Company.

For the last thirty years, Don Barden has been in the media, cable television, and real estate businesses, and he now owns five radio stations as well as a computer-based education service that is used in over five hundred school districts. His wife, Bella, is an attorney who graduated from the University of Michigan Law School, sits on the board of the Museum of African American History, and runs their real estate company, Waycor Development. Among other structures, Waycor has built apartments, a medical clinic, and a detention center. One of its recent projects was a joint venture with New York developer Donald Trump to develop a riverboat casino business on Lake Michigan in Gary, Indiana, which serves over eight thousand people a day.

"But people around here started getting a little jealous of Barden's power and money," says a former *Michigan Chronicle* reporter, "and they kept him from opening a casino within Detroit. They gave the green light for white casino operators like MGM, but not to him." Although Barden was blocked from establishing the first black-controlled casino in Detroit, he remains an influential player in both black and white circles.

Other prominent Detroit blacks in business today include William Pickard, chairman of Regal Plastics, a $40 million injection molding company; Donald Davis, head of the $100 million First Independence National Bank; Jon Barfield, a trustee of Princeton University and head of $62 million Bartech, an employment staffing company; and Donald Coleman, owner of Don Coleman Advertising, a $110 million advertising agency with one hundred employees based in Southfield. A graduate of the University of Michigan, Coleman serves General Mills, Kmart, Ameritech, and Shell Oil among his client base.

Just as the wealthy have followed a certain path from one group of professions to another, they have carved a specific path through certain residential neighborhoods as they have become wealthier.

"When I first arrived in Detroit, the two top addresses for well-to-do blacks were on Chicago Boulevard and Boston Boulevard," says Robin Hamilton Sowell, who moved to Detroit in the 1940s and now lives in nearby Southfield with her husband, Myzell, an attorney who was born in the city. "Most of those large homes had originally been occupied by whites," explains Sowell, "but as neighborhoods ended their restrictions on selling to us, many of the black physicians and attorneys moved onto those two streets." Over time, as blacks became even more affluent, they moved west on those same two streets, into even more prestigious neighborhoods. So important were these two streets to a black person's status that when the society columnists of the *Michigan Chronicle* reported on black elite gatherings or family celebrations, they would regularly list the specific Chicago Boulevard or Boston Boulevard address of the host in the newspaper.

A member of the Detroit Links remembers the parties that socialites Andrew and Beatrice Fleming Jackson used to host forty years ago in their home on Boston Boulevard. "We didn't show up at their Christmas parties if we weren't driving a Cadillac or wearing a mink. That was our version of Park Avenue, and we were unapologetic about the display of excess back then."

The movement of many of the affluent blacks can be tracked by observing where middle-class Jews had moved. "At first, most blacks were in the southeast," says an attorney who lives in the Rosedale Park area. "Then we moved into Jewish homes that were vacated in the southwest. Then, as the Jews started moving farther north on the west side, we followed them—always just a few years behind."

But even as affluent blacks spread out farther away from their core neighborhood, which began in the Black Bottom at Hastings, they inevitably came back to shop, eat, and socialize at black-owned establishments in the old neighborhood. What put an end to the bustling black neighborhood along Hastings was the construction of Interstate 75—the Walter P. Chrysler Freeway—which was built right through Hastings and all of its commercial establishments in the 1960s. Just as the city planners had done in other cities like Atlanta and Houston, Detroit officials dealt a lethal blow to a thriving black neighborhood when they razed businesses and constructed the four-lane highway.

Another major change to the black neighborhoods and the few upscale integrated sections occurred in 1967 after the police raided a black social club known as the Blind Pig. What began as a police raid and altercation between white officers and black patrons soon turned into the now-famous Detroit riot that led to mass burning and looting of property throughout the downtown area. The 1967 riot created dramatic consequences for whites and affluent blacks. Black middle-class families had already entered the mostly Jewish west-side Russell Woods neighborhoods in the early 1960s. But now whites who were, at first, a little uncomfortable about living near blacks were downright scared of living near blacks of any class. Whites fled to the suburbs, and the liberal, affluent whites ran to the northwestern edge of the city. As these whites left their homes—many of them still carrying restrictive covenants in their deeds—the residents sold them to blacks, who were now finally permitted to live in the best city neighborhoods. Prior to that time, most of the wealthiest of blacks were living in completely segregated black neighborhoods.

Although many members of the black elite stayed within the city limits after the riot, they moved farther northwest, completely populating Russell Woods and venturing into the more upscale Rosedale Park.

And just as the black elite preferred certain new neighborhoods, they favored certain schools for their kids. In the 1960s and early 1970s, the preferred public schools were Mumford High and Cass Technical High School downtown. "I entered MacKenzie High in 1963," says Michael Goodin, who grew up in Russell Woods, "but there was already a caste system that said that the high-yellow, penny loafer and khaki-wearing black doctors' kids were at Mumford, while the jitterbug-Temptations-listening and conk-wearing blacks were at my school." Some people called the Mumford blacks "e-lights" because they made up the light-skinned elite.

My wife, who never attended the city's public schools, recalls that by the time she reached high school in the late 1970s, Mumford had been displaced by Detroit Country Day and Kingswood-Cranbrook, two private schools outside the city, as the preferred school for old-guard children. "Some Detroiters, though," explains her cousin, Leslie Hosey Robinson, "try to find good schools within the city limits—places like University of Detroit High School, Renaissance, or Gesu—because they are usually more racially diverse." Robinson, a psychologist who attended the Gesu School in the early 1970s before going on to Mount Holyoke College, recalls that her father, a Detroit dentist, saw great value in sending her brothers and sisters to good integrated private schools *within* the city limits.

Damon Keith's daughter, Debbie, recalls that by the mid-1970s it was rather common for black Detroiters to attend private schools outside the city. "I was living in Detroit, but I went to Kingswood-Cranbrook in Bloomfield Hills," says the media executive, who returned to her hometown after graduating from Princeton University. Like the pricey Kingswood-Cranbrook, Country Day and Mercy are also outside the city boundaries; the Friends School is one of the few remaining prestigious private schools within Detroit. Palmer Woods resident Deborah Copeland says that the emphasis today is definitely on private schools.

"When I was in high school in the sixties, the popular schools were Central, Mumford, and Cass, but I would say that two-thirds of my children's acquaintances are in private or parochial schools."

The even more prestigious neighborhoods of Greenacres, Sherwood Forest, and Palmer Woods opened up for the black elite slightly later, in the late 1960s and early 1970s.

"There is no doubt that the greatest concentration of Detroit's black elite today live in either Sherwood Forest or Palmer Woods," explains a real estate broker who lived in northwest Detroit before moving to Birmingham, an upper-middle-class suburb just outside of the city's boundaries. "While the three neighborhoods are almost indistinguishable," says the broker, "Palmer Woods is slightly more prestigious, with larger homes." Laid out on winding roads with names like Canterbury, Strathcona, Lincolnshire, and Gloucester, the 1920s Tudor and colonial houses are built on spacious lots. The exclusive northwest neighborhoods are buffered by the greenery of Palmer Park, Detroit Country Club, Evergreen Cemetery, and the Palmer Park Golf Course. The most famous street in the area is Fairway, which runs along the Detroit Country Club and includes the home of millionaire entrepreneur Don Barden. Located on the northern edge of the city limits, like the gold-coast upper Sixteenth Street neighborhood in Washington, these streets still have white residents who form the more liberal white coalition still living in the city.

Psychologist Leslie Hosey Robinson lives in a Greenacres colonial with her husband and children and is an unabashed supporter of the city's residential areas. "Many people don't realize how many choices there are when they are looking for nice neighborhoods within the city limits," says Robinson.

In addition to the northwestern section of the city, there is increasing interest among the black elite in east-side and downtown locations. "There are many professionals who want to live closer to downtown, so they move to Indian Village, the Marina, Lafayette Towers, or Harbortown," says Robinson.

Indian Village, a neighborhood of large older homes built in the early 1900s, does have some black residents. But it is one of the mostly white city neighborhoods that help to make up the city's population of 250,000 whites.

When well-to-do blacks first had the opportunity to move beyond the city limits in the mid-1960s and early 1970s, the choices were ini-

the auto industry. The act was so blatant and insulting to Roy Roberts, en head of the General Motors Truck Division, that it drew public critism from Jack Smith, chairman of General Motors. Smith and others ropped their memberships and denounced the blatant racism of the club's decision.

Although Grosse Pointe is much closer to Detroit—bordering the city on its eastern edge—it is a community that was even more aggressive about keeping blacks out of its neighborhoods. It is often referred to as "the Pointes," and there are actually five communities—Grosse Pointe, Grosse Pointe Farms, Grosse Pointe Shores, Grosse Pointe Woods, and Grosse Pointe Park—that make up the twenty-square-mile suburban area. If Bloomfield Hills is "new money," then Grosse Pointe is clearly "old money." This is where the automotive barons built their estates along Lake St. Clair during the early 1920s. There are many other old and new million-dollar homes throughout Grosse Pointe.

"There are a lot of middle-class white families scamming over there in Grosse Pointe Park," explained a Bloomfield Hills mother who tells of her difficulty in getting a broker to even drive her and her husband through parts of Grosse Pointe in the early 1970s. "When I saw the brokers refuse to even show me the middle-class housing in Grosse Pointe Park—homes that were well below my intended price range—I realized that the Pointes were very different from Southfield and Bloomfield." As the woman coasts through the tree-lined streets in her car, it becomes obvious that there is a pecking order among the five sections of Grosse Pointe. They are not all as uniformly elite as one would expect. "Yes, they've got multimillion dollar homes in the Shores, the Farms, and the Woods," says the woman with annoyance in her voice as she drives down a street named Kercheval, "but it just burns me up that they wouldn't even let me bid on this stuff in the Park! And not that I would ever live in a house as small as those."

Given Grosse Point's close proximity to the city, many of its realtors had reportedly established a sophisticated method by which they kept out the nearby Jews, blacks, and other minorities for many years. This was a well-known point system for scoring potential buyers by evaluating their ethnicity, skin complexion, accent, and other characteristics. This formal system, which continued until the 1960s, is just a hint at this community's determination to keep Detroit's black population outside. Even the public parks are closely restricted: you must be a resident with a permit.

tially quite slim, but they most often settled on Southfi
included the popular Northland Mall and that borders
edge of Detroit. Similar to the New York suburb of New
ethnic and economic makeup, it had a relatively subs
educated Jewish population that was more racially open minc
other nearby suburban communities. In fact, today Southfield,
is often listed as among the more welcoming towns in the Unit
for professional blacks. Sociology professor Andrew Bever
Queens College in New York, conducted a study in the early 1990s
identified cities where the black median household income exceede
income of white households in the same area. With their large nun
of doctor- and lawyer-headed black families, Southfield blacks earned,
average, 20 percent more than their fellow white residents.

The downriver working-class white towns, or the "bubba suburbs"—as many members of the black elite refer to communities like Lincoln Park, Allen Park, Melvindale, and Trenton—have never been desirable for black professionals. The other middle-class and upper-class suburbs each have their own interesting story and thus reveal why Southfield had for so many years remained a favorite of blacks with money.

Dearborn, the home of Ford Motor Company, for example, is a much more welcoming place today, but throughout the 1950s, 1960s, and early 1970s, the city had a mayor, Orville Hubbard, who very openly expressed his views about keeping Dearborn a white community—no matter how many blacks moved into neighboring Detroit. As he pointed out in the late 1960s, "I don't believe in integration. It leads to socializing with whites, intermarriage, and then mongrelization." So strong was the antiblack sentiment that had carried over from the Dearborn mayor's thirty-six years of vitriol that blacks were even uneasy about shopping in the town's Fairlane Mall when it opened in the late 1970s. Today, the town is far more accepting of black residents.

The other two suburbs that would naturally have attracted the interest of affluent blacks are Bloomfield Hills and the Grosse Pointes. Bloomfield Hills, a community that is due north of Detroit, is considered to be the more progressive of the two. A very affluent suburb, it is also quite far from downtown and is considered a haven for "new money" as compared with the more aristocratic and Waspy Grosse Pointe. Bloomfield Hills received an ugly bit of national publicity in the early 1990s when the Bloomfield Hills Country Club refused to accept a black applicant who also happened to be the highest-ranking African American

While many of these suburbs have a handful of black families, there are white residents who remember how Detroit was transformed from a 70 percent white population in the early 1960s into a 70 percent black population by the 1980s. Some worry that suburban demographics could change too.

Many city dwellers who did not escape to the suburbs did choose to escape to a resort in northern Michigan called Idlewild. In response to the black elite's inability to vacation at white resort areas, many well-to-do Detroit blacks helped to establish this midwestern version of Oak Bluffs. Beginning in the early 1900s, blacks such as the intellectual W. E. B. Du Bois made the three-hour trek up to Idlewild and built vacation homes around the lakes in the area. "My parents started bringing me there when I was four years old," recalls Detroit attorney Joseph Brown, who remembers seeing celebrities such as Sammy Davis Jr. and Dinah Washington perform at the Paradise and El Morocco Clubs. "During the forties, people went not just for the quiet fishing and the horseback riding," Brown explains, "but also for the nightclubs and the bars."

Robin Sowell says the area is not as popular as it used to be. "When people started going to integrated resorts or east coast vacation areas like Martha's Vineyard and Sag Harbor," explains the Southfield resident, "Idlewild lost its popularity. Now, it's mostly a retirement area. My husband does continue to go in August just to reunite with old friends." Like Sowell's husband, others attend a special Idlewild Week each August where special parties and groups assemble from Chicago, St. Louis, and other midwestern cities.

Since Detroit did not offer the same variety of entertainment as cities such as Chicago and New York, private parties and social gatherings were more sought-after—and because of this, black society in the city was reported on with unusual frequency.

"If Myrtle or Grace didn't report on it, then it wasn't worth attending," explains a lifelong resident of Detroit, as she recalls the social events covered by Myrtle Gaskill and Grace Sadler, society columnists for the *Michigan Chronicle,* Detroit's black weekly. "But God help you if you snubbed them or certain elements in the black community that they protected as their own. I'll never forget what Grace did to us when we hosted a national conference of Deltas and used the Sheraton Cadillac hotel, a white-owned building, instead of the black-owned Gotham

Hotel." The woman pulled out a yellowed clipping from a bright-red photo album with Delta Sigma Theta emblazoned across the cover. The article had the date "January 1957" written on the bottom.

It was one of Grace Sadler's columns, and in between its high-society terms of endearment, Sadler had woven in a tough problack message that was rare in the society columns of other black city papers at that time. She chastised members of Detroit's black elite sorority for not supporting the city's only black-owned luxury hotel. It began with the words, "'Will you love me in December as you did in May?' but alas the response was in the negative as far as the Delta sorority convention was concerned. Nearly one thousand Deltas met in Detroit last week . . . but no one dreamed that a Negro-owned hotel of Gotham caliber could have been so completely snubbed by any local or national body of Negroes—in spite of integration!"

With less of a political edge, throughout the 1950s Myrtle Gaskill was there, writing about the annual one-hundred-dollar-a-plate NAACP Freedom Fund Dinner at the Latin Quarter or other locations and attended by people like Judge Elvin Davenport, the J. J. McClendons, Dr. and Mrs. Alf E. Thomas, and Dr. Wendell and Iris Bell Cox.

The biggest event of its kind anywhere in the country, the NAACP Freedom Fund Dinner is still a major event among the black elite each spring. "The event has grown so large with politicians and famous people," said Robin Sowell, who has been a frequent guest at the formal affair in recent years, "they moved it to Cobo Hall. This is a must-attend evening for the governor, the mayor, and other state officials. It's one of the Detroit events where blacks and whites come out in equal numbers." Nancy Glover agrees, "I think we raise more money at our NAACP dinner than at any other NAACP in the country. We get more than two thousand people here."

Other popular gatherings include the annual Barristers Ball, a black-tie affair hosted each February by some of the top lawyers, judges, and members of the local Wolverine Bar Association. An even more select crowd from the black elite attend the Cotillion Club Debutante Ball each year, which is sponsored by the Detroit Cotillion Club, a group of prominent businessmen who sponsor two or three dozen debutantes. The extravagant affair features two runways for the debutantes and escorts, who are initiated and rehearsed during a ten-week preparation period. A feature of the televised event includes the selection of a Miss Cotillion Club. Those selected in the past include Dykema Gossett attorney Nicole

Lamb-Hale, who attended Detroit's Gesu School, the University of Michigan, and Harvard Law School. A Jack and Jill alum and member of the Deltas, Lamb-Hale was typical of the bright and socially prominent debutantes to be presented.

"When we started the Cotillion Club in 1950," explains Judge Damon Keith, whose daughter debuted while in high school, "we did it because there were no businesspeople sponsoring debutante balls where black girls could debut. We wanted our daughters to be taken seriously too."

"I don't know if I really like these cotillions, even if they raise money for charity," remarks a father of two adult daughters during a dinner I am attending in northwest Detroit. "I grew up here and I think it's a throwback to those high-yellow blacks that were all a part of that Marion and Alf Thomas crowd."

A woman looks over at him skeptically. She is there with two Delta sorors who also grew up in the city. "Oh, you're still smarting because you didn't get invited to the wedding."

I look at the woman as she drinks from her wineglass. "What wedding?"

The woman looks at her two girlfriends and says, "He knows what wedding. I mean *the* wedding."

It seems that there has been virtually no black Detroit party that created more excitement (for those invited) and more dissension (for those not invited) than the 1958 wedding reception for Patsy Stubbs and Harold Fleming, which had nearly one thousand guests. Thomas's mother, Marion Turner Stubbs Thomas, and her stepfather, Dr. Alf Thomas, were stylish, high-profile, and well respected among the Detroit elite. Marion, who had made her name twenty years earlier by founding the elite children's group Jack and Jill, was the daughter of a Philadelphia physician and had been left a widow in the 1940s when her first husband, Dr. Douglas Stubbs, died suddenly.

Patsy, a Vassar graduate and a member of the Deltas, was toasted at a wedding reception that took place just outside Detroit on a nearby island in Canada. "Patsy's stepfather and his brother, Sam, both owned an island and I think they even named it after Marion," recalls Mary-Agnes Davis, who attended the wedding.

"That wedding got a whole lot of people upset," said the woman as she sympathized with the father sitting at the head of the oval dining room table. "My husband knew both of those girls and he still didn't get invited. Of course I was there, but we weren't married yet. And I'll tell you, for every one person who got an invitation—"

"There were ten more who wanted one and didn't get it," interrupted her fellow Delta soror.

"Sam and Alf's father owned a group of convalescent homes and they were the only black millionaires in this town in the forties and fifties," the offended dinner host added. "The whitest-looking black people you ever saw. As high-yellow as they come."

A medium-complexioned woman at the table glanced at the Cartier watch on her slender wrist. "Yeah, but that didn't last. I can tell you that," sniped the woman with near satisfaction.

"What didn't last?" I asked.

"The money *or* the marriage," two of them added, almost in unison.

The man looked across at me. "The family lost their money by the early seventies, and Harold and Patsy split up too. Damn shame." He paused and reached for a plate of corn bread.

Just as certain parties and annual events played an important role in maintaining the status of certain black families, so did membership in certain churches. Plymouth Congregational, St. Matthews, Bethel Baptist, Hartford Avenue Baptist, Tabernacle, and Little Rock Baptist were favored by the city's black elite.

Plymouth, located on East Warren Avenue, was long considered the intellectuals' church and it ranked at the top. It was not uncommon during the 1950s to see Plymouth's Rev. Horace White in the local papers, for which he would occasionally write thoughtful letters and editorials on the dangers of segregation and bigotry in Detroit institutions. He took on many of the passive local white ministers, like Rev. Malcolm Sylvester of St. Andrews Presbyterian Church, and chastised them for not being more aggressive in the effort to bring about racial harmony. Robin Hamilton Sowell and her husband, Myzell, an attorney, have been members of Plymouth since the 1950s. "We got married at Plymouth in 1951, when the church was located at Beaubien and Garfield," says Mrs. Sowell, whose wedding was conducted by Reverend White. "Then we moved the church to Warren when the medical center began eating up the residential neighborhoods around us." In addition to the Sowells, Plymouth Congregational is known for such families as the Bells and the Hoods. Dr. Hailey Bell was a dentist whose family also owned radio stations WCHB and the former WJZZ. The Hoods have been ministers at the church for two generations. Nicholas Hood Sr. was head minister as well as a member of the city council before his son, Nicholas junior, held the same two positions. Nicholas junior's wife, Denise Page Hood, a member of the Links, is a federal judge.

St. Matthew's Episcopal Church is another church that is popular among Detroit's old guard. Established in 1851, it served free blacks and runaway slaves who had been brought north by the Underground Railroad during the 1850s and 1860s. Most respected because its former priest, Rev. James Theodore Holly, became the first black bishop in the Episcopal Church, St. Matthew's attracted many prominent Detroiters. "Lots of our Co-Ette Club members belonged to St. Matthews," says Mary-Agnes Miller Davis. "If they didn't go to Plymouth, they were at St. Matthews." The church moved from St. Antoine and Congress in 1881 to St. Antoine and Elizabeth—all east-side locations. Finally, after Interstate 75 was built in the late 1960s, destroying many of the black east-side establishments, the church merged with St. Joseph's at Woodward and Holbrook in 1971. Many artifacts from the 1881 building have been placed on permanent exhibit at the Museum of African American History downtown.

Hartford is well known for its Yale-educated minister Rev. Charles Adams, and the selling point for many in the Tabernacle membership was the presence of fellow congregant Damon Keith.

Detroit's new black elite is becoming a much more visible segment of the city now that it has a sophisticated and well-educated mayor who very much has ties to this world. As Mayor Dennis Archer's profile has risen, he has brought many accomplished people with him. C. Beth DunCombe, a member of the Detroit Girl Friends and sister-in-law to Mayor Archer, is considered one of the most influential members of the new black elite. In addition to being a partner in the law firm of Dickinson Wright, where she negotiates major development deals and important bankruptcy cases, DunCombe is also president of the Detroit Economic Growth Corporation and chairwoman of the city's casino advisory committee.

Because of Archer's skill in working with white government officials and business leaders—a talent that was not visible or valued in the Coleman Young administration—he has improved business opportunities for black entrepreneurs and professionals. In fact, *Industry Week* magazine recently listed Detroit as the number-one city for manufacturing businesses, and the city also received a $100 million grant as an empowerment zone that the federal government recognized as worthy of special investments.

Mary-Agnes Miller Davis doesn't know why, but as she sees more blacks move to suburban areas that were once closed to them, she notices that whites are coming back to pockets of the city. "I've watched Detroit come full circle," the longtime resident says. "At first, whites held on to parts of the west side, and now I see them moving back to the Boston-Edison neighborhood. I hope black people don't give up on this city. We have a great history here."

Black Elite in Atlanta

There is no major metropolitan area that has a better-organized black upper class than the city of Atlanta. Exerting its power in the worlds of politics, business and academia, Atlanta's black elite sets the gold standard for its counterparts in other cities.

"We've had three black mayors with national reputations," says my friend Janice White Sikes of the city's Auburn Avenue Research Library on African American Culture and History, the nation's best collection of black Atlanta history documents. "We are home to the best-known historically black colleges. And in addition to hosting the Olympics we have some black-owned companies that are the oldest of their kind in the country."

Although she has spent most of her career researching and writing about an older, more rural Georgia, it is obvious that what excites Sikes most as we sit in the dining room of the Atlanta Ritz-Carlton is talking about the new Atlanta and how the black community has played a role in making it one of the most popular destinations for elite blacks in search of a city where they are in control.

"This city produced older civil rights leaders like Dr. Martin Luther

King Jr., Julian Bond, and Congressman John Lewis," she adds while looking over some notes describing her uncle, a black class-of-1933 Harvard graduate, "but Atlanta has also elevated people like Andrew Young, Maynard Jackson, and Johnetta Cole to national standing in recent years."

Unlike other cities of its size and sophistication, Atlanta has seen a black elite forge strong enough ties between blacks, whites, and the business communities of both groups to elect three consecutive black mayors. What is also interesting is that Maynard Jackson, Andrew Young, and current mayor William Campbell are solidly representative of the black upper class—a characteristic that historically has not been welcome in black electoral candidates in cities like Washington, Chicago, or Detroit. In fact, when Marion Barry and Coleman Young of Washington and Detroit, respectively, were campaigning in mayoral races, they bragged about their ties to the urban working-class community. In Atlanta, good lineage, money, and top school credentials are appreciated by the black mainstream.

In addition to excelling in political clout, black Atlantans outstrip other cities' elite in the area of college ties. Atlanta's black academic community is larger than any other city's because of prestigious schools like Spelman, Morehouse, Morris Brown, and Clark Atlanta. When former Spelman College president Johnetta Cole received a $20 million gift from Bill and Camille Cosby (she is a Spelman alumnus) in 1993, other cities and their black colleges took notice of the strong black university consortium that was growing on the southwest side of Atlanta.

And further reinforcing the role and place of the black elite in the city are its black-owned businesses. While it does not outnumber New York or Chicago in black entrepreneurs, the city does claim the nation's largest black-owned insurance company (Atlanta Life), the largest black-owned real estate development firm (H. J. Russell), and some of the country's top black-controlled investment firms, law firms, auto dealerships, and food service companies.

With all of Atlanta's high-profile black schools, political leaders, and businesses, as well as its overall economic and population growth, it is not a surprise that many young black professionals from the North and elsewhere are relocating to Atlanta. I recall when my brother was in his last year of dental school in Boston—a city where the black elite often attend school, but rarely settle—he announced his desire to move to Atlanta. To him and to my family, the city and its new black elite looked so happy and so successful the idea made good sense. We called friends that we'd grown up with in Jack and Jill and others from the Links and other groups who had relocated to the southern city.

Unfortunately, though, it took almost no time to discover that relocation would have been a major mistake. It took no time to uncover a dirty little secret about Atlanta and its black elite.

One childhood friend, who had relocated to Atlanta after practicing medicine in New York, told us, "You can make a million dollars a year and live in the nicest house in Buckhead, but you'll never be accepted by the old black elite in this town."

"These people will stare you in the face after you've told them of your great accomplishments," explained our friend, "and the only thing that will matter to them is who in your family went to Morehouse, and for how many generations your family has lived in Atlanta."

This was not the only opinion we heard of the old-guard blacks in the city, but we heard enough similar stories to convince us that my brother should reconsider his plans and remain in New York, where family lineage mattered less.

With all the success that exists in the black community in Atlanta, there is a hard line drawn between the new and old members of the black elite. The two groups seem to work together to empower the larger black community during election time, but the old elite seem to be quick to exclude the newcomers from the social circles that define the city's original black upper class. And they seem to do this with greater ease than old-guard blacks in any other city—including Washington, D.C.

Before the population and economic boom began in the 1970s, Atlanta's privileged blacks had a rather firm definition of who they were and where they stood in relation to whites and to other blacks. Ella Gaines Yates has mixed emotions about the way Atlanta has changed over the last twenty years. Part of her time is spent among old Atlanta and another part is spent interacting with the new city that is emerging. "There are parts of this city that remind you of a New Jersey suburb," she says, noting the number of people who have relocated to Atlanta from New York and New Jersey. "Many of the new people are accomplished and extremely pleasant," she explains while sitting in her home in the southwest Cascade Heights neighborhood, "but some of them have no knowledge of, or interest in, the history that preceded them in Atlanta."

Many members of the city's old guard—people like Ella Yates and the men and women she knows—would probably acknowledge that part of Atlanta has become known for its ability to remake itself: to tear up its historical roots every few years and erect a shiny new facade that reflects little of what preceded it. Yes, as a result of the changes there have been many improvements, but much of the past has also gotten lost through these changes. It happened when the busy Interstates 75 and 85 were run through the black Auburn Avenue neighborhood, and it happened when black sections of the city were demolished in preparation for the 1996 Olympics.

Yates is old Atlanta. She's fourth-generation. She's Spelman College. She lives in the right neighborhood. She's earned the right graduate degrees, held the right jobs, and married into the right family. She even belongs to the right literary club—one that includes a small and elite group of black Atlanta's old-guard women who have met once a month, without fail, for eighty-nine consecutive years. Yates is the real thing, and she knows where the bones are buried.

But there are many new wealthy blacks in Atlanta who wouldn't know that. Some of the new black money that has landed in Stone Mountain, in Guilford Forest, in Buckhead, or in Dunwoody wouldn't know about the Yates family, or the bank or the drugstore they founded. They wouldn't know about the book party that was just given for Julian Bond's mother, Julia, or about the recent funeral services held for Eloise Murphy Milton, or about the medical contributions of Asa Yancey or the business accomplishments of the Blayton family. They wouldn't know that the Herndon mansion, which overlooks the city of Atlanta, was built in 1909 by a black millionaire in the insurance business. They wouldn't know the old names of Rucker, Aikens, Harper, Cooper, Dobbs, and Scott. In fact, many members of the new elite don't live within the city limits. They are not among the 260,000 blacks that make the city only one-third white. The new black elite are on the outskirts of old Atlanta—both geographically and socially.

According to a third-generation Jack and Jiller who grew up in the town, "The only old names that mean anything to the new transplants are Martin Luther King and Andrew Young—and we consider both of them to be *new*."

"It's not that Atlanta natives don't appreciate outsiders," explains Ella Yates, a native Atlantan who graduated from Spelman College before later going on to earn a law degree as well as a master's from Clark-Atlanta University. "It's just that we want people to express an interest in what so many generations contributed to the community." After returning from the funeral of a family friend, Eloise Murphy Milton, at First Congregational Church, Yates remarked how much the church, the mourners, and Mrs. Milton's passing had reminded her of old black Atlanta. "Having been the first black director of the Atlanta Fulton Library," she explains, "I am very conscious of the city's black history." Her husband's father, Clayton Yates Sr., was a major part of the city's black history, and a lot of his work was performed in concert with Eloise Milton's husband. They were involved in some of the most important business deals of early-twentieth-century black Atlanta.

Several months before she died at age ninety-nine, I had the chance to meet Mrs. Milton. She talked about the Atlanta Links, which she helped charter in 1953, as well as the businesses and projects that her husband had launched with Clayton Yates, Ella's father-in-law, in the 1920s and 1930s, "My husband, Lorimer, and Clayton Yates were partners in almost every project they pursued, including the reorganization of Citizens Trust Bank in the early thirties and the opening of the Top Hat Club," said Milton, pointing out that the bank still operates with several branches in the city. "They also opened the Yates and Milton Drugstore when there was a growing number of black businessmen in the city."

"Although I was clearly more involved with social organizations, I grew up in a family that was always in business," explained Milton, who attended Atlanta University and Oberlin College. Two prior generations of her family had already run businesses. A city school was even named after one of her ancestors. "My grandfather, David T. Howard, was an undertaker, and my parents owned a grocery store on Decatur Street. None of them were major businesses, but they were successful."

Like many in this group of elite women, Milton vastly understates her family's success and contributions. Although her grandfather was born a slave, his business was actually successful enough to make him among the six richest blacks in the city by 1896. He was also a major benefactor of Big Bethel A. M. E. Church.

Milton lived a life that was different from that of many Atlanta blacks, and she remembers spending some of her years on her grandfather's farm in north Atlanta. "He owned a great deal of land in what we considered the country. It was nowhere near Auburn Avenue or where the rest of the black community lived at the time."

It is not until I speak to my Aunt Phyllis (the AKA die-hard), who is close friend of Mrs. Milton's daughter Eleanor, that I learn how unusual Milton's grandfather's wealth really was. I discover that the David T. Howard farm was located in what is now called Buckhead (the whitest and most affluent section of Atlanta) and it abutted the land of the Woodruffs, the family that founded Coca-Cola.

"Remember, we're talking about the early 1900s. My mother grew up in a completely different environment than most black Atlantans," explains Eleanor Milton Johnson, who is one of Milton's two daughters. "Because of the great exposure that she and her family gained through their wealth, Mama made many progressive decisions that would not have been considered by other blacks. And my father, Lorimer, who graduated from Brown University in 1920, was just like her in that regard. So it was not a surprise that I ended up at Mount Holyoke College in the 1940s when most of my Atlanta friends were attending Spelman and Morehouse. My sister, Betty, graduated from Mount Holyoke too—but only after first going away to the Northfield boarding school in New England."

The Howard-Milton family is only one example of black success in the community. Another representative of this group is the Jackson-Wiltz family. Yvonne and Teresa are both daughters of prominent Atlanta physicians. Both of their mothers joined the Atlanta Links and the Inquirers Literary Club. Both of their fathers joined the Atlanta Boulé. Both women were raised in Jack and Jill by parents who attended Friendship Baptist Church and who socialized with other black professionals who lived in southwest Atlanta.

Yvonne Jackson Wiltz and Teresa Wiltz are mother and daughter, and in many ways their life experiences appear to be identical, a common pattern among old-guard Atlanta families who value certain traditions. But experiences that appear to be repeated generation after generation in these families are actually being changed greatly by a city that has grown from a segregated small town to a thriving metropolis where the old black families are now interfacing with families who are not from their race, their class, or their city.

"When I was growing up in the Atlanta Jack and Jill, it was much different from the Jack and Jill my daughters were raised in," says Yvonne Jackson Wiltz who recalls that the Jack and Jill of the 1940s was operating in a segregated city where the activities remained in the black community and where everyone knew everyone else's family. "All of our activities were focused on people's homes because few public facilities were available to us. We swam in each other's backyard pools or played in each other's playrooms and living rooms. By the time my daughters joined, we were exposing them to the High Museum and Fox Theater downtown—places black people were not welcomed before. And what was also changing was that we were finally meeting new families who had not grown up in Atlanta."

Even though Yvonne and Teresa's world in Atlanta was primarily a black one, socially, their academic experiences also reflected the changes that integration brought about. Yvonne, as well as her parents, Ruth and Marque, went to black colleges. "My mother went to the private laboratory high school on the grounds of Atlanta University before she went to Atlanta University for college, and my father went nearby at Morehouse," explains Yvonne, who attended Spelman College. She met her husband, Philip, before he entered Howard Medical School.

Teresa, on the other hand, attended the Westminster Schools, Dartmouth College, and Northwestern University School of Journalism—all predominately white institutions. When she was attending the expensive private Westminster Schools in the Buckhead section of Atlanta with the governor's son and other children of wealthy white families, the school's racial makeup was about the same as the population in its surrounding north-end neighborhood. "Out of two hundred students, only about four of us in my class were black. And we all knew each other," says Teresa, who is now a features writer at the *Chicago Tribune*. Not surprisingly, the black students in her class were all children of doctors—including Shanta Sullivan, the daughter of Dr. Louis Sullivan, who was President Bush's secretary of health and human services.

And while Teresa's Jack and Jill experience was, in fact, different from her mother's, her Jack and Jill playmates came from the same famous Atlanta families: former mayors Maynard Jackson and Andrew Young both had daughters in Teresa's group. "I remember when a lot of us in the chapter helped Paula campaign for her father when he ran for Congress," adds Teresa. Such activities were commonplace among Teresa's crowd.

But as old and established as the Jackson-Wiltz clan is in Atlanta, there are still families among the elite that would say they are not quite old enough. Like Washington, D.C., Atlanta has people in its old-guard black community who scrutinize neighbors, marriages, and family credentials so closely that even some who have been in the city for four generations are still considered new.

"Black Atlanta is a very closed society if your family has been here less than three generations," says a Harvard-educated attorney who arrived in the city over a decade ago. "I grew up in Jack and Jill, and my parents are professionals, but there's a group here that wants nothing to do with new people like me. When I first got here, I'd go to these business events where I'd meet guys my own age, and it was like they were talking in code," says the attorney. "I'd be hearing last names of people I didn't know. And when I asked who they were, people would look at me like I was from outer space. And the women were even worse. It got to the point where I had to start reading the society columns and go to the library to get history books on Atlanta."

A New Yorker who is a friend of the attorney agrees. "I just keep my mouth shut at these events. I was at First Congregational Church a couple of years ago and this group of older women I met scared me to death. When I mentioned that my mother had gone to Spelman around the same time as they were there, they all looked at each other as if to say, 'How can that be?' Then one of them asked me, 'Why don't we know her?' When I told them she moved to New York and pledged Zeta after graduating, they completely lost interest in our conversation.

"And when people around here ask you, 'So, who *are* you?' or 'Who are your parents?' they want more than just a name. They want to know how many generations of Atlanta you represent, what year your grandparents graduated from Spelman and Morehouse, which literary club your great-grandmother belonged to, what street in Collier Heights your parents lived on in the sixties, and who in your family goes to Friendship Baptist Church."

Ella Gaines Yates believes that this characterization is not entirely true. She belongs to one of the old families that appear in the Atlanta history books, but she also has the experience of living in places outside of Atlanta. This has provided her with the perspective that is often missing in old families who have stayed rooted in one place.

"Atlanta has certainly changed since my parents' generation and my grandparents' generation," says Yates, who spent time working in Virginia after being appointed as the state librarian by the governor. "There is a certain worldliness that is brought to a community when respectful outsiders relocate there. I think people in this city can appreciate that. I just think that the old guard wants to be assured that Atlanta's black history is not being forgotten or ignored. After all, many of our older neighbors are doing some fascinating things here."

Among the neighbors Yates refers to are the Yanceys, a family that truly personifies the black doctor elite in Atlanta. Accomplished, affluent, well educated, and philanthropic, the Yancey family has ties to Morehouse, Spelman, the Atlanta School Board, Howard University School of Medicine, the Tuskegee Veterans Administration Hospital, Meharry Medical School, Delta Sigma Theta, the Boulé, Jack and Jill, Links, and the elite First Congregational Church.

Dr. Asa G. Yancey Sr. was born in Atlanta, and like his two brothers he also became a physician. Educated in the Atlanta public schools, he was valedictorian of the city's Booker T. Washington High School. He graduated from Morehouse College in 1937 and then from the medical school at the University of Michigan in 1941. In spite of his and his two brothers' accomplishments in the medical field, and the fact that three of his four children are physicians, Yancey abhors using the term "elite" to describe him or his family. "We do not refer to ourselves as upper-class or middle-class. We consider ourselves to be sound, hardworking African Americans, and we are not unique in Atlanta. This city has lots of accomplished families who have distinguished themselves in different professions."

"Asa Yancey is a very modest man, but what people should know is that there are a lot of black doctors in the South who would never have become surgeons if it were not for the programs that he started for us," says an Atlanta surgeon who was trained in the first black surgery program in Georgia.

Modesty notwithstanding, Yancey is similar to another group of successful blacks of his generation whom I interviewed in other cities, and whom I found to dislike the "upper class" label. Many of these individuals—most of them male, southern, and over the age of seventy—spoke out strongly when I suggested that they were members of an elite or of a successful minority. Having grown up around my paternal grandfather, who was equally disdainful of accepting such a label or of other blacks who accepted it, I came to understand that there were at least a couple of reasons for this response.

For one, like Yancey and my grandfather, all of these men had grown up as witnesses to the worst examples of southern racism. They had grown up in settings where, despite their educational or career accomplishments, they were obliged to display false modesty and obsequious behavior in the company of whites. Even when my grandfather was an adult in Memphis operating a successful trucking and hauling company and transporting products for Kellogg's and other companies during the 1940s, 1950s, and 1960s, young and old white business clients called him by his first name while he was required to call them "Mr." or "sir." Even when I was a child growing up in the 1970s, I would see him sometimes accidentally slip and say "Yes, sir" or "Thank you, sir" to a white restaurant waiter or gas station attendant. Taking on this inferior role in a white-dominated Jim Crow South required one to remain humble, hide one's intelligence, and disguise one's academic and career success because much of it could be stripped away if one was perceived by whites as acting "equal" or "uppity."

And from the black community's perspective, it was easy for black men in the South—particularly successful ones—to be thought of as "selling out" or becoming an "Uncle Tom." So the natural response for any black man who wanted to maintain respect in the larger black community was to eschew the labels or signs of success. Just as my grandfather did, and as older men in Atlanta seem to do, Dr. Asa Yancey makes consistent references to his ties to the ordinary, more mainstream black community. It is clear that he wants it known that even if others consider him to be part of an elite, he is not one who will ever embrace the term—even in a city where so many of the newer well-to-do blacks are proud to show what they have accomplished.

But in spite of his reticence, most blacks—and many whites—in Atlanta already know of his accomplishments and his position among the elite. After completing his own surgery residency at the Howard Medical School under Dr. Charles Richard Drew, the well-known black physician who developed blood plasma storage and blood banking, Yancey spent two years in a gynecological surgery residency at Meharry and then became chief of surgery at the Tuskegee Veterans Administration Hospital. It was there that he established the first accredited surgical graduate medical education program for black surgeons in the state of Alabama.

"Before he came to Tuskegee," says a former student of Yancey's, "there was no place in the whole state where blacks were allowed to train for surgery."

Yancey then returned to Atlanta in the late 1950s. "After I got back to Atlanta and became chief of surgery at Grady Memorial Hospital, I realized that there were just too many black doctors here who had no way of getting training for surgery, so I helped start Georgia's first accredited graduate program for the training of black surgeons." Several years later, he began an affiliation with Emory University School of Medicine, ultimately becoming a professor of surgery and a partner in the Emory University Clinic. In 1984, he became clinical professor of surgery at Morehouse Medical School. His new programs in Tuskegee and Atlanta affected a whole generation of black doctors.

For the last thirty years, the Yancey family has also had a significant involvement in public and private education in the city as well as the state. They are shy about telling me this, but Asa and his wife, Marge, have both held seats on the Atlanta Board of Education. She has been a Spelman College trustee and was the first black woman to sit on the board of regents of the University System of Georgia. A native of Detroit and a graduate of Wayne State University, she had been a schoolteacher before moving to Atlanta. "I first joined the board of education in 1982," says Mrs. Yancey, who is also a member of the Atlanta Links, "and then Governor Harris appointed me in 1985 to the board of regents. It's important that black people be represented on these boards because our children's education affects their future and our community's future."

The Yanceys are so well respected in the Atlanta community that it is not unusual to find photos of their various family celebrations on the front page of the *Atlanta Daily World*, as happened when Asa senior and Marge recently celebrated their fiftieth wedding anniversary.

Another family of physicians are the Browns. Dr. Calvin Brown and his wife Joy San Walker Brown met each other while she was a biology major at Spelman and he was a premed student at Morehouse in the 1950s. His grandfather had been a missionary as well as a physician. "I came to the city for college, but my husband grew up here. Other than the time that we moved to Nashville when my husband went to Meharry," explains Mrs. Brown, "we have remained in Atlanta. We raised our daughters here and they stayed to attend Spelman College."

Today, both of the Browns' daughters, Joisanne and Sannagai, are physicians in Atlanta, following their father, Calvin, a member of the Boulé, who is in general practice. Joy, a former teacher in the Atlanta Schools, says that it was not a surprise that her daughters pursued challenging careers in medicine. "They grew up surrounded by people like the Yanceys and other friends of ours, and they went to a college where young women were told they could excel in the sciences," explains Mrs. Brown, who volunteers much of her time through organizations like the Girl Friends and the Links, which she had previously served as chapter president.

Laying the foundation for people like the Yanceys were other black doctors such as Henry Butler and Thomas Slater, who founded the National Medical Association; Dr. Homer Nash, Dr. Hamilton Holmes, and Dr. James Porter, who began their practices before World War I; Earl McLendon, who opened an Atlanta medical clinic in the 1940s; as well as many others, including Dr. Albert Cooper, Dr. Waymond Reeves, and Dr. Antoine Graves.

In addition to the city's physicians, other contributors to Atlanta's early history were its intellectuals, who were tied to various churches and institutions. They included people like Henry Proctor of First Congregational Church; George Towns of Atlanta University; Benjamin Davis of the *Atlanta Independent*; Peter Bryant of Wheat Street Baptist Church; Edwin Driskell, head of Union Mutual Publishing Company; John Hope, president of Atlanta University; and many individuals involved with the leadership of Friendship Baptist Church.

As well known as these names and institutions are among the black community, they have almost no place in the collective memory of Atlanta whites. In fact, because Atlanta was a city that did very little to preserve historic buildings and settings, it is easy to live here as a member of the black or white community and have little sense of the other. Although the city came out of segregation with greater ease than cities like Memphis or Montgomery, the black and white communities place very different value on the various elements of Atlanta history that have actually survived destruction: There is "their" history and "our" history.

For example, while well-to-do whites in the city revere the history of Margaret Mitchell, author of *Gone With the Wind;* Robert E. Lee; and the many other symbols that accompany the Confederate side of the Civil War, blacks pay tribute to the black civil rights movement that was advanced by hometown sons like Dr. Martin Luther King and Ralph Abernathy. Many whites say that the different focus on history has little to do with race. But blacks say that race is the primary factor that separates the two groups' histories and experiences.

"White people don't respect us here," says a retired Spelman professor. "They know that we Atlanta blacks have a lot going for us—powerful elected leaders, strong colleges, and the hard-to-ignore legacy of Martin Luther King. They claim that celebrating the Confederacy has nothing to do with race. Let's not forget that the Confederacy supported our enslavement, and that these same supporters later fought against integration." The professor points to a more modern example of that bigotry. "One way in which the white Atlanta elites continue to show their disrespect is by supporting a state flag that reflects the Confederate logo."

While some wealthy white Atlantans dismiss their poorer white counterparts in the outlying "bubba counties" as rednecks or bigots, many of them claim to accept the local black elite and insist that they embrace the state flag and Robert E. Lee only because the Confederacy is a part of Georgia's history.

"I don't buy it," says Atlanta resident and author Janice White Sikes. "Let's remember that the Georgia flag didn't even have the Confederate symbols on it until they were added in the late 1950s. The flag was actually changed as a sign of defiance after the federal government forced integration on our schools and public accommodations." Sikes is correct when she points out that many blacks are justifiably offended by a flag that was specifically redesigned to underscore the antiblack sentiments of white citizens.

But how does the black upper class in this city of 3.5 million people deal with the slights that blacks don't quite have the power to completely erase? Some say they work hard to preserve the surviving remnants of the old black middle and upper classes. The city's black leaders have invested a great deal in maintaining Auburn Street, the locale for many of Atlanta's first black businesses. These same individuals were the ones who called for the establishment of the Auburn Avenue Research Library, which opened in 1994 and serves as an important archive on black Atlantans who have excelled in scholarship, business, and civic life.

Because even museums and libraries have their limitations in terms of being able to promote the history of the black elite in Atlanta, current members of the elite take it upon themselves to look back at their own families and share the personal histories of some of their relatives. In my conversations with people like Eloise Milton, I learn that these family stories are truly the best way to learn about the city's earliest black elite.

"Although my grandfather, David T. Howard, was eventually a very well-to-do man," said Eloise Murphy Milton, describing one of the first successful black businessmen in Atlanta, "he was born a slave—not a free black like some of the other blacks who became successful in the North in the late 1800s."

As Milton explains, her grandfather had been born a slave to a white master who decided to give him a small inheritance upon his freedom. Very few blacks in nineteenth-century Atlanta received their freedom before the end of the Civil War. In fact, of the twelve thousand people who lived in the city in 1860, three thousand were enslaved blacks and only twenty-five were free blacks. So, when one looks at the speed with which Atlanta developed an educated and well-to-do class of blacks, one cannot credit early generations of free blacks such as one finds in the history of cities like Washington. Instead, the basis for Atlanta's black elite is found among former slaves who gained modest benefits either from their white masters or, more often, through the education and physical support that they received from Atlanta's early established black university community.

During the late nineteenth century and the early part of the twentieth century, Atlanta's black society grew out of Spelman College, Morehouse College, Atlanta University, Morris Brown College, and Clark University.

The first of these to open was Atlanta University, in 1869, and like the other colleges, this campus served as a place of employment as well as a place of learning for many of the city's black families. "These schools turned this city into an oasis for black people," says Spelman graduate Joy San Walker Brown. "Not only did many of us have the chance to attend these schools, but the campus programs gave blacks who didn't attend the schools the opportunity to see musical performances, dramatic groups, and national leaders that were not accessible to them anywhere else. Remember that this was a segregated city that was not concerned with exposing black children and adults to intellectual and artistic forums."

Many of Atlanta's original and current black elite are affiliated with the schools either through their alumni status, like Brown and her husband, who attended Morehouse, or by virtue of their employment at one of the schools.

"Most of the families who have been here for a generation or more will have some kind of close tie to at least one of our colleges," says Morehouse graduate Dr. Asa Yancey; he notes that Mayor Maynard Jackson's mother and five aunts all went to Spelman.

In addition to the support that the colleges gave them, another factor that contributed to early black success in the city is that there was an unusually large and concentrated black customer base to support black entrepreneurism. Unlike Chicago or New York, where blacks made up only a small percentage of the population, Atlanta was already more than 40 percent black by the 1890s. Because of this, black-owned businesses had no difficulty finding patrons to support them.

Between the universities growing up on the southwestern edge of the city and the black owned-businesses growing up on Auburn Avenue, just north of downtown, the black community had relatively good odds for success.

Of those business leaders who found their success outside the university community, several also used politics to bolster their position in the community. One person who succeeded at business and political brokering was Henry Rucker, who served in 1880 as a Georgia delegate to the Republican Convention in Chicago. He was also an appointee of President William McKinley in 1897, when he was named collector of internal revenue. In fact, by 1904, Rucker had become so successful in the barbershop business that he built a five-story office building, which became the city's first building that would rent space to blacks.

A family that looms even larger than the Ruckers and the Howards in the worlds of business and civic affairs is the Herndons: "The Herndon family is still thought of as the first family of Atlanta's black elite," says Atlanta resident Keith Chaplin, who graduated from Morehouse and now works at CNN. "Not only did Herndon become a rich man in the 1890s, but he also left an estate and a business that several generations of blacks have been able to visit and patronize since his death."

As you stand in the backyard of the Herndon mansion, you can scan the entire Atlanta city skyline. The massive 1910 Beaux Arts classical-style building is a testament to the wealth that Alonzo Herndon amassed by the turn of the century through a variety of businesses.

"Although he eventually founded Atlanta Life Insurance Company in the early 1900s, his earlier successful business was an upscale barbershop that served an affluent white clientele at 66 Peachtree Street," explains Dr. Carole Merritt, an Atlanta historian and the curator of the Herndon home.

Like many of the black-owned upscale barbershops in the country, Herndon's shop catered to affluent white customers who wanted haircuts, shaves, manicures, and skin treatments. They referred to the shop as the "Crystal Palace," and it featured black barbers dressed in white suits who worked in an ornate room lit by crystal chandeliers. Each day, white businessmen arrived for appointments and waited on leather sofas in the center of a marble floor. "That was the largest and most formal barbershop I'd ever seen. It had at least twenty-five leather barber chairs in it," recalls Chicago Boulé member Truman Gibson, who remembers visiting the shop as a child while his father, Truman senior, was an executive at Atlanta Life.

"Mr. Herndon's wealth supported many institutions in this city, including First Congregational Church, which he attended," says Dr. Merritt. "And his son, Norris, gave generously to many groups like the NAACP, the Butler Street YMCA, and Morris Brown College, to which he gave $500,000 in 1948."

Although Herndon was to turn over his mansion, businesses, and riches to his son, Norris, the latter never married or had children.

"But fortunately," says Morehouse graduate Keith Chaplin as he looks up at the crystal chandelier hanging in the Herndon home's music room, "the family established a foundation that preserved the mansion and its contents so that generations of black people could see what one of their own had accomplished."

Although Herndon and other members of Atlanta's black elite had much greater interaction with middle- and upper-class whites than blacks in other southern towns that lacked the city's sophistication, the quality of black life and race relations had a serious setback in the early 1900s when a series of Jim Crow laws were adopted by the Atlanta city council and other agencies in order to further separate blacks and whites. It restricted blacks in their business dealings and their housing options, as well as their educational and job opportunities, at a time when the first generation of successful blacks was just becoming established. For example, at a time when black-owned eating establishments were flourishing and attracting both black and white customers, the Atlanta city council passed, in 1910, an ordinance that required restaurants to become single-race establishments—either "white only" or "colored only." In 1913, the city created legal boundaries for the segregation of black and white residential neighborhoods. And in 1920, the council limited the success of black-owned beauty salons and barbershops by making it illegal for them to serve white women or white children.

Alonzo Herndon's salon and his other business enterprises survived in spite of these new laws—eventually allowing him to build the mammoth Atlanta Life Insurance Company, which has remained one of the largest black-owned insurance companies in the nation. Upon his death in 1927, his son continued with the company, and today the downtown area is bordered by the insurance company's large modern headquarters building. And leading down Auburn Avenue onto the main downtown thoroughfare of Peachtree Street are banners honoring the Herndon family for their role in the city's history.

Another prominent family that also established a successful enterprise on Auburn Avenue was the Scott family. They are best known as the founders and owners of the newspaper *Atlanta Daily World*, as well as landowners who purchased a great deal of Atlanta property during the depression. "I am the third generation of my family that has worked in publishing," explains Portia Scott, editor of the family-owned paper. "My grandfather William A. Scott Sr. was a publisher and printer of church bulletins while he was living in Mississippi. During the time that he was working with Booker T. Washington in the Business League, he bought his own printing presses."

On August 5, 1928, twenty-eight years after her grandfather bought his first presses, Portia's uncle William A. Scott II launched what was then called the *Atlanta World*. "My father, Cornelius Adolphus Scott, his five brothers, and three sisters had all come to Atlanta because this was where blacks could find good high schools and universities," explains Portia, who grew up in the Atlanta chapter of Jack and Jill and remains an active member of the Atlanta Links. "Most of the brothers went to either Morehouse or Morris Brown, and all three of my dad's sisters—Ruth, Vashti, and Esther—went to Spelman."

As Portia Scott points out for her own family, there were many who benefited from and made use of the same schools and universities—an advantage that black Atlantans had over many other segregated cities. They were a people who were not going to miss out on opportunities just because the city refused to provide them an equal opportunity in education.

When the wife of John Hope, Morehouse's president, joined the wives of David T. Howard and Alonzo Herndon to establish a kindergarten for black children in 1905, there began a tradition whereby the old-guard families worked together to fill the void that Atlanta's segregated schools left in the black community. This tradition of turning to privately funded schools and programs was one that would continue to link the city's black elite.

For example, beginning in the 1930s, a core group of old elite families sent their children to a private laboratory elementary school they had helped to create at Atlanta University. "My parents and several other families like the Coopers and the Yateses decided they wanted to establish a school with top teachers and students," explains Eleanor Milton Johnson, who attended the Atlanta University Laboratory School from first through twelfth grades before going to Spelman and Mount Holyoke colleges. "The elementary division was set up on the Morris

Brown campus," Johnson says, "and the high school was located on Spelman's campus." Although the school was clearly populated by many children from elite families, Johnson says the intent was for the school to have a cross section of children, "so the parents established a scholarship program to make it accessible for any gifted student who wanted to attend." There have also been a number of private nurseries and kindergartens that grew out of the churches that some of the old-guard families belonged to.

The largest segment of old black Atlanta, however, has its roots at Booker T. Washington High School, the first public high school built for blacks in the state of Georgia. "Before Booker T. Washington was built in the 1920s," says Ella Gaines Yates, a graduate who went on to Spelman, "blacks had to rely on the small private schools that were run at Spelman or Atlanta University, or they were sent to black boarding schools like Palmer Memorial in North Carolina. A small group of kids with wealthy parents were sent up to the white boarding schools in the Northeast."

Some Atlantans remember when state legislator Grace Towns Hamilton sent her daughter, Eleanor, to a Vermont boarding school in the 1940s. "She was the only black in the entire school," says a man who knew Eleanor Hamilton as a child, when her parents were big names in the city. "Eleanor grew up with her own nanny because her mother was this busy politician, and her father was busy with the university," says the childhood friend. "I guess they figured a boarding school would be better for her, since they were always out of town, but I just can't imagine growing up in this city and missing the entire black school experience. Those of us who went to Booker T. had the chance to learn from black teachers and black administrators who had master's degrees and Ph.D.'s from the best schools. That's not something you'd get in the white prep schools back then—or today, for that matter."

Ella Gaines Yates often runs into her Booker T. classmates at the meetings that her graduating class holds on the first Sunday of each month. "There were around 425 students in my class, and each month, we have as many as fifty alumni from the class joining us at meetings."

That same loyalty is found at the colleges in the city as well.

"When I first got to Atlanta in the early fifties, what I found was a real oasis," says Joy San Brown, who came from Houston to attend Spelman College. "My family did not like the idea of my coming to Georgia for school because they saw it as such a bigoted state, but the Atlanta community that surrounded Spelman and the other colleges was more enriching than anything we could have expected. We had one of the most dynamic faculties, which not only informed the students but also informed the community and helped to transform the city into a metropolis that attracted talented students and adults."

Brown, who belongs to the Atlanta Links and Atlanta Girl Friends, was such an unabashed Spelman supporter she sent both of her daughters there. "My daughters, who are both physicians now, like their father, feel that they received the kind of foundation that only a college specializing in training the nation's smartest black women could provide," says Brown.

One cannot overestimate the influence that was wielded in the city by the five prestigious colleges located there. Among the many alumni who graduated from these schools are Dr. Martin Luther King Jr.; Children's Defense Fund founder Marian Wright Edelman; former U.S. surgeon general and current Spelman president Audrey Manley; former secretary of health and human services Dr. Louis Sullivan; *Ebony* magazine editor Lerone Bennett; and sociologist E. Franklin Frazier.

Many of the people who graduated from these schools remained in the city as its future leaders. Among them are the city's first black mayor, Maynard Jackson, who graduated from Morehouse. Many other prominent political and civic leaders have ties to the colleges there. One of them was Grace Towns Hamilton, a politician and civic leader who came from a family of achievers in both politics and education. Elected in 1965 as the first black woman in the Georgia legislature, she had previously been named as head of the Atlanta Urban League in the 1940s. Hamilton's father, George, had been a professor at Atlanta University and an active member of the NAACP. Her grandfather had served in the state legislature and her husband, Henry, was a member of the faculty at Morehouse.

Another prominent Atlanta native who claimed ties to the city's universities was NAACP executive secretary and *Crisis* magazine editor Walter White. A 1916 graduate of Atlanta University, White eventually became an adviser to presidents Roosevelt and Truman. His father was a graduate of Atlanta University and his family was a respected name in the black community. "Walter's sister was the maid of honor at my mother's wedding," remarks Eleanor Milton Johnson.

Clearly the most famous man to come from this community was Dr. Martin Luther King Jr., who grew up on historic Auburn Avenue in the

1930s and 1940s. His family's church, Ebenezer Baptist, also located on Auburn, was pastored by members of his family for nearly one hundred years. A graduate of Morehouse, King was a member of the Boulé, and at age thirty-five he was the youngest person ever to win the Nobel Peace Prize. After his death, his wife, Coretta Scott King, remained in Atlanta, where she and their children have established foundations and programs to commemorate his life and work in civil rights.

Although he was not a native Atlantan, one of the city's most prominent intellectuals was Dr. William Edward Burghardt Du Bois, a longtime member of the Atlanta University faculty. A native of Massachusetts and a 1895 Ph.D. graduate of Harvard, W. E. B. Du Bois quickly established himself among the elite when he was teaching at Atlanta University from 1897 to 1910 and then again from 1933 to 1944. It was he who used the term "talented tenth" to describe the black intellectuals who were graduating from top schools and entering professional life. Although he was an early member of the Boulé and a friend of many of the most accomplished blacks in the nation, Du Bois had conflicts with a large segment of the NAACP leadership, causing him to break with the organization during the late 1940s.

Included among the oldest and most respected grandes dames of Atlanta society are Ruth Thomas Jackson, Ann Murray Cooper, Marge Harper, and the recently deceased Eloise Murphy Milton. Not coincidentally, they are all founding members of the Atlanta chapter of the Links. "This is the original chapter of the Links," explains a woman who stood in the lobby of the Ritz-Carlton, as her husband gave the keys to his car to the valet. "There must be two or three other Links chapters in this town," the woman adds after describing the significance of these four women and the contributions they had made through the Links and their other activities, "but we always make sure to come to the Atlanta chapter's events." As the woman glides across the carpet with her husband, headed to the silent auction to be held at the chapter's annual brunch, she comments on how jealous she is of Ann Cooper's legs; she greets a friend from One Hundred Black Men; she blows a kiss to Portia Scott, Juanita Baranco, and two or three other attractive women who are all members of Atlanta's young elite crowd; and she gives a big hug to Dr. Hugh Gloster, former president of Morehouse College.

Other grandes dames and female members of the establishment in Atlanta have included Harriet Chisholm Nash, Julia Bond, Alice Holmes

Washington, Belle Brooks Dennard, and Grace Towns Hamilton. These are people my Atlanta friends have met over the years, and they have always been the names that people mention at old-guard events. Many of them have belonged to the Links; Girl Friends; Jack and Jill; the sororities; the literary groups such as the Utopian Literary Club, the Chautauqua Circle, and the Inquirers; or the many bridge clubs.

"My mother was very active in the Chautauqua Circle for decades," remarks Eleanor Milton Johnson, referring to her mother, Eloise Murphy Milton, and the monthly literary group that gathered at each other's homes or at restaurants. "They invite speakers to discuss news or social issues in addition to literary subjects."

What link all of these individuals more than anything else are the old neighborhoods that once served as the core of the black community. Auburn Avenue is probably the historic center of the old black Atlanta neighborhood. Affectionately referred to as "sweet Auburn," this northeast street is a national historic site district and includes commercial establishments, as well as the Martin Luther King Jr. Center, Big Bethel A. M. E. Church, Ebenezer Baptist Church, the *Atlanta Daily World* office, Mutual Federal Savings and Loan, the Herndon Building, the Henry Rucker Building, the headquarters of Atlanta Life Insurance, and many other businesses and historic sites. The Auburn area is located on the eastern edge of downtown Atlanta. By 1918, there were nearly one hundred black-owned businesses and professional firm offices located on the street.

Hunter Street (now known as Martin Luther King Drive), which is located on the west side of Atlanta, is another important street in black Atlanta's history. A nine-mile street that runs from downtown and goes out west past the Atlanta University center and into the suburbs, Hunter became the main part of middle-class black Atlanta's commercial establishments—with certain parts of it also serving as a beautiful residential section for the black elite. Beginning in 1922, black businesses such as Amos Drug Store, Parks Shoe Rebuilders, the Crystal Theater, and the Broadnax Building opened on Hunter Street. Other businesses that established themselves there were Sellers Brothers Funeral Home and Citizens Trust Company.

In these black neighborhoods, the black establishment found the institutions which they would anoint as their favorites. For example, the old guard chose three favorite funeral homes for their families. They were Ivey Brothers, Sellers Brothers, and Haugabrook's. Their favored cemetery was Lincoln Memorial in northwest Atlanta. Their favorite college hangout for years had been the Waluhaje. Their favorite place for outdoor gatherings was on the golf course of the New Lincoln Country Club, which was established in the 1930s.

The residential section of Hunter Street was a prestigious address for blacks in the 1940s and 1950s. In fact, the old Atlanta families who live there refer to it as Hunter far more than by its new name, Martin Luther King Drive. Among its residents have been Theodore M. Alexander, founder of the insurance firm Alexander and Company, and many members of the Scott family, as well as many doctors like the Jacksons and the Coopers. "One of the reasons why so many members of our family happened to live on Hunter," explains Portia Scott, "is that my uncle bought many of the lots during the depression. After years went by, he either gave or sold different lots to our family members." Portia remembers growing up on the street along with friends who were in her Jack and Jill chapter. "Since we were growing up during segregation in the 1950s and couldn't go to certain public recreational facilities, I specifically remember the kids on our street who had big backyards and swimming pools—like the Alexander family—because this is where we as kids were socializing."

The old black elite residential neighborhoods include the Hunter Hills section along the old Hunter Street, as well as the Collier Heights section, which is situated in the northwest part of the city. Homes on streets like Engle Road, Skipper Drive, Waterford Road, Old Know Drive, and Woodmere Drive were built during the late 1950s and early 1960s and are located between Collier Road and Bankhead Highway. Over the years, these homes, many of which were designed by black architects, have attracted people like Dr. and Mrs. William Shropshire, Dr. and Mrs. Asa Yancey, real estate developer Herman Russell and his wife Otelia, and many other prominent Atlanta families.

Even though Atlanta had an overwhelming number of well-educated, well-to-do black families, the city drew clear lines around where blacks lived and conducted their business. As much as these families did to give their children a feeling of confidence and pride, it was difficult to shield them from the fact that whites gave them second-class status when it came to living and socializing where they wanted. Of course it helped that the five colleges provided access to educational and cultural activities on the weekends and that groups like Jack and Jill ensured an introduction to a core group of professional parents and their children, but it was impossible to avoid the obvious racial divisions that the city's leaders allowed and encouraged.

There was no division as obvious as the Peyton Wall, which was erected in 1962 around the borders of a well-to-do southwest neighborhood in Atlanta. After middle- and upper-class blacks had begun moving to the nicer residential neighborhoods at the far western end of Hunter Street in the late 1950s and early 1960s, white residents began discussions on ways to keep the blacks from buying in or even driving through their residential subdivisions. White residents living in a neighborhood known as Cascade Heights were able to get the city to approve the building of a brick wall at Peyton Road and Harlan Road as a way to establish a literal line that blacks were not to cross.

"The Peyton Wall was one of the most blatantly racist and offensive gestures that was offered to us in the sixties in Atlanta," remarks Ella Gaines Yates, who lives in the neighborhood that the wall had been built around. "And what was so mean-spirited about it is that it was backed by the city's officials."

Eventually, blacks were able to buy in the Cascade Heights area established around the Peyton Forest neighborhood in the early 1960s.

"It's hard for people to imagine today how separated blacks and whites had been in this town," says Portia Scott, "We were not even allowed to join the Girl Scouts when I was a child. They only allowed white girls to join. I'll never forget when Philippa Brisbane's mother, Kathryn, integrated the Girl Scouts by organizing a troop for the black girls in the fifties."

As the 1960s brought blacks to the city's southwest neighborhoods, many whites left for the suburbs. And just as the residential housing patterns changed racially, so did the public schools and the commercial shopping areas. Lisa Cooper, an Atlanta attorney who grew up in the southwest section of the city, remembers how the racial makeup of her high school changed during the 1970s. "When I entered the eighth grade at Southwest High School in 1969, the senior class was approximately 90 percent white," recalls the Duke University graduate, "but my eighth-grade class was 95 percent black. The grades in between further demonstrated that the school was getting blacker with each grade, as the years progressed. You could see it in the yearbooks and in the classrooms. Whites were leaving that part of the city to such an extent that by the time I was a senior in 1974, all the grades were almost completely black."

Cooper, who worked for the state attorney general's office before her current position with a federal judge, grew up as a typical child of the Atlanta black professional class. Her father, Dr. George Cooper, is a Kappa who attended Morehouse and Meharry Dental School. Her mother, Carolyn Cooper, is a graduate of Fisk and belonged to one of the old-guard bridge clubs. Lisa grew up in the Cascade Heights neighborhood and, after pledging AKA at Duke, attended law school at the University of Georgia. In spite of her sheltered upbringing, the separation of the races was always evident to her at Southwest High School.

"While there were never any major conflicts," explains Cooper, "everybody knew where the lines were drawn. My class did not see the problems in the way the class of 1969 and class of 1970 do. Those two classes often have separate reunions based on race. Since those two classes had large racial splits, I've heard that they have black class reunions and white class reunions. My class was interesting because although there were almost no whites left by my senior year, it was a white classmate, Michele Belloire, who was voted 'Miss Southwest' for the year. It got a lot of attention in the press, but I think it showed that the black students were open-minded enough to see beyond color and vote for someone they actually liked."

Cooper also remembers that during her years in high school, the popular place to shop in the southwest part of the city was Greenbriar Mall. "By my generation, Auburn Avenue and Hunter were no longer popular shopping districts for young people," says Cooper, "but going to Greenbriar or going downtown was commonplace for blacks who lived in the southwest part of the city."

By the time Teresa Wiltz's generation came along six years later, the racial balance in the southwest end of the city had changed even more: There were virtually no white residents, students, or shoppers on that side of town. "And the good stores that used to be in the southwest were all gone by the time I was in high school," explains Wiltz, whose family has been in the city for several generations. "Many of us went north to shop at one of those malls, or we went downtown."

During the last twenty years, a younger generation of black professionals have been moving to the city's suburban areas—places like Stone Mountain, East Point, College Park, and, if they are particularly wealthy and want to remain inside the city, Buckhead. They have also been attracted to such upscale developments as Niskey Lake and Guilford Forest.

While the suburbs don't seem to offer the black cultural activities that exist around the Atlanta University center, there are chapters of the Links and Jack and Jill serving these growing communities. And as the black families continue to move there, the racial attitudes awaiting them are changing. "It's amazing when I think of how much things have evolved even on the outskirts of this city," says Joy San Brown, who lives inside the city. "I remember when they were burning crosses on people's lawns in Stone Mountain. And now, it has become a very popular place for black families."

Like many other cities with a black elite in its urban or suburban neighborhoods, Atlanta has a number of churches that the wealthy and well-connected call their own. One of these has long been the First Congregational Church, where one finds many of the black Atlanta doctors. One of its most famous ministers was Henry H. Proctor, an 1894 graduate of Yale School of Divinity. Given its popularity among the black doctor crowd, the church was the site of the founding of the National Medical Association. President Taft's visit to the church in 1898 was another event that enhanced the prestige of the congregation.

With members such as former mayor Andrew Young, one still finds that Atlanta's elite has a special fondness for First Congregational Church. Other popular houses of worship are Friendship Baptist Church, which was founded in the 1860s and is considered the oldest church in black Atlanta; St. Paul of the Cross; Ebenezer Baptist; Wheat Street Baptist; and Big Bethel A. M. E. It is at these churches that the old guard meets the new members of Atlanta's black elite.

As my brother discovered when he was graduating from dental school in the 1980s, Atlanta's black community can look very different if you're new in town. Regardless of your academic and business credentials, the old elite is pretty much set in stone. Where there are greater possibilities for social mobility is among the new elite that has been establishing itself in great numbers since the 1970s. The best-known link between old and new was brought about by the city's first black mayor.

Before Maynard Jackson was elected vice mayor and mayor of Atlanta in the 1970s and 1980s, he and his family were well known in Atlanta society. Jackson's mother, Irene Dobbs Jackson, was from a respected family that had been deeply involved political activism. The mother of the city's first black mayor, Irene was one of six sisters, all of whom attended Spelman College. Her father, John Wesley Dobbs, was a founder of Mutual Federal Savings, a grand master of Georgia's Prince Hall Masonic Lodge, and an organizer of many political groups in the city. One of her sisters, Mattiwilda Dobbs, became an opera star and performed throughout Europe and the United States.

"The Dobbs family was already on the map here," says a third-generation Atlantan who remembers when Maynard entered politics, "but it was really Maynard who really shook up the establishment and made them realize he could bring the old and the new together. When he ran for vice mayor in 1969, it was before the old politicos were ready to anoint him. Even with his family's background, there were still people who turned on him. But as history has proved, he showed them all."

Although he is from an old New Orleans family, Andrew Young is another important force in Atlanta political circles. As a former congressman from the district (elected in 1972) and a former mayor of the city, he has been a prominent voice in the city since his early days of working with Dr. Martin Luther King Jr. He was credited with managing a large part of the 1996 Atlanta Olympics. His brother, Walter, is a well-known dentist in the city.

Another prominent Atlanta resident is Jesse Hill, who arrived in the late 1940s and ultimately became a permanent fixture when he was named president of Atlanta Life Insurance in the 1970s. "Everyone knows that Jesse Hill supports black Atlanta," says Keith Chaplin, a Morehouse graduate who is one of the younger members of the social set. Having just run into Hill at a Links brunch and auction at the Ritz-Carlton, Chaplin points out that the businessman was the first African American to serve as president of the city's chamber of commerce. "People like him and the Baranco family are great role models in both the business and the social world."

The Baranco family is best known in recent years because it is virtually impossible to buy an automobile in Atlanta without passing through a showroom that is controlled by the Baranco Automotive Group. Headed by Gregory and Juanita Baranco, the nineteen-year-old company owns dealerships in Atlanta and Tallahassee, Florida, that sell Lincoln, Mercury, Pontiac, and Acura cars and General Motors trucks. Juanita, an attorney, is an active member of the Atlanta Links. Gregory is a member of the Boulé and also serves as chairman of the board of Atlanta Life Insurance.

Another popular family of today are the Russells, who run a major construction company that did a great deal of the building at the Atlanta airport and other important sites and has annual sales in excess of $150 million. Herman Russell, founder of the forty-six-year-old H. J. Russell Construction Company, and his wife, Otelia Hackney Russell, a former teacher, have three grown children, who are also in the business. Otelia is in the Girl Friends. Still considered members of new Atlanta, they have broken down some of the barriers to the oldest groups and families.

The Scott family became well known to black Atlantans when William A. Scott II established the newspaper *Atlanta World*, now known as the *Atlanta Daily World*. His brother, C. A. Scott, was publisher until recently. Today, family members Alexis Scott Reeves, Ruth Scott Simmons, and Portia A. Scott still operate the popular publication. Eleanor Milton Johnson grew up with some of the Scotts and remembers C. A. Scott throughout her life in the city. "Mr. Scott was 'Mr. Atlanta.' He ran that paper for sixty-two years and he knew all of us in that town like we were family," remembers Johnson.

Many of the important names in Atlanta are individuals tied to Atlanta Life, Citizens Trust, Mutual Federal, and the colleges. Atlanta Life's key people have been Jesse Hill and Don Royster, as well as Gregory Baranco. During recent years, important names at Citizens Trust have been William Gibbs, Owen Funderburg, Charles Reynolds, and Johnnie Clark. At Mutual Federal, important names are Al Whitfield, Hamilton Glover, and Boulé member Fletcher Coombs. Among the fast-moving entrepreneurs are Nathaniel Goldston, who operates a food service company; Felker Ward, who heads his ten-year-old investment firm; and Floyd Thacker, who founded a $30 million construction and engineering firm. Morehouse's Hugh Gloster, Spelman's Johnnetta Cole, and Clark-Atlanta's Thomas Cole have been among the well-known names in the university community. Among the political and civic leaders have been Shirley Franklin, a key adviser to mayors Young and Jackson as well as to the Olympic organizers; Henry and Billye Aaron, who have contributed their names and resources to causes; politicians Michael Lomax and Marvin Arrington; Dr. Alonzo Crim; Links national president Patricia Russell-McCloud; Bishop John Adams; and Friendship Baptist's William Guy. And there are many others, such as Donald Hollowell, Asa Hilliard, Tim Cobb, Thomas Sampson, Charles Johnson, Madeline Adams Cobb, Herman "Skip" Mason, and Xernona Clayton.

Perhaps better than any other city, Atlanta has maintained a black upper class that is committed to holding on to the political and economic power of its own community while also participating actively in the decisions made for the surrounding 3.5 million black and white residents in metropolitan Atlanta. It has included some of the most famous political figures in the nation—most of whom came out of the professional class.

Spelman graduate Joy San Walker Brown is proud that so many of the black institutions in Atlanta survive generation after generation, and she believes it is because of the city's black residents and leaders, who support them. "Whether it's sending our daughters to Spelman, or maintaining a bank account at Citizens Trust, or making a contribution to the Butler Street Y, or attending the Links' annual auction," she says, "we will support each other in this town."

"Many people compare us to the blacks in New York, Chicago, and Los Angeles," says a longtime Boulé member, "but our black elite is much more cohesive. Those cities haven't been able to elect one black mayor after another. They can't walk down a street like Sweet Auburn and know that old black businesses like Citizens Trust, Mutual Life, the *Atlanta Daily World*, and Atlanta Life are still thriving. They can't point to people like Maynard Jackson, Julian Bond, Andrew Young, Mayor Campbell, and John Lewis and say, 'Look at the people we produced locally and on the national scene.' And there is no other city in the country that can claim elite individuals like this, or black universities of the level that exists in Atlanta."

Black professionals in this city are justifiably proud of and unapologetic about their community's success. They realize the importance of acknowledging and celebrating the fact that they have had an elite for many generations during the city's development. As Atlanta historian Dr. Carole Merritt says, "You cannot understand the history of black Atlanta or black America if you don't include its upper class. Only then will the story be complete."

Other Cities for the Black Elite: Nashville, New Orleans, Tuskegee, Los Angeles, and Philadelphia

Large metropolitan areas, like New York, Atlanta, Washington, and Chicago, are not the only cities that have seen the development of black elite communities since the 1800s. Although these four are clearly the largest and best known, there are smaller cities as well as smaller black communities within equally large cities that have interesting and unique stories to tell. Although their number of well-to-do black families may not reach the hundreds that exist in these other cities, these towns still have certain neighborhoods, churches, social clubs, businesses, institutions, and family names that are specifically old-guard. They can be found in cities such as Nashville, New Orleans, Cleveland, Los Angeles, and Philadelphia, and in towns as small as Columbia, South Carolina, and Tuskegee, Alabama.

I have always found that when I visit friends or family members residing in moderately sized or small towns with old-guard families, there is an almost claustrophobic atmosphere immediately enveloping our every activity. Because the towns are smaller, the black elite appear to have a disproportionately large presence, and you see them everywhere: at the one church everyone belongs to, at the one Labor Day party everyone has been invited to, at the one cotillion everyone is supposed to support. Moving into a community like this and attempting to live outside its narrowly defined institutions can be extremely difficult.

In places like Columbia, South Carolina, one can see the black elite's story told in one neighborhood, or around just six or seven old businesses or families. For example, only a few families loom larger than the one that runs the Manigault-Hurley Funeral Home, which has seen four generations of the family run the same business since 1911. With degrees from Morehouse, Atlanta University, and Howard University, the Manigault-Hurley family are decidedly establishment, but as the founders of the Columbia Urban League, they continue the tradition of elite families with purpose.

"All three members of my generation graduated from Howard, including my brother, Brian, who graduated from its Medical School before working under former secretary of health and human services Dr. Louis Sullivan," says Michelle Manigault-Hurley, who is the fourth generation to join the Columbia-based funeral home. Like most old-guard families, her parents have ties to all the right groups. Her father, Anthony, a Morehouse graduate, is a thirty-second-degree Mason and, along with his wife, Alice Wyche Hurley, founded the Poinsettia Cotillion.

"We raised our children in Columbia's chapter of Jack and Jill," says Alice Hurley, "but we also wanted our daughters to have the debutante experience, so we started the formal affair in 1980." The cotillion, which is a white-tie formal ball held during the Christmas season, brings together many of the old-guard families in Columbia and nearby Orangeburg—many of whom belong to old-guard churches like St. Luke's Episcopal, Bethel A. M. E., and First Calvary.

"We're not a big city like Washington, but we have families like the Nances, the Richburgs, the Coopers," says a member of the Columbia Boulé with a hint of self-satisfaction. "Since we're small, we have our own special social core, and there's not many who try to upset us or compete with us."

Alice Hurley, who stays busy as a member of the Deltas, the Moles, the Links, and the National Smart Set, is also a graduate of Boston University and Atlanta University Graduate School of Social Work. In addition to being the coordinator of social workers in the local school district and a licensed funeral director like the other members of her family, she assumes her role in community service by giving back through her annual Poinsettia Cotillion and Columbia's Urban League Guild fund-raisers. "This is really a small town for us," she says, "and my family can have an impact through the groups and activities we've started. It's hard to do that in the larger cities."

NASHVILLE

While a bit larger than Columbia, Nashville is another southern city with a black elite contingent that seems disproportionately large for its surrounding working-class community. In that city, many of my black friends tell me that the high-profile black residents are often embraced by both the black and the white communities. One of those high-profile people is Patsy Campbell Petway.

When she pulled up in a sleek black Cadillac at the Swan Ball last April with her escort, Lewis "Bill" McKissack, photographers from *Town & Country* magazine, the *New York Times,* and several other national publications were waiting alongside the driveway to capture her arrival at the annual Nashville black-tie event.

"We had no idea that people outside this town knew about this ball," says Petway as she recalls the evening. For thirty-five years, the virtually all-white, all-rich crowd from the Belle Meade neighborhood and other sections of the affluent west end of Nashville have paid exorbitant ticket prices—now one thousand dollars—to dance and dine the night away at the Cheekwood estate and garden where the black-tie charity ball takes place each year.

Even though the Campbell, Petway, and McKissack names are a big deal in the Nashville black community—a community that centers on Fisk, Tennessee State, and Meharry Medical College—the Swan Ball was one of those pivotal moments when a handful of upper-class blacks actually interact with upper-class whites in this southern city. "I'm on the board of Cheekwood with Leatrice McKissack," explains Petway as she points out the need for blacks to involve themselves in charities that benefit the black community as well as the larger community. "We need to make our presence known wherever we live," says the Tennessee State graduate, who belongs to the Nashville Links and the Girl Friends. And this is something that all three of these families have done in Nashville for the last fifty years. Their role in the black community there has been virtually unparalleled.

A lifelong resident of Nashville, Patsy has been an important figure in the city's civic activities—as was her husband, Carlton, who died of a heart attack in 1991. He was a city councilman and a wealthy, high-profile litigator who taught at Vanderbilt Law School and received an appointment to the U.S. attorney's office as the first black to serve in the Middle District of Tennessee.

Bill and Leatrice McKissack's family is best known for founding McKissack & McKissack, the architectural and engineering firm that designed and built Carnegie Library at Fisk in 1908, the Tuskegee Squadron Air Base and Pilot Training School in 1942, the Morris Memorial building for the National Baptist Convention, the National Civil Rights Museum in Memphis, and Nashville's Capers C. M. E. Church, as well as many other churches, homes, and buildings on such campuses as Howard University, Lane College, Texas State University, and Tuskegee University.

"Our family members were among the first registered architects in the state of Tennessee," says Leatrice Buchanan McKissack, who heads the firm, which has also opened offices in Memphis, Philadelphia, Harrisburg, Washington, and New York since Moses McKissack Jr. first opened a business under the name McKissack Contractors in the 1870s. "My husband, William, was the grandson of Moses junior, and he graduated from Howard's School of Architecture and Engineering. In fact, all five of my husband's brothers earned degrees in architecture or engineering so they could enter the business. And I was happy to see our three daughters, Andrea, Cheryl, and Deryl, do the same."

"I remember back in 1942, when Calvin and Moses McKissack went up to the White House to get an award from Franklin Roosevelt when the company was named the best black business in America," says a proud Nashville doctor as he sits in a front pew at the old-guard Holy Trinity Episcopal Church in south Nashville. He says he often runs into Leatrice and her daughters near the firm's Broadway office. "I knew Leatrice's stepfather, Dr. Alrutheus Taylor, who got a Ph.D. from Harvard and was the dean at Fisk," adds the doctor.

Leatrice sees her friend Patsy Petway at board meetings for Cheekwood, or for other activities in Nashville, and their circle of black friends is an interesting one. It includes Leatrice's friends from her years at Fisk, as well as the associates they both knew from their years at Tennessee State. Leatrice knew such people as writers Langston Hughes and Arna Bontemps. In fact, on the evening when Petway first introduced me to her friend, Leatrice was dining with screenwriter Alice

Randall, who is married to sixth-generation Nashvillian David Ewing. As I am quickly reminded by one of the dinner guests, Alice was originally from Washington but was adopted by Nashville society during her first marriage—into the historic Bontemps family.

"Among the three of them," says an animated member of the Nashville Links, referring to Patsy, Leatrice, and Alice, "they know everybody across three generations. In a small town like Nashville, there are certain people you have to know and you have to get their background right." The woman explains that newcomers mix up facts on important but unrelated people who have the same last names—people like Helena Patton Perry, the pediatrician, who was married to two doctors, and the unrelated Rosetta Perry, who is quite rich and owns the weekly newspaper. Or they are confused about which families belong to which Links chapter.

"Like the other week," explained the woman. "There was this young upstart trying to convince me that Corinne Schuster was the founder of the newest Links chapter! I instantly knew this girl was new in town. There are like four Links chapters in Nashville and everybody knows Corinne is in the original one—the real Links chapter. Her chapter has been around for thirty-five years, and she used to be president of it!"

"That's the kind of mistake people shouldn't be making," adds the woman's luncheon companion. "After all, the original Links chapter has the big names: Alberta Bontemps, St. Clair Foster, and Edwina Hefner—her husband, James, is president of Tennessee State."

And not surprisingly, it's also the chapter to which Patsy Campbell Petway belongs. Her sister Doris belonged to it, and her niece Pamela Campbell Busby is a member today. These are facts that the old guard keeps straight.

Patsy's other niece, Gail Campbell Busby Schuster, points out how fortunate they are to live in a city that is sophisticated, yet still small enough for families to know and recognize each other. Gail's grandfather—Patsy's dad—Emmett Campbell, had done the electrical wiring on the old-guard Holy Trinity Church. "When my son, Evan, was baptized here," says Gail, "it gave me such pride to know that my family's history was in this church and in this town. It makes you feel rooted to know your own history is surrounding you."

Gail remarks that her mother, along with Aunt Patsy and several of her other aunts, attended Tennessee State. "In fact," adds Patsy, "three of my sisters—Irene, Alberta, and Kathryn—went to a small private demonstration elementary school on the grounds of the college in the 1920s because President W. J. Hale had young children that he wanted educated nearby." And Patsy's uncle, John Galloway, also a member of Holy Trinity, was principal of Pearl High School, the public school that three generations of the city's black elite have attended. Her sister, Phyllis Campbell Alexander, says that life in what was then the segregated north side of town was quite insulated from the problems blacks found elsewhere. "When we were growing up in a black community that had its own universities, our world was insulated from the bigotry we might have faced at segregated institutions run by whites," explains Phyllis, who belongs to the Links in Los Angeles. "When we went to a ballet or an opera as children, the performances we saw were special ones, put on at our universities' campuses, by our own people. Because black Nashville had these resources, we were in a fortunate situation."

But now, many old-guard families who grew up and established themselves in north Nashville, near the three historic black campuses, are seeing the next generation of black professionals venture to the newer, whiter, and richer west side of town. "My father and mother moved to the western side of Nashville in 1974," says Gail Campbell Busby Schuster, "so I missed the whole Pearl High School scene. Along with my sisters, Suzanne and Pamela, I had a much more integrated school experience than my parents' generation." Gail's father, George Busby, a dentist on the faculty at Meharry, belongs to the Boulé, where he socializes with Dr. Henry Foster—President Clinton's controversial surgeon general nominee, as well as a crowd that is beginning to reflect the new generation to which Gail refers: young adults who grew up in Jack and Jill, but who were raised at integrated schools, in integrated neighborhoods, far removed from Jefferson Street, the Ritz Theater, the Bijou, and Meharry Boulevard, which the black elite favored in the 1930s and 1940s. They still go back to churches like Holy Trinity, St. Luke's, Park Memorial, and Capers Memorial. They still know the history of the Boyd family, who founded Citizens Savings Bank. They even still call Patton Brothers or Kossie Gardner to arrange services at the Gardner family funeral home when one of their loved ones dies. But things are not the same as they used to be, because they live and work around whites more than ever before.

"Some things have changed here," admits Patsy Campbell Petway, "but we can pretty much find the people we grew up with. Even if places like Brown's Hotel and Dr. Price's pharmacy are gone, there are enough institutions and events that have remained constant—like my AKA chapter, which still has its debutante ball at Christmas. Of course, now it's grown so big that we have it at the Opryland Hotel."

But Patsy Campbell Petway doesn't get caught in the past. She doesn't let the changes slow her down. Whether it's dinner with sister-in-law Rose Busby, a fund-raiser that's she's attending with Bill McKissack and her two sons, or an event that her Girl Friends chapter is arranging, she's moving. On the afternoon that we spoke she was on her way to Aspen, Colorado, for the wedding of her godchild, Kimberly Webb, an attorney who grew up in Houston and was marrying Douglas Qualls of Memphis. "Kim is the daughter of Hildred Webb, one of my best friends from Tennessee State," she explains as she reminds me that I had dined with Hildred and Kim in New Orleans, along with Patsy's childhood friend Aaronetta Pierce, at the last Links convention that my parents had brought me to.

"You remember Kim from Wellesley. You met her when you were at Harvard. And her brother, Johnny, went to church with you last Easter."

I nod with immediate comprehension as I recall that Hildred's husband, John—a Houston surgeon—is in my fraternity, Sigma Pi Phi Boulé, and he had hosted me during a recent tour of Texas.

Once again, Patsy reminds me of how small the circle really is.

New Orleans

"There are the *uptown* Dejoies and the *downtown* Dejoies, and from what I hear, the only time they speak to each other is at funerals," explains a member of the New Orleans chapter of Jack and Jill after I meet her in the dining room of Dooky Chase, a black-owned restaurant popular with the older families of New Orleans. The woman has spent the last twenty minutes attempting to tell me how complicated it gets when you try to rank the impact of certain black families in New Orleans.

"Many families—including my own—aren't as *in* as the Dejoies, the Haydels, or the Tureauds," she says, "but even when you do find people from an important family, you have to determine which branch they are from because there are so many ways to label family relatives here: by the ward they come from, by their black or Creole heritage, by the whole downtown-uptown-back-o'-town distinctions. In some of the families, there is a black group as well as a white group. It's hard to keep it all straight. But if you don't, people get very offended."

This is the kind of pressure I was feeling as I was on my way to have lunch at the Martha's Vineyard summer home of Ellie Dejoie Jenkins. Although she doesn't use such terms to identify herself, Ellie Dejoie Jenkins is clearly a "downtown Dejoie," and I didn't want to make the mistake of forgetting that fact. Her family is old New Orleans, and she has all the right old-guard credentials.

"I don't like to get caught up in all those ridiculous distinctions," says Jenkins. "If others do, that's their business."

Ellie is a slender, small-boned, light-complexioned black woman. It seems to be well-established that the wealthy Creoles and the aristocratic blacks who come from her background live in downtown New Orleans, while the working-class blacks and upper-class whites reside uptown and the working-class Creoles stay in "back-o'-town."

"You have never seen class warfare until you've come to New Orleans," says a member of the Boulé who grew up in the city and has known the downtown Dejoie family for years. "There are the rich blacks and the blacks who worked in service for the whites, then there are the rich whites, and then there are the Creoles—who might be very, very poor and uneducated, or there's the group of Creoles who are rich and who may be either white-looking or black-looking. It was not uncommon to find three branches of the same family living as three different racial or ethnic groups. Historically, there was so much mixing in that town, you find people with relatives of every shade."

Ellie Dejoie Jenkins's grandfather was a physician, and her father, Paul Hipple Vitale Dejoie, was a successful businessman and a member of the elite Old Men's Illinois Club. When Ellie was growing up in the elite downtown section of New Orleans, it was rather obvious that there were two branches of the Dejoie family. Although the branches did not always get along, they were all people with money and influence. "My family owned the Louisiana Life Insurance Company, the Louisiana Funeral Home, and the Dejoie Flower Shop, so we grew up knowing many of the businesspeople in New Orleans," explains the attractive woman as she sits with a couple of friends and her husband, Judge Norman Jenkins, on the deck of their summer home, a contemporary masterpiece that is surrounded by woods. "There was another Dejoie that owned the *Louisiana Weekly* newspaper—but that's a long story because my immediate family had ties to that before as well."

Knowing the uneasy relationship that many old New Orleans families have with various branches, I know enough not to press Ellie. She is gracious and diplomatic about "the other Dejoies" even though there are others who are not when the subject is raised. I already know about one other Dejoie—C. C. Dejoie, who had been publisher of a black newspaper, the *Louisiana Weekly*. In black New Orleans society, it is not in good taste to ask people about the ancillary branches of their families. It is also not in good taste to ask people who are mixed with black Creole or white ancestors to tell you whom they most closely identify with.

As her small white poodle, Precious, hops into my lap, I ask Ellie about some acquaintances I had known in her native city. I also ask her about the city's first black mayor. "What about Ernest Morial's family? They were contemporaries and in the same crowd as your parents, right?"

One of Ellie's visitors tries to answer the question first. "Hardly. Ernest just happened to marry well."

Ellie, who is very protective of the important families of New Orleans, raises her hand to stop the flow of negative remarks before they get started.

I realize I will have to get my information elsewhere, because Ellie is an unqualified fan of the Morials.

"Ernest Morial was a brilliant judge and politician, but he was from a working-class family," says a third-generation Dillard alumnus who admires the family that produced the city's first two black mayors. "But 'Dutch,' as we called him, got great connections by marrying into the Haydel family when he met Sybil." Ernest Morial, who was elected mayor in 1977, according to this self-described fan of the Morials, was actually *not* from the elite downtown section. "He was from the working-class 'back-o'-town' section."

The Haydels have long been a powerful, well-to-do family in New Orleans. "I grew up with Sybil Haydel and knew her quite well when she married Ernest Morial," explains Ellie Dejoie Jenkins. "Her father, C. C. Haydel, was a physician, and his wife was a good friend of my mother, Thelma, and my aunt Pearl, who were all in the Gloomchaser's Club together. Everyone in our crowd was proud when he and his son got elected mayor. We feel like an extended family."

While Ellie is talking, I overhear one of her friends remark that Atlanta mayor Andrew Young was originally from New Orleans, and that his mother was also a Gloomchaser—as was Mrs. Alfred Dent, wife of the president of Dillard University.

"Well, Sybil's family were all great supporters of Ernest," continues Ellie, "and he served on the court of appeals, then as mayor, and we were all thrilled to see his and Sybil's son get elected mayor a few years ago. They are a talented family."

Although Ellie now lives in Philadelphia, her ties to New Orleans are strong and old. Like many of the elite blacks and Creoles of color, she went to a Catholic private school, Xavier Prep, and then to Xavier University. She also has a master's degree from Columbia. She is a member of the Girl Friends. Her childhood friend, Sybil Haydel Morial, is a member of the Links. Sybil's brother, C. C. Haydel Jr., is a physician who belongs to the same New Orleans Boulé chapter as their late father.

Following in the footsteps of their parents, many of the middle-aged men and women in the old guard belong to the national black groups as well as to the groups that are uniquely New Orleans. New Orleans Link Viola King belongs to a group called the Merrymakers. "We are a bridge group of thirteen women that started fifty-seven years ago," says King who graduated from Southern University and includes Sybil Haydel Morial among her friends.

"Many of us belonged to clubs that grew up around the Mardi Gras festivities," says Harold Doley, who grew up in New Orleans and can trace his paternal ancestors back to 1720 in Louisiana. "My father's group, the Illinois Club, was started in the 1880s, and one of its big activities was putting on an annual cotillion for the daughters of many of the old families."

Like Ellie Dejoie Jenkins, Doley also attended Xavier University and the private, all-black Xavier Prep before that. Growing up in a relatively small city, where even the Catholic schools were racially segregated, they both felt that their community was a small and intimate one. They both knew the doctors who worked at Flint-Goodridge Hospital, they both ran into friends on St. Bernard Avenue downtown or in the neighborhood surrounding elite Eastover Drive and in the "Sugar Hill" section near Dillard, and they both were part of the elite who had ties to St. Luke's Episcopal Church and Central Congregational Church.

The first black individual to buy a seat on the New York Stock Exchange, Doley now runs an investment firm that has offices in New York and New Orleans. He is also the owner of Villa Lewaro, the twenty-thousand-square-foot mansion that was built by Madam C. J. Walker after she became the first woman millionaire in 1908. He is one of the many black New Orleans success stories that became famous elsewhere but still holds tightly to his native city. "Few people realize that Andrew Young and Bryant Gumbel are from New Orleans," says Doley, who met

Gumbel many years ago at a childhood birthday party and remembers when Young was a student at Dillard and later joined Central Congregational Church as assistant pastor.

"Even if we later move to other cities," explains Doley, "the culture of black New Orleans is so strong that we end up returning to the community or finding others around us who share these roots."

TUSKEGEE

Although Tuskegee, Alabama, is a relatively small city with a population of less than two hundred thousand, the community played an important role in the development of the black elite. Not only did the increasingly well-endowed college, Tuskegee Institute, attract a talented faculty, but the area also served as the training ground for thousands of black dentists and physicians. The town remains a place with fond memories for many of the people whose family names, like Dibble, Tilden, and Branche, are associated with it.

"The Veterans Hospital in Tuskegee was one of the few places in this country where black physicians could receive training in surgery," says Dr. George Branche, who grew up in Tuskegee while his father, Dr. George Branche Sr., worked as a neuropsychiatrist at the hospital. "My father started working there in 1924 after he had attended Lincoln University and Boston University Medical School," recalls Branche, as he and I are leaving our monthly meeting of the Westchester Clubmen, a forty-year-old social organization of black professional men.

Branche remembers growing up on the grounds of the hospital with his sister, Martie, and his brother, Matt, who was eventually to attend Boston University Medical School. His brother Matt is in my Boulé chapter and his son, George, grew up with me in Jack and Jill before going off to Princeton.

"There were several Tuskegee families who lived on the grounds in an area we called 'the circle,' and they were all doctors tied to the Hospital," recalls Branche as he remembers how much that rarefied environment protected him and other children from the very segregated world of Tuskegee's town outside these gates. "We were protected from the bigotry of the town in so many ways. We went to an elementary school on the grounds and had a swimming pool and tennis courts there as well. We also grew up seeing many famous black people like George Washington Carver and General Benjamin Davis and others."

Among the other families living on "the circle" with the Branches were the Tildons and the Dibbles. Toussaint Tildon, a psychiatrist and the director of the hospital during the 1930s, 1940s, and 1950s, was a class-of-1923 graduate of Harvard Medical School. More than two generations of black physicians point to Tildon as their role model and mentor; they either trained under him or were invited to become a part of the growing institution.

"Everybody knew Dr. Tildon because of the stature he brought to the hospital," says Dr. James Norris, a New York plastic surgeon who was in charge of surgery at the V.A. hospital in the late 1960s and early 1970s. "He, Dr. Dibble, Dr. Yancey, and Dr. Branche were all important figures in both the medical and the social communities."

Tildon's four children were among the kids that George and Matt Branche remember from their childhood. "Hortense Tildon, Ann Dibble, and others of us still run into each other," says George, who was later to avoid the segregated Alabama schools when his parents sent him north to attend Boston Latin School.

Hortense's daughter, Margaret Calhoun Williams, recalls growing up in Tuskegee but not being fully aware of her family's history in the community. "I knew that a lot of people knew my grandfather, but it wasn't until I was much older and moved away that I got a sense of the impact he had on so many doctor's careers," says Williams, who remembers that although she did not grow up on the campus like her mother, she did not really encounter the town's segregation in the way that others might have. She was in Jack and Jill as a child and went away to boarding school at Northfield Mount Hermon in the ninth grade. "Several of the Dibble children also went to my school," Williams recalls.

The Dibble family's history at the V.A. hospital began with Dr. Eugene Dibble, a class-of-1919 graduate of Howard Medical School who was chief of the hospital surgical section. He and his family also lived on the circle. His daughter, Ann Dibble, is a graduate of Fisk and sits on the board of several major corporations, as does her husband, Washington lawyer Vernon Jordan. Like their childhood neighbor Hortense Tildon, Ann Dibble and her brother Eugene Dibble sent their children to Northfield Mount Hermon in Massachusetts.

"I was in the Tuskegee Links with Mrs. Branche," says a former Tuskegee resident who knew many of the families with ties to the hospital, "and I have to say that place produced some amazing histories.

Not that Ann Dibble's background wasn't interesting on its own, but her first husband was Mercer Cook in Chicago, and his lineage was also interesting. His father, Mercer senior, graduated from Amherst in the twenties, and got a Ph.D. from Brown in the thirties. He was a professor at Howard and ambassador to Senegal. You can't beat families like that." The former Tuskegee resident also points out that although they are living in different parts of the country, virtually all of the Branches, Tildons, and Dibbles are members of the Boulé and the Links.

Another magnet for talented blacks in the small community was Tuskegee Institute, which had grown increasingly wealthy through the connections that its founder, Booker T. Washington, had with well-connected white philanthropists at the turn of the century. With his growing endowment, he was able to hire the best professors for his university. "Booker T. Washington asked my father-in-law, Robert Taylor, to come to Tuskegee and design the campus buildings," says Marian Taylor of Manhattan. "Robert had graduated from MIT in 1892 as the first black graduate and the valedictorian in the area of architecture, and Mr. Washington had seen some of his work." Marian is married to Taylor's son, Edward, who belonged to the Boulé, and she points out to me photographs of the historic buildings that he designed. In total, he is responsible for forty-five structures on the campus, including the chapel, the science building, and many dormitories.

New York surgeon Dr. James Norris points out that the John Andrew Hospital, which is affiliated with Tuskegee Institute (now called Tuskegee University), was also a major center for black doctors. His father, Dr. Morgan Norris, was an intern at the hospital after graduating from Howard Medical School in 1916. "Although the surrounding town was extremely segregated, the campus of the V.A. hospital, John Andrew Hospital, and the college was an oasis for the black intelligentsia," remarks Norris. "People like Asa Yancey went there and set up important internship programs that allowed blacks to enter fields of medicine that were closed to blacks in hospitals throughout the rest of the country, including the North."

Norris also recalls that even after his own father left Tuskegee and was later serving as a trustee at Hampton, he would return to Tuskegee's medical community for regular medical seminars. "It is remarkable to think of how much the institutions in that small town changed black America."

LOS ANGELES

There is a city park named after his father. Dozens of local charities have been supported by three generations of his family. Others in his community point to him when they talk about one of the nation's oldest black insurance firms.

Ivan J. Houston is a member of one of Los Angeles's most prominent black families. A lifelong resident of the city, he is an accountant and the former chairman of Golden State Mutual Life Insurance Company, a firm which his father founded in 1925, and which is now the nation's third largest black-owned insurance company. As a well-respected business leader and adviser to L.A.'s corporate elite, Houston has sat on the boards of such companies as Metromedia, Pacific Bell, Kaiser Aluminum, and the First Interstate Bank of California.

Although the Houstons are considered one of the old-guard families, L.A.'s black elite history is among the youngest of all the cities in the United States. This is because Los Angeles developed long after eastern cities like Washington and New York, and also because the black population here did not grow out of the early eighteenth- and nineteenth-century slave populations of the East Coast or out of the 1930s black southern immigration to the Midwest.

"Even though blacks were a rather small percentage of the Los Angeles population at the time I was growing up in the 1930s," says Houston, who—like his father—belongs to the city's Boulé chapter, "there were many professionals—physicians, dentists, attorneys, and entrepreneurs—working and living around us in the Central Avenue neighborhood and on Adams and in the Sugar Hill section. Many of them went to the same churches—First A. M. E., Second Baptist, and later Holman United Methodist at Adams and Crenshaw."

Since the black L.A. community was far newer than those in the Midwest and on the East Coast—only a few hundred blacks lived in the city at the turn of the century—the old guard remained small for years. Thomas Shropshire, a fellow member in Ivan Houston's Boulé chapter, points out that some of the old black organizations were very late arriving in the city. "I organized our Guardsmen chapter here only a few years ago," says Shropshire, a retired Philip Morris executive who lives in the city and has sat on the board of Howard. "The Guardsmen didn't move to the West Coast until recently, even though it was started in the 1930s on the East Coast."

Among the prominent black Los Angeles families—the Houstons, the Dunnings, the Hudsons, the Groveses, the Weekeses, the Lindsays, the Blodgetts, and the Somervilles—one finds almost no ties to Hollywood. Instead, their professions and their successes were in the fields of banking, medicine, dentistry, insurance, real estate, hotels, and engineering. "When my father, Norman, founded Golden West, we lived next door to the Hollywood actress Louise Beavers," says Houston, "but my father's friends were mostly businesspeople or doctors." Houston points out that the racially restrictive covenants in the city's neighborhoods placed all the black people in close proximity to each other, but this did not bring together black entertainers and their black professional neighbors. They were two entirely different social groups.

Another phenomenon that distinguished the old L.A. black elite from the groups in other cities was the academic experience of its members. Many of them had no ties to prominent black colleges like Howard, Morehouse, or Spelman. Ivan Houston, for example, went to the University of California at Berkeley—like his father. "Blacks that grew up in Los Angeles during the 1920s and 1930s had a much different experience than those who grew up in southern cities or even eastern cities. Many of us went to predominately white California state universities."

Similarly, blacks in Los Angeles were more likely to have had integrated grammar-school and high-school experiences than their counterparts in places such as Atlanta or Washington. While blacks in those cities grew up in segregated elementary and secondary schools, L.A. blacks attended schools that were predominately white with a small number of Asian and Latino students. "In fact," Houston says, "many of us had been educated our entire lives with white classmates. Our neighborhoods might have been all black, or mostly black, but our early school experiences were integrated ones. So it was not a surprise that I would end up in a school like Berkeley or my brother Norman would end up attending UCLA."

Among the elite families, there were people like the dentist Dr. John Somerville, who built the Somerville Hotel, the precursor to the black Dunbar Hotel on Central Avenue. There was also another dentist, Claude Hudson, who was the beginning of several generations of Hudsons who excelled in the banking business. Joe Dunning was a successful engineer who attended MIT. "My father and the Blodgett family worked to start Liberty Savings Bank," says Houston.

The generation of black professionals who followed Ivan Houston moved further south than the South Central neighborhood that had so long been associated with the black community. They moved, principally, to three upscale suburban areas within the city: View Park, Ladera Heights, and Baldwin Hills. At the foot of these three neighborhoods is a park named after Ivan's father.

"Since we don't typically include these black entertainers or the people who are tied to the music or film business as a part of our crowd," says a Los Angeles Jack and Jill mother whose children graduated from the organization in the 1960s, "there are really two sets of affluent black people in this town: the ones who live in our neighborhoods and join our organizations, and the ones who make a lot of fast money in music or TV but are not very educated or rooted in the community. Some of them want to attend our affairs so they can check off a box and say they 'belong.' Back in the sixties and seventies, they wanted to break into this crowd, get their name in Jessie Mae Brown Beavers's columns in the *Los Angeles Sentinel,* and then move on."

The Jack and Jill mother remembers the temptation that she and others have had to invite industry or celebrity types into their affairs, only to be reminded that the old guard and the industry people have little in common. "I remember working on fund-raisers," the woman recalls, "and all of us pretty much lived up here in View Park or Baldwin Hills, and we thought we should include a celebrity or some record executives to broaden the guest list. Big mistake. They have no class and no patience for people who won't give them the star treatment. That's why I am so glad that the Links cotillion hasn't turned the list of girls into a pageant full of celebrity daughters. The only one who had any class was the singer Marilyn McCoo, who debuted in 1959." The woman pauses for a moment. "But then, of course, she was Jack and Jill and came from a good family."

Of all the annual old-guard events in the black community, the Links cotillion is one of the favorites. "I have gone to many of those cotillions, and it's one of the nice southern traditions that Los Angeles has adopted even though it no longer exists in some cities," says Links member Phyllis Campbell Alexander, whose husband, Joseph, is a surgeon in the city. "For the last fifteen years, my Links chapter has been known for an annual awards luncheon for our achiever program that takes place at either the Beverly Hilton or Century Plaza," explains Alexander, "where we raise around one hundred thousand dollars and offer college scholarships to students in the L.A. schools." Other major events have included the Omega Starlight Ball and the AKA's annual Fantasy in Pink Ball, which typically attracted as many as fifteen hundred people at the Biltmore Hotel in its heyday in the 1970s.

"Laura Anderson, Marjie Davis, Nancy Graves, Laura Hunter, Angela LaMotte, Laurie Marine, and Cynthia Wilson." It's been two decades since they graduated, but Los Angeles native Teresa Clarke can recall their names as if it were yesterday.

"There were seven other blacks in my class of sixty-nine girls," says Clarke as she recalls the class of 1980 at the Marlborough School, the most exclusive girls' school in Los Angeles. Located in the conservative, blue-blooded neighborhood of Hancock Park, the prestigious school sent a hefty percentage of its students to Stanford, Berkeley, UCLA, and the private colleges on the East Coast. Although Clarke was later to attend school with my wife at Harvard College, Harvard Law School, and Harvard Business School, she has very clear memories of what it was like growing up among the black elite in the Los Angeles of the 1970s.

"Although Marlborough attracted smart blacks from top neighborhoods in L.A.," remarks Clarke, a former Goldman Sachs investment banker who grew up in the city's upscale View Park neighborhood, "it was obvious that our white classmates who lived in Hancock Park or in other West Side areas like Bel Air, Beverly Hills, Pacific Palisades, or Brentwood had absolutely no idea of where we came from. They had no idea where Baldwin Hills, View Park, and Ladera Heights were—and since whites didn't live there, they really didn't care."

The three neighborhoods that Clarke refers to are the sections where one finds the majority of today's old-guard black L.A. residents. Clarke's experience demonstrates that although smart, well-to-do blacks and whites came together and befriended each other in the very best schools, they still remained worlds apart in the diverse city of 3.5 million.

A few years younger than Clarke, James Bond, a Los Angeles native, agrees that most whites on the West Side would have no ties to these affluent black neighborhoods, which were far away geographically and in ethnic culture. "While a lot of my classmates would have heard of Baldwin Hills and View Park," says the class-of-1990 graduate of the Bel Air Prep School, "most white students would never have been there before."

Even though she grew up in sprawling Los Angeles and had many white friends, Clarke's world was well defined. "I used to give an annual

Christmas party that was attended by students from the private schools and parochial schools around L.A.," she explains. Besides Marlborough students, there were kids from Westlake, Harvard, Buckley, and Crossroads, and a couple from University High. But what is interesting is that while they spent their schooldays with white students, the evening parties around the pool at home were exclusively black.

Clarke grew up in a professional family that includes a grandfather who graduated from Lincoln University, an uncle who graduated in 1944 from Yale, a grandmother who graduated from Howard, and a mother who graduated from Howard and holds a doctorate from UCLA. Her circle of black neighbors and classmates are from very similar families who live in beautiful homes in the neighborhoods of View Park and Baldwin Hills. She stays in touch with her old girlfriend and former neighbor Shaun Biggers, who was a class behind her at Marlborough and Harvard. The daughter of a prominent black surgeon—also in the L.A. Boulé—Shaun is now an obstetrician in New York. When I see Clarke or Biggers today, they reminisce about their experiences of growing up in a tightly knit black Los Angeles. Interestingly, like the older generations of the L.A. elite, this group still does not include entertainers or other blacks in the music, television, or film industries.

Spelman graduate Heather Bond Bryant remembers when her family moved from Baldwin Hills to the affluent and mostly white Laurel Canyon neighborhood in 1971, and then a few years later to an even wealthier West Side neighborhood abutting the Beverly Hills Hotel. "I have always had close friendships with whites wherever I lived," says Bryant, who recalls that the only black neighbors she had on the West Side were Billy Dee Williams and Rae Dawn Chong's family, "but what kept me truly connected to my black world were my family and those early friendships I had formed with kids in View Park and Baldwin Hills."

Bryant's father, a residential developer, had built their prior home in 1961 on the highest point of Baldwin Hills—on Don Carlos Drive, where the north side of the house had a view all the way across the city. The family could see the Hollywood Hills twelve miles north of them. "And we had a lot of glass on the west side of the house," remembers Bryant, "so that on a clear day, you could see Santa Monica and the ocean. A lot of the houses in those two neighborhoods have incredible views."

The families in those neighborhoods were the envy of some of the blacks living in the older South Central neighborhoods. "That was the place to move back in the seventies," says a Los Angeles attorney in her thirties who grew up on Crenshaw in the South Central neighborhood. "There was this big doctor-lawyer-CPA crowd back then—people like the Groveses, Feasters, Hunters, Littles, Paxtons, Moultries, Gibsons, LaMotts—and a lot of them lived up in those View Park–Baldwin Hills neighborhoods and went to St. Bernadette's Church. I went to one of those open parties when I was in junior high, and it was the first time I'd ever seen a black person with a Rolls Royce."

Today, there is even a growing number of black professionals moving into the previously all-white West Side areas of Beverly Hills, Pacific Palisades, Bel Air, and Brentwood. "My wife, my son, and I moved to Pacific Palisades thirteen years ago," says Los Angeles Boulé member Bernard Kinsey, "and we see blacks living in every area around the city. Many have moved to San Bernardino and Riverside counties, but what we have to remember is that no matter where we settle, it is still necessary to give our black children a positive self-image. This is why we had our son, Khalil, participate in Jack and Jill activities." Kinsey, who cochaired Rebuild L.A. with Peter Ueberroth after the riots in 1992, says that L.A. is not very different from other cities when it comes to successful blacks.

"Ultimately, our success is going to be found in starting our own businesses," says Kinsey, who has made a successful career as a developer of multimillion-dollar homes in the Pacific Palisades and Bel Air neighborhoods. His own home, which has a 300-degree view of the Pacific Ocean, is in a town where very few blacks live, but Kinsey points out that like most blacks who live in integrated, affluent settings, he always remembers who he is and how he got there. "My wife and I recently appeared on the *Tony Brown Show*," says Kinsey, "and I thought about what a shame it is to see what's going on out here with affirmative action. There are many blacks who have succeeded because of it, and now they are turning against other blacks who could benefit from it."

As Kinsey and others of L.A.'s black elite are beginning to see, blacks who wield clout in the area are no longer simply those who feel rooted to the old South Central community that grew up in the 1920s with an Afrocentric outlook.

PHILADELPHIA

Many members of black society point out that while it is no longer a city which leads in its number of black elite families, Philadelphia remains an important town because it is where black society's three most important selective clubs were founded. Native Philadelphian Dr. Nellie Gordon Roulhac does not seem fazed by the common thread. "The Links, the Boulé, and Jack and Jill shared many of the same members because they were all black people looking for opportunities to socialize with blacks who had similar backgrounds and similar desires to improve the black community." A member of the Philadelphia Links, she is a former national president of Jack and Jill and was married to the recently deceased Dr. Christopher Roulhac, who was a member of the original Boulé chapter.

"My husband, our children, and I all gained immensely by belonging to these groups," says Roulhac, "because they further rooted us to the community." Although she was later to spend part of her adulthood in Memphis and in Georgia, Roulhac's family story is primarily rooted in the genteel setting of historic black Philadelphia. Her father, Dr. Levi Preston Morton Gordon, was a dentist who graduated from Howard University and its dental school. Her uncle, Dr. Chester Gordon, was a physician who graduated from the University of Pennsylvania; and her grandfather, Dr. Alexander Gordon, was pastor at the city's Monumental Baptist Church.

Even though she has degrees from Cheyney University, Columbia, and the University of Sarasota, and has a stellar social background and family history, Dr. Roulhac insists that she does not represent old-guard Philadelphia. "There are many distinguished families who have a much longer history in this city than my own."

In fact, although the city's overall population is considerably smaller than New York's and Chicago's, Philadelphia had a prominent black upper class long before these other northern communities. Many members of the old guard in the North have family histories that seem almost new when they are matched with those with Philadelphia lineage.

One of the women in Roulhac's crowd, Emilie Montier Brown Pickens, came from one of those old families. Born in Philadelphia, Emilie was also a Link and a national president of Jack and Jill. "My mother's family was responsible for building the oldest African American home in Pennsylvania, just outside the city in an area known as Glenside," says William Pickens, who now lives in New York.

John Montier, who erected his home in 1770, was a free black man and was eventually included in a census that was taken in the 1790s by Benjamin Franklin. "My mother's family also had a family burial ground across the street from the 220-year-old home," explains Pickens, "and there were more than seventy of our relatives buried there during the 1800s."

While most members of Philadelphia's black society cannot trace their local roots to the 1700s as the Pickens-Montier family can, there are, nevertheless, many family names that have been around for generations.

"When I was growing up in Philadelphia, the town was small enough that you knew the names of the old families," says Boulé member Boyd Carney Johnson, whose uncle and father founded the city's chapter of the Alphas when they were at the University of Pennsylvania in the 1920s. In addition to names like Alexander, Purvis, Adger, Forten, and McKee, Johnson was particularly familiar with the Minton family name. "The Mintons were Philadelphia royalty—especially among the doctor crowd."

After having attended Phillips Exeter Academy and the Philadelphia College of Pharmacy, Dr. Henry M. Minton went on to graduate from Jefferson Medical College in 1906 and was the founder of the Boulé and of Douglass Hospital, Philadelphia's first black medical center.

"A lot of my parents' friends at Mother Bethel Church used to wonder where Minton got all his money because my dad and his friends were all doctors, and none of them had what he had," says an elderly physician. "He had lots more than regular old black-doctor money."

"And more than white-doctor money too," adds a fellow physician who remembers Henry and his protégé, nephew Russell Minton. "They had the kind of money and lineage that would normally make somebody bored by us ordinary black doctor types."

The phrase "black doctor crowd" does not seem to be an unusual one among Philadelphia society. Whereas the crowds or cliques in Atlanta and Nashville were formed around the various black colleges, Philadelphia had no black universities, thus leaving the divisions to be drawn around certain professions—mostly the legal and medical professions. And there were also entrepreneurs. The Mintons fit into all of the groups.

Minton's grandfather, also named Henry, had arrived in Philadelphia from Virginia in the 1830s and opened a catering firm that exclusively served wealthy whites. This was a business that many of the old black families engaged in during the 1800s. "Many of the most successful nineteenth-century black entrepreneurs, like Norris Herndon of Atlanta, opened businesses that exclusively served a rich white clientele, and these lessons that were learned in the South were copied in the North," says Atlanta's Janice White Sikes of the Auburn Avenue Research Library on African American Culture. These patterns were copied in the North for many occupations. It happened with barbershops, catering, and tailoring. Then, with the money earned from such businesses, entrepreneurs invested in more respected and profitable ventures like banking, real estate, newspaper publishing, and insurance.

And, as in Grandfather Minton's case, they merged their businesses and families with others in their small elite circle. One such merger was between Minton and John McKee, a multimillionaire who was also in the catering and real estate businesses, when McKee's daughter Martha married Henry's son, Theophilius Minton, an attorney. By the time their son, Dr. Henry McKee Minton, graduated from college, he was already part of a wealthy dynasty.

"But they're almost all gone now," says one of the physicians who recently attended the funeral of Henry's nephew, Dr. Russell F. Minton Sr. "It's strange, given all the history they had in this town. Russell went to medical school with a lot of us at Howard. He was chief of radiology and head of the hospital here, and his kids grew up here. Way back, they used to be the toast of the Philadelphia cotillion—the big fund-raiser for Mercy-Douglass Hospital. But now, you meet some of these new people moving into Chestnut Hill that don't even recognize the name. Especially since the kids moved away."

When one does hear the Minton name today, it's usually in one of the neighborhoods that are favored by the old guard—somewhere in the Mount Airy or Germantown neighborhoods of North Philadelphia. "Those neighborhoods first opened up for black professionals in the late 1940s," says Dr. Melvin Jackson Chisum, a physician and member of the Boulé.

As Chisum points out, blacks have, over time, lived all over the city: in the north, west, and south ends. "In the 1930s and 1940s, South Street was a major commercial street when blacks populated South Philadelphia," says Chisum, who recalls that a major segment of the black elite had ties to South Philadelphia's Douglass Hospital on Lombard Street. "My mother graduated from Douglass's nursing school in 1916, during the time of that first major group of black physicians. One of them was Dr. Nathan Mossell, who had been among the original physicians at Douglass Hospital and had been the first black to graduate from the medical school at the University of Pennsylvania," says Chisum.

As Chisum points out, the Mossell name was an important one at that time. Nathan's brother, George Mossell, was one of the first blacks to graduate from the university's law school; and George's daughter, Sadie Tanner Mossell, became a trailblazer in the legal profession as the first women, of any color, to practice law in the city.

As the south side became more congested, blacks expanded into the west across the Schuylkill River and into the north. The growing "doctor crowd" was found at Mercy Hospital, which was opened before World War II in West Philadelphia. "They had the big-time doctors," recalls a retired dentist who now lives in the Mount Airy neighborhood. "People like Minton, Eugene Hinson, Nolan Atkinson—practically the whole Boulé crowd."

"But a lot of my parents' friends felt that life was really rooted on the south side," says Boyd Carney Johnson as he flips through his mother's old class-of-1911 yearbook from the Philadelphia High School for Girls. "That's where many of their institutions and favorite places were—places like Christian Street, Chew Funeral Home, St. Simons Episcopal, and Mother Bethel. Everyone had some kind of tie to South Philadelphia." Among Johnson's family friends were such names as Abele, Upshur, Webb, Sewell, Chew, Anderson, Clifford, and Christmas. "Anderson" was Marian Anderson's family, who lived nearby during the 1920s and 1930s. "Clifford" was Patricia Johnson Clifford, who was the daughter of Charles Johnson, the first black president of Fisk, and the wife of Dr. Maurice Clifford, who was the first black at Philadelphia General Hospital. "Christmas" was Dr. Lawrence Christmas, a dentist who was big in the Philadelphia chapter of the Alphas. "Patricia Clifford was in a woman's group called the Stork Club with my sister Pauline. Dr. Christmas was everybody's family dentist, and his wife, Alice, was a good friend of my mother's," explains Boyd Johnson. Two of the Christmases' daughters, Alice Christmas Mason and Marie Christmas Rhone, went on to become well known in New York society. Alice is a well-regarded socialite and real estate dealer on New York's Upper East Side who includes Barbara Walters, Connie Chung, and the Trumps among her friends. Her sister, Marie, became national president of the black women's group the National Smart Set in the 1980s. Today, Marie's daughter, Sylvia, is the highest-ranking black woman executive in the record industry: She is a graduate of Wharton and the chairman and CEO of Elektra Records, a $300 million division of Time Warner.

Many old Philadelphians feel that the black community's heyday was in the 1940s, when Raymond Pace Alexander and Sadie Tanner Mossell Alexander were a young power couple. "Everybody knew Ray and Sadie," says Boyd Johnson. "Ray Pace and Sadie were not just society people—they were activists," he explains. "Two incredible lawyers. He graduated from Penn in 1920 and Harvard Law School in 1923."

"I worked with Sadie when she acted as legal counsel to our sorority," says Nellie Roulhac. "Not only was she the first black Ph.D. from the University of Pennsylvania; she was also the first woman to practice law in the state. The two of them touched so many of us."

Boyd Johnson recalls that Ray Pace Alexander's sister, Dr. Virginia Alexander, was the obstetrician who delivered him. Dr. Chisum recalls being initiated into Ray's Boulé chapter in 1966. "Although it was populated by my mentors from the hospital—doctors like Edward Holloway, Russell Minton, Ed Cooper, Lancess McKnight—I knew it was Ray Pace Alexander's chapter. He was Mr. Boulé and Mr. Philadelphia back then."

Alexander, who served on the city council, went on to become senior judge of the court of common pleas. "Ray blazed the trail for me in Philadelphia," says Norman Jenkins, who was named a judge on the same bench several years later.

Other big names in the Philadelphia social circles in the 1940s were Dorothy and Emanuel Crogman Wright. After attending the University of Pennsylvania, she was one of the charter members of the Links and Jack and Jill, and he was the president of the Citizens and Southern Bank and Trust Company. Their daughter, Gwynne Wright, grew up in the Philadelphia Jack and Jill and is a member of the city's Links chapter. "When I was growing up, this was a very tight-knit community where people knew each other because they went to the same church, belonged to the same organizations, or lived in the same neighborhood," says Wright, who notes that her Jack and Jill chapter included the daughters of the Alexanders as well as many other children of old Philadelphia.

Jetta Norris Jones remembers growing up in South Philadelphia when her father, Austin Norris, was practicing law and serving as the Philadelphia editor of the newspaper *Pittsburgh Courier*. "My father came to Philadelphia in 1919 after graduating from Yale Law School, and he quickly began practicing law while simultaneously launching a paper called the *Philadelphia American*," says Jones, who lives in Chicago and is a class-of-1950 Yale Law School graduate. "My dad was always certain that he knew everyone in this town. It was that small a community." A prominent litigator in the city, Norris represented many famous individuals including Marcus Garvey and Father Divine. He, too, knew Sadie and Ray Pace Alexander. "I just joined my husband, Jimmy, at a Guardsmen Weekend in Chicago," remarks Jones, who graduated from Philadelphia's Girls High before attending Mount Holyoke, "and we ran into their daughter, Mary. It shows you how tight the Philadelphia circles really were. It was a nice reunion for both of us."

During the 1950s, a large contingent of the Jack and Jill, Links, Girl Friends, and Boulé crowd had moved north to the neighborhoods of Germantown and Mount Airy. Germantown High School was already a popular high school by that time. Now that West Philadelphia and the older South Philadelphia neighborhoods were no longer the only available places for blacks to live, some of the black institutions began to pass away. "Mercy-Douglass Hospital closed once there was no longer a concentrated black audience in West Philadelphia," remembers Dr. Melvin Chisum, who had served on the staff before practicing downtown.

By the 1970s, many members of the elite were completely rooted in North Philadelphia with no ties whatsoever on the south side. Living in Mount Airy, Germantown, and in many of the surrounding suburbs, they began to regularly send their kids to the expensive, mostly white private schools like Germantown Friends School, Chestnut Hill Academy, and William Penn Charter.

"We sent our three children to Germantown Friends and Chestnut Hill," says Ellie Dejoie Jenkins, who belongs to the Girl Friends and lives in the Mount Airy section, "but the city also now has some great magnet schools at Central High and Girls High. So there's really no reason why we should feel we need to leave the city in order to find good schools." Jenkins describes some of the historic parts of black Philadelphia as she sits on the deck of her Martha's Vineyard summer home with her husband Norman—a Philadelphia judge—and a few friends, including Dr. Valaida Smith Walker, who also belongs to the Philadelphia Girl Friends.

Norman, who remembers practicing law in the same offices as Sadie Alexander, graduated from Columbia University before later being appointed to the court of common pleas. He, Ellie, and Valaida note that most of their friends have roots that go beyond just South Philadelphia today. "Although Mother Bethel is the best-remembered church," explains Norman, "St. Thomas Episcopal, St. Luke's Episcopal in

Germantown, Mount Carmel Baptist, Grace Baptist, and Zion Baptist are also popular churches among our friends and colleagues." Valaida also points out that many blacks are discovering that the suburbs are not as alien as they originally thought. "I was born in Darby, where my father was a physician, and he was not in the least bit outside of the black circle," explains Valaida, who became a tenured professor and vice president at Temple University after earning degrees at Howard and Temple. "We do not lose our connection to the black world just because we live outside the city."

Valaida Walker's point is discussed more than one would think. "I was feeling like we were the oddballs when we took our son out of Central High and put him in Penn Charter with a school full of white kids," says a retired attorney who lives in North Philadelphia, "but then my wife and I started hearing that there were even a couple of black kids at Shipley and the Haverford School out there on the Main Line. We had to draw the line someplace, and we just couldn't go any whiter than the private schools in the city."

Today, more and more black Philadelphians are trying to balance their ties to the black community with their desire to move into more spacious areas outside the city. The Main Line suburbs seem to be the final frontier. "Of course, there are many of us living in the suburbs now," says a retired dentist. "In fact, Nolan Atkinson has been out on the Main Line for years."

Today, black Philadelphia's borders regularly extend into the Main Line—old Waspy towns that were shut off not only to blacks but to Jews and other nonwhite Protestants. Today one finds blacks in Bryn Mawr, Haverford, and Upper Merion.

"Dr. Atkinson has been in Bryn Mawr for years," says an elderly woman to a group of friends who have gathered in her living room in Mount Airy. "You know his wife, Frances, who started the Links. Well, her great-grandfather was the first black to graduate from Oberlin College—in the 1840s."

The group nods with approval.

"That's right, George Vashon. He was also the first black lawyer in New York."

The group has a phenomenal grasp of the facts surrounding the people in its circle.

"Yes, the Atkinsons were in Bryn Mawr before even Jews got there," remarks one of Dr. Atkinson's Boulé brothers. "In fact, the Atkinson girl, Carolyn, grew up there in Jack and Jill and went to Vassar."

"That's right, she's married to that Thornell man who went to Fisk and Yale Law School. Isn't he in the Boulé?" asks the hostess.

"Whatever happened to Carolyn and Richard?"

The group looks around the room—certain that somebody has kept tabs on a family like the Atkinson-Thornells.

"Well, I can tell you," says the lone Washingtonian in the group, "that they live in D.C. He's teaching at Howard, she heads a school in the district, and all three of their kids went to Sidwell. One of them—Paul—works on the Hill. He went to Penn."

Several people in the room nod with approval. I happen to know Paul Thornell, and from what he has told me in the past, these facts sound quite familiar. They've wrapped up another multigenerational story with no loose ends. It's a common occurrence in cities that very quickly come together with extended families.

Passing for White: When the "Brown Paper Bag Test" Isn't Enough

"Excuse me, my name is Sara Lewis. By any chance, are you black?"

"No, I'm not."

Sara took a good look at the thick lips of the white-looking twenty-two-year-old as a few of us stood there with her in the Harkness Commons student center. There was something about the shape of his head that made her wonder.

"Really?" She asked. "Not at all?"

The student shrugged uncomfortably and hurried past her to the dining hall.

When I was a first-year student at Harvard Law School, there was a male classmate who several of us blacks thought was passing for white. Although most of us would never have found the need—or the courage—to confront the student in the way that our black classmate, Sara, did, this student remained a focus of our attention. For the first few months of our first semester, "Bob" (not his real name) was a subject of conversation when we, as a black group, got together for BLSA (Black Law Student Association) events, or when we saw him entering a classroom or dining hall with a white woman or a group of whites. Although he'd long ago insisted on his white racial background, the rest of us had just taken it as fact that he was a black man in denial. We didn't need further confirmation.

"What I'd like to know is whether his girlfriend knows," Clarence asked no one in particular as we sat at an all-black table in the Harvard Law School dining hall.

Several of us glanced over at the "suspect." He was holding the hand of a slender white woman with blond hair and blue eyes.

Sara rolled her eyes. "Of course she doesn't know. They never know."

"Tell me about it," added Henry, another black classmate sitting at our table. "All white people see is skin and hair."

A white person sitting at this table would have been mortified to discover how adept and obsessed black people—particularly affluent black people—are at identifying black characteristics and black physical features in people who claim to be white. It is not something that blacks obsessed about as a sort of sport. It is because of our family experiences of knowing relatives or friends of relatives who have made the once-too-common decision to pass for white.

Even though we were never able to get "Bob" to acknowledge his racial background, we did all feel vindicated when a news article proved what we all suspected. Just before graduation, a national magazine ran a story on high-ranking blacks in corporate America. Among the individuals profiled was a high-achieving, self-identified black man who looked very much like a darker version of someone we all knew. He identified himself with surprisingly similar credentials, the same hometown, and the same last name, and we realized what our "white" classmate really was.

Skin color has always played an important role in determining one's popularity, prestige, and mobility within the black elite. It is hard to find an upper-class black American family that has been well-to-do since before the 1950s that has not endured family conversations on the virtues of "good hair, sharp features, and a nice complexion." These code words for having less Negroid features have been exchanged over time for more politically correct ones, but it is a fact that the black upper class thinks about these things more than most. This is not to say that affluent blacks want to be white, but it certainly suggests that they have seen the benefits accorded to lighter-skinned blacks with "whiter features"—who are hired more often, given better jobs, and perceived as less threatening.

A study funded by the Russell Sage Foundation in 1995 concluded that whites feel more comfortable around light-skinned blacks than they do around dark-skinned blacks, and hence light-skinned blacks receive better job opportunities from white employers.

It should not be a surprise to learn that some light-skinned blacks have availed themselves of these opportunities. And history shows that generations of blacks with parents in both the South and the North have taken this skin color issue to such an extreme that they have used it not only to dupe employers and landlords but friends and spouses as well. They have internalized the rhyme that some cynical black elite kids have used for generations:

If you're light, you're all right
If you're brown, stick around
But if you're black, get back.

Having been introduced to disturbing attitudes like this from an early age, it is probably not a surprise that I decided to undergo a nose job soon after graduating from law school. In fact, some of my black childhood friends wondered why I had waited so long.

But telling jokes and altering one's physical appearance are not the only manner in which some members of the black elite respond to the ongoing fixation on the skin color and physical features of black Americans. Another response has been to escape the black experience altogether by "passing."

When I talk to friends, relatives, and colleagues who share my social groups and clubs, it is unusual to meet individuals who *don't* have some stories or experiences of black family members who have passed. Since neither of my parents were born of light-complexioned parents, I have to look back at least three generations in order to identify family members who might have had such an option.

Charlotte Schuster Price, the widow of a prominent Washington physician, remembers well the day that one of her older brothers made the decision to stop living as a black man and to enter the world of whites.

"It was in the 1920s and my family was all in Washington for my brother Ernie's graduation from medical school at Howard University," says Price as she recalls the event. "The whole family was proud of him. Then all of a sudden, Ernie turns to my father, hands him the diploma and says, 'I hope you like this diploma because it's yours now. I'll never be able to use it.' That's when we finally realized that Ernie was going to start living his life as a white man."

The day that Price's brother decided to walk away from his ties to the predominately black medical school was also the day that he walked away from his family. "That was more than sixty years ago, and that was one of the last times we saw him." Price pauses for a moment. "Actually, he did come to my father's funeral, but he kept to himself and avoided telling us about his life. We got the sense he was living somewhere in Westchester, because after the funeral he asked me to drop him off in a parking lot in Westchester—I think it was just off the Cross County Parkway."

That her brother felt he could not succeed or be happy as a black man seems ironic to Price, particularly since she ended up marrying a successful black urologist who had also graduated from Howard Medical School and because she raised two successful sons: One of the sons, Kline Price, is a prominent Washington physician; and the other, Hugh Price, is one of the nation's most important black civil rights leaders in his position as head of the National Urban League.

An elegant and well-educated woman who served as a Howard University library archivist after receiving degrees from Howard and Catholic University, Price saw the split between certain of her light-skinned black family members who decided to live among a separate race. Of her nine black siblings, three chose to live as whites. So schizophrenic were the racial choices of that generation of her family that Charlotte often distinguishes between them in our conversation by referring to them in terms of their chosen race: "My black sister who lives in Boston . . . my white brother who moved to Canada . . . my white sister who married a doctor . . ."

A former resident of Connecticut, Price was married in the elite St. Philip's Episcopal Church in Harlem and recalls that her "white" brother, Ernie, once dated Vicky Bishop, the daughter of St. Philip's aristocratic black rector, Shelton Bishop. "At the time, Ernie was living as a black man," explains Price. "They eventually broke up when he decided to become white. I remember Vicky coming to our house in this chauffeur-driven limousine, visiting my mother, trying to ask why Ernie wouldn't see her anymore. It was so sad." Suddenly Price adds an ironic twist. "Later, the Bishop girl ended up dating my other 'white' brother, Gus."

Despite the abandonment that families like Charlotte Schuster Price's have suffered, people like her have a remarkable ability to maintain a sense of humor and a positive black identity. "Gus didn't keep in touch with us, but Ernie did at least stay in touch with our mother, and I'll never forget the time I found out that the two of them ended up accidentally putting their daughters in the same New York private school. Both girls were named Barbara, both had the same last name, and both were passing as white children. At the time, I don't even think they knew each other."

Charlotte, who belongs to the Deltas (her husband was an Alpha) and raised her sons in Jack and Jill, has all the credentials, the affluence, and the friendships that qualify her as a grande dame of black society, but she maintains a modesty that is disarming. She even laughs when she refers to the antics of some of her "white" relatives, who work hard to disguise their black identity. "Some of them call me up and want to know all about their black cousins," she says, "but the minute I ask them questions about their own lives or relationships, they clam up. One of my white nephews whom I have never met is actually coming to visit me this weekend."

When I ask her where Ernie and Gus are now, she pauses. "You know, I really couldn't tell you. They could be dead by now, but I really don't know."

During conversations with many light-complexioned blacks among the elite, one hears a wide range of family secrets and experiences. An Atlanta attorney who was born into a family where many relatives passed shared some conclusions after observing the way he saw black cousins and siblings pass into the white race. Although he used no such label, I have heard a number of "tips" repeated by other light-skinned blacks. For lack of a better term, I call this disturbing litany of tips "The Rules of Passing":

1. Passing will be easier if you attempt it while away at college—preferably on a campus that is predominantly white and is located in a small rural town.
2. Change your last name to one that is not associated with black family names. Avoid such surnames as Jones, Jackson, Johnson, Williams, Thomas, and Brown.
3. Re-create your family tree by describing yourself as an only child born of parents who died years ago, and who were also only children.

4. Relocate to a new community that insulates you from interacting with blacks and that is at least a few hundred miles from your family's home. Avoid cities like New Orleans and Charleston, South Carolina, where whites are adept at spotting light-skinned blacks who are passing.
5. Think of some manner in which to "kill yourself off" in the minds of black people who know you and your family. If your parents or siblings are willing participants in assisting you, they can say that you now live outside the country, that you have entered a cult or religious order, or even that you have died.
6. Realize that blacks—and not whites—are the ones who can threaten your security as a black person living a lie. Avoid any meaningful interaction with black people. Affluent blacks who understand the "passing" phenomenon and may try to "out" you are particularly dangerous.
7. Develop associations with organizations and institutions that will buttress your new white résumé. Convert to the Presbyterian Church or the Republican Party. Contribute to charities like the Junior League or the Daughters of the American Revolution.
8. Recognize the physical features that can undermine your new identity. Avoid getting tanned at the beach. If your hair is not straight, keep it short, wear a hairpiece, or maintain weekly touchups.
9. Enhance those physical features that can support your new identity. Lightening your hair color, narrowing your nose, thinning your lips, and adopting a more conservative style in clothing and speech are all simple steps that can aid your transition.
10. Realize that no one in your life (including a spouse) should be trusted with your secret, except for your adult-aged child, who, presumably, will maintain secrecy because of his or her own self-interest in living as a white person.
11. Avoid applying for high-profile positions or admission into selective clubs or lineage-obsessed institutions like secret

societies or prestigious co-op boards, so that you will not be subjected to probing questions and searches.

12. Avoid the appearance of being secretive about your racial identity. If your physical appearance makes it possible, claim to be of white European background. If you have a darker complexion, claim to be a mixture that includes a white European background (e.g., Irish, Dutch, German, Polish) and a darker European or Middle Eastern background (e.g., Greek, Cuban, Lebanese, Portuguese). Never claim any ethnic group from continental Africa or Mexico or Central America.
13. Avoid sitting with or being photographed with black people because if you have any vaguely black features, those characteristics will be exaggerated and suddenly make you seem quite similar to "real blacks" standing near you. The similarities will quickly become obvious to all.
14. If the members of the black family you have "divorced" are willing to support your efforts to "pass," always meet them on neutral territory where neither you nor they live, work, or socialize. Never meet them at your home, and never meet them in settings that are predominately white or that are places your white acquaintances might frequent.
15. If the black relatives you have "divorced" are unwilling to support your efforts, make a complete break from them, because they can too easily undo the facade you have created in your new community and new life.
16. To avoid the risk of giving birth to a "throwback child" with black features, consider adopting a white child.
17. If having your own child is a priority to you, you will be better able to explain your child's dark features if your spouse is a member of a dark-skinned ethnic group. Southern Italians, Greeks, Armenians, Brazilians, and Cubans are among the groups that fit this category.

"The Rules of Passing" have pretty much remained unchanged for families today, even though they were first established by northern and southern families who were avoiding the harsh discriminatory practices in existence during the slavery and early postslavery periods.

"Mother, I'm moving to Los Angeles and I will stay in touch, but I don't want you and Daddy to visit me there."

"Sheila, don't do this. Don't do this. It's a terrible life." Mrs. Harrison sat at her makeup table with the phone to her ear.

As her husband sat on a chair, he buried his head in his hands.

"Mother, I'm sorry, but I've made my mind up."

"But, Sheila, sweetheart, just come home and let's talk about this. We can figure it out; we can help you."

As she tells me the story of how she "lost" her child, Varnelle Harrison recalls how she attempted to talk her daughter out of passing. As she stares at the black-and-white photo of Sheila and her two other daughters standing in front of a summer beach cottage, she clenches a tissue in her left hand. Angry and somewhat embarrassed by her daughter's choice, she asks me not to use her or her daughter's real name (or identifying characteristics), even though she knows that her daughter has long since found a new identity.

"We had sort of expected that this call might come one day," says Harrison, a woman who, ironically, has strong ties to her black sorority, which she joined more than sixty years ago. "Some people don't understand why a black person who was born with a good background of educated and well-to-do parents would want to pass, but I think it's more likely that *we* would try to pass rather than a poor black person because we actually get to see what the most privileged white person has in life. We have the same education, the same money, and the same potential. In a way, we get so close that it becomes an awful temptation." For that reason, she wasn't terribly surprised that Sheila, a smart and ambitious child, would one day fall prey to that temptation.

Varnelle says it started becoming obvious when Sheila spent four years in college with only three visits to her parents' home. "She came home Christmas of freshman and sophomore year—and the summer in between," says Harrison as she recalls. "She told my husband and me that she was doubling up on coursework in order to graduate in three years, so she was working through vacations and doing research with professors during the summers."

It was all a lie, and Varnelle and her now-deceased husband, Roger, had an inkling of what was going on, particularly when Sheila had planned two visits with them and insisted that they stay at an inn that was forty-five minutes from campus.

When Sheila went away to college in the early 1950s, she picked a small women's college in the Northeast against the advice of her parents, who had both attended a black southern college.

"I told her she should be going to Howard, where she could meet some nice friends from good families," explains Varnelle, who is now almost ninety and has had limited contact with her daughter since that phone call. "But she came up with this college in New England that none of us had ever heard of. I remember asking her why she would ever want to go to a school in a place like that. And the minute I asked it, I knew the answer. It was devastating to my husband. And I think I just got angry."

Although she had been raised in a black neighborhood with an entire circle of well-to-do sophisticated black friends, Sheila had intentionally picked a white college in a rural community. Such an environment would allow her an easy transition out of a black culture. Her mother now concludes that it was Sheila's testing ground to see if she could live a life of passing.

"It was like she killed herself off as a black person, and then reemerged with an entirely new identity," says a childhood friend who knew Sheila as a black kid in their southern hometown.

"She almost never came home for summers during college—always telling her parents that she was doing extra papers and research with the hope of graduating early," says the childhood playmate. "Around her junior year, we started to hear rumors that she had suffered a breakdown and was institutionalized. Somebody else said she'd left the country and settled in India somewhere. And a couple others—friends of Sheila's parents—told my parents that she'd killed herself. All these crazy stories to explain her disappearance. It was just a nontopic for Sheila's parents. They just never discussed her anymore."

According to a white classmate who claims to have known Sheila only as a white person, Sheila got married, unbeknownst to her parents, during the late part of her junior year to a local white high school graduate. "I think his last name was Masters," says the classmate. According to the former classmate, Masters was a quiet, rather simple, unsophisticated man who worked in a grocery store. Presumably, he was somebody that she knew would care little about her background and would be suitably impressed by the fact that she was a college student.

"I later got the sense," says Varnelle, who says it was years before she ever heard about a marriage—and to this day, she is not certain it took place—"that she told this boy that we were deceased and that she was an only child with no other family. I never met him or saw him, but he probably wasn't too concerned about her background. After all, unless you're in places like certain parts of South Carolina or Virginia, whites don't think about blacks passing that much. It's foreign to most white people."

Evidently, according to a former friend of Harrison's daughter, Masters was also somebody Sheila would not stay married to. Making the calculated decision that her husband's value was only in his name and white family heritage, she made no efforts to meet his friends or to build a life together. One black childhood friend learned that Sheila had taken off the next year from college and guessed that her intent was to gain a different graduation year—further altering her original school records. With a new last name, he concluded, her plan was to graduate and enter the real world with a new name and a new identity. When Mrs. Harrison hears this pieced together, she says, "Before my husband died, we pretty much figured out what she'd done, but I long ago stopped trying to understand the convoluted decisions our daughter made. I'm not even sure she married this man. She may have just taken his name. It's like trying to figure one of those murder mystery novels."

At age twenty-two, only months before graduation, Sheila evidently parted from the white man who was believed to be her husband. Soon after graduating, she left town for Los Angeles with a completely new name and identity.

Today, Sheila has two different identities for two different communities. The black community in her hometown knew a Sheila Harrison, a black woman, who they believe, for the most part, is either dead, institutionalized, or living in some other country. The white community in her adoptive city and surrounding environs know a Sheila Masters, a white socialite, whose Vermont doctor father and Greek mother died when Sheila was a child.

"I knew Sheila as a black person," says an elderly black physician who belonged to the same fraternity as her father. "And what's so amazing is that none of us have interacted with her since she re-created herself as a white woman. I have a pretty good idea of where she lives, what she does, and what she looks like. I had the opportunity to meet a white colleague who had actually been to Sheila's new home fifteen years ago. My colleague knows her only as a white person. From what I understand, when people ask her about her background, she says that her maiden name is Sheila Masters and that she is part Greek."

It seems that Sheila never mentions her former husband, but on the rare occasion when someone learns that she has previously been married, she will quite matter-of-factly offer the incredible story that "my name was originally Masters, but I ended up marrying a very distant cousin—also named Masters." With the dexterity of a double agent, Sheila has developed a clever way to guard her true identity and steer even the most curious genealogist directly into a white family tree. Of course, it was her former husband's family tree, but by claiming him as a distant cousin, she suddenly made it her family history too.

"I tried to rekindle my friendship with her," says a retired college professor who had grown up with Sheila and had been a friend of her father's. "I ran into her and confronted her in an airport several years ago, but the lies were so outrageous and so well-rehearsed that I couldn't get through to her. It was so ridiculous to be keeping that story going now that she was successful and living a great life. It's not as if this was still the 1950s or 1960s anymore. She kept insisting that I was mistaking her for somebody else. Here she was with the same face, the same voice, and the same first name—and she's telling me that I'm confusing her with somebody else. It absolutely amazes me that white people can't see the black in her. She even has a southern black twang. But I guess the whites she socializes with have absolutely no ties to black people. If she's gone through that much trouble to live in the white race, all I can say is good riddance. They can have her."

A black secretary who worked for one of Sheila's white friends says, "When I first saw Mrs. Masters stop by to see my boss, I immediately assumed she might be black. Even with her white skin and the straight hair, in my eyes, there was nothing else she *could* be. But then I could tell the way she wouldn't meet my eyes when I greeted her that something was up. She was always direct with the white secretaries but not with me. She wasn't mean to me, but it was almost as if she was scared of me. And here she was, this rich, confident lady with all these rich white friends. I guess she was afraid they'd all dump her if they knew she was black." The secretary paused and shook her head in disgust.

"I hope you enjoy your new scarf," said the saleswoman as she slid the change across the counter and placed the large Goldsmith's shopping bag on the counter for Erma Clanton.

"I'm sure I will, thank you." As Clanton walked back past the silk scarves that she had been perusing, she stopped in front of one of the mirrors to adjust the lapel on her jacket. In the reflection, she caught a glimpse of someone in the background who looked familiar.

Erma turned around and made her way to the next department—ladies' hats. She looked around her and saw the glass countertops and small racks lined with the latest fashions from New York and abroad. What Erma had always liked about Goldsmith's was that it brought a big-city, northeastern flair to her hometown of Memphis. A large, sophisticated New York kind of store without the New York hustle and bustle.

"Sadie!" she called out, recognizing the profile that she had only glanced at earlier.

"Oh, my," responded a woman who was trying on a navy-blue hat with a spray of silk flowers in the back.

Erma had not seen Sadie (not her real name) since they had graduated from Booker T. Washington High School and gone off to college in the early 1950s. Almost a decade had passed, but she would have recognized her friend's bright-red hair and sharp olive features anywhere.

Sadie smiled faintly as the saleswoman removed the navy-blue hat and replaced it with a brown one. She spun around and rushed over to her friend, stopping Erma short.

"Girl, it is so good to see you." Erma put down her bag and leaned over to hug her old friend.

"No," the friend whispered briskly. "Erma, please don't hug me. Don't touch me."

Erma's outstretched arms froze strangely.

"I'm sorry, Erma. I'll grab your hands and shake them." The olive-complexioned hands very tenderly grabbed Erma's deep-brown wrists and brought them down to Erma's sides with a whisper, "It's very nice to see you, Erma, but not here."

As Erma glanced over Sadie's shoulder to the arched eyebrows of the white saleswoman, she was confused. Then she glanced up at the price tag that hung from the new hat on Sadie's head. She suddenly understood what was going on. Sadie was passing for white.

"Yes, Sadie, some other time."

"It had not occurred to me why Sadie wouldn't let me hug her until I saw the unpurchased hat on her head," explains Clanton, who is a retired professor from Memphis State University and who shares this memory with me during a recent visit to my aunt's home in Memphis. "At that time, blacks were not allowed to try hats on in department stores in Memphis. We had to either buy it or just look at it without touching it. So, when I realized she was trying a hat on with the assistance of the saleswoman, I realized she was passing. And if she'd hugged me, I would have blown her cover."

Erma shook her head—not with disgust over Sadie's actions, but more over the fact that she had not immediately recognized the situation.

One of the reasons many blacks have historically allowed their black brothers and sisters to pass without exposing them is either that it caught them off guard or that they were afraid of subjecting these individuals to consequences from the law or the general white public. Others neglected to expose these passing blacks because it sometimes gave them satisfaction to see one of their own outsmart white people and the outrageous Jim Crow treatment of the black community. My father, who is the lightest-complexioned member of my immediate family but by no means light enough to pass, remembers a childhood acquaintance who used to pass whenever she went to the movie theater in Memphis.

"For Charlene, it was a big joke. A group of us would all walk over to the theater together, and about a block away, all the black-looking kids would drop behind and let her go ahead of us. When we got to the theater on Beale Street, we'd all see Charlene standing on line with the white people at the front door. She'd slyly wink at us, and then we'd go around to the side alley door where black patrons had to come in. Up in the balcony where blacks had to sit, we'd all be able to look down and see Charlene fooling the white kids that sat around her. When the show was over, we'd meet up about two blocks away and Charlene would be laughing about her prank."

The idea of "sticking it to white people" or beating them at their own game of racial segregation and favoritism has been an issue for many generations in the black community. While children might have done it as a prank, many parents and adults, like Sheila Harrison and Sadie, did it before—and continue it today—because of the economic advantages that are afforded to whites in the area of housing, employment, or treatment in public facilities. With the exception of Congressman Adam Clayton Powell, who had passed as white and dated white girls during his early college years at Colgate University in upstate New York, one rarely finds high-profile blacks attempting to lead double lives, so any advantages being gotten by black people who pass are usually individual and generally affect only the people who are being duped.

In her book *Ambiguous Lives,* George Washington University history professor Adele Logan Alexander looks at the generations of her family living as free blacks in the eighteenth and nineteenth centuries and points out that of her paternal grandmother's eight siblings—all of whom were light enough to pass—only one did. And among her father's siblings, she speculates that only one—a brother—passed, and that after he chose to do so, he created such selective rules about who could not meet his white wife that "the rest of my father's family simply got fed up with it."

"Just like my own relatives living during that time and later periods," says Alexander, "most educated or privileged blacks feel a sense of obligation to acknowledge their own black community and to give back to others who may not have the same advantages. Most are simply not willing to abandon an entire community."

Unfortunately, there is a high price to be paid by family members—particularly the children of blacks who pass. There are many stories of families who have been polarized on the issue of racial passing. I have at least a half-dozen cousins who are often mistaken for white and could pass if they chose. I feel grateful not to have lost them or other close relatives to the practice, but I can't help wondering what it must be like to be privy to remarks made by white people on the issue of race when they have no idea who is present.

There have been many blacks who have managed to divorce their parents or fool their friends and colleagues, but the passing blacks create an even more complex situation when they raise children. The passing black who marries a white will sometimes tell the spouse and sometimes not. If the white spouse objects, it's something that he or she can avoid or dismiss through divorce. But what happens to the child of passing blacks? How are they affected by the lies and by the fact that their racial makeup is permanent and will always be called into question?

"Kids used to think I was adopted when I was in college. They'd see the photo of my parents in my dormitory room and say, 'Who's that?'" says Loretta Josephs, a fifty-two-year-old light-brown-complexioned woman who now wears her graying brown hair in small braids, as she tells the story of how she grew up as a dark-skinned child in a family that lived as white people. It took her twenty years to come to terms with the fact that her parents had developed all kinds of lies to avoid confronting the family's true racial makeup. And when she asks me to disguise her name (and certain identifying characteristics), it becomes obvious that she is still not ready to fully accept her family's situation in a public way. She admits that her parents went so far as hiring a black nanny who would pose as Loretta's mother in situations that might prove awkward for the other family members.

"My mother was white and my father was black—but passing—so I was a real problem. My two older brothers came out real light, but I came out dark. I was a *throwback child*."

As Josephs opens a large four-ring photo album, her point becomes immediately obvious. She sat peacefully in the dining room of her spacious colonial home while pointing out yellowed family photos taken in the 1940s. One photo showed a white couple on a picnic blanket with two white boys sitting in front of them. The boys, with clean-shaven heads, looked to be no more than four or five years old. They smiled broadly at the camera with complexions as milky-white as their parents'. The baby, dressed in a striped jumper, was also smiling—but she didn't seem to fit into the scene. She looked strangely out of place, even while being held in her mother's lap. The baby was black.

"That was me when I was ten months old. And as you can see, I was the darkest thing in that picture." The woman showed a succession of photos—all family gatherings—from birthday parties to outings at the park, Christmas tree poses, backyard cookouts. And in each one of them, Loretta was immediately identifiable. She was always the only black-looking child in the group. In a few of the photos, there was a black woman in a white uniform standing at the back.

"My parents hoped I would get lighter, because, as you can see from the pictures, we lived in a white world—went to a white church, lived in a white neighborhood. My mother used to scrub me twice a day—hoping that my skin would lighten up. She would make up a bath mixture in the tub using up a quart of milk, two squeezed lemons, and a teaspoon of liquid bleach. When she was done, she'd rub my knees and elbows with the halves of the lemon, all the while saying to me, 'Now, if you stay off your knees, they'll lighten up.'"

And there was something else Loretta remembers that was always kept near her bathroom tub.

"And she always kept a jar of Nadinola Bleaching Cream within reach. It was in the bedroom, in the kitchen cabinet, in the glove compartment of our car. Nadinola Cream—for clear complexions. That was a popular thing back then, but I remember you could only find it in the black neighborhoods, so my mother always had our maid get it for her. Of course I had no idea that all kids weren't scrubbed this way every morning and night. I thought it was normal."

Like many very young children of color-conscious parents, it was a long time before Loretta even noticed her own color difference. It was a long time before she noticed that she was considerably darker than her parents and brothers. Like other children, she saw size and gender as the primary differences between herself and her brothers and parents. The distinctions her parents drew and the rules they established seemed to be logical and fair when they were issued to her.

"When I was told by my parents not to play with the kids in the neighborhood, I thought it was because I was a girl and they were worried that I'd get hurt. When we went to the beach and they kept me fully clothed with a visored hat pulled tightly over my head and ears, I thought it was because it was unladylike to get tanned. As they held me under umbrellas, protecting me from what Mom called 'the sun's harsh rays,' they offered an innocent explanation for everything. They wouldn't allow me ever to pick up the telephone because they said little girls didn't do that. It wasn't until I was about five years old that I sensed real differences and started to realize there was more behind the special rules and special treatment. One example was when my mother used to hot-comb my hair with a blazing iron comb every morning. Once again, I thought all ladies got their hair hot-combed."

"One day, my brother Jimmy came in the room and asked, 'Why do you do that to Retty's hair?' My oldest brother, Sammy—who was eight or nine then said, 'Because she's got nigger hair when she wakes up every day.'"

"Even though Sammy didn't speak the words in a mean tone—and though I wasn't quite sure what 'nigger hair' was, I could immediately tell it got my mother mad. That night, my brothers and I heard my mother tell my father, 'You know, Sam, Retty's going to be a problem for us when she starts school in September.' All we could hear then was my father say, 'I'm sorry.' Then we heard mom start crying."

"About a month later, to our complete surprise," explained Josephs, "my mother announced that she and Dad had hired a maid."

"'Your father and I decided that with Retty starting school in two months,' my mother said gaily, 'I will need more help around the house.

You boys are old enough to do your own things and get yourselves ready for school. But Retty will need extra help, since she's young. So the new maid will be helping Retty with her things.'"

"'So the maid will really be Retty's maid,' my father added—picking up almost exactly where mother had left off—as if with a script. 'She will walk Retty to school at St. Catherine's, take her to the playground on Saturdays, bring Retty to the doctor, and so forth.'"

"'St. Catherine's?' Jimmy asked. 'Isn't she going to *our* school?'"

"'No, Retty will not be going to public school. As a girl, it's better for her in a nice Catholic school.'"

Loretta then remembers her father interrupting the discussion to explain how she and her brothers were to address the new maid. "'You are each to call her Mam when she gets here.'"

"'Mam?' my younger brother asked. 'What's her name?'"

"'Mam—that's her name and that's what you will call her.' My parents then got up from the dining-room table with their plates and went into the kitchen."

Loretta remembers her brother staring at her. A new school, a new maid, and new rules were all elicited by her coming of age. She admits she was confused. "All three of us were surprised. But my surprise soon turned into a feeling of superiority. My brothers glared at me and left the table."

The following month, when "Mam" joined her family and moved into a bedroom over the garage, Loretta was moved into a bedroom that allowed her to share Mam's bathroom. As her mother walked her brothers to school, Mam took Loretta by bus to an integrated Catholic school that was just on the other side of her town's border.

"By the time I was in the fifth or sixth grade, I realized a lot of things. I realized that neighbors were whispering things about me being a Negro, or that I was a half-breed, or that I was Mam's child. I realized also that my parents had kept me from answering the phone because people had been calling and saying racist things into the phone. But most devastating of all for me was when I realized that the reason why Mam was hired as 'my maid' was that she was, in a sense, acting as my 'mother' for the people who were watching from the outside world."

By the time Loretta was sent away to boarding school at age twelve while her brothers remained at home, she had stopped believing all of her parents' special rules about what young ladies were supposed to do. She realized that she was standing in the way of her family's ability to live as a white family in a white world. Today, living very obviously as a black woman, she says that her parents are deceased and that she has little contact with her brothers. Married to a black attorney who collects African art, she has no white friends and fully embraces her black identity. "I intentionally wear my hair in cornrows and get black in the sun," Josephs says with a slight chuckle that is belied by more than a few tears in her eyes. "I feel that I've got to make up for the years of blackness that my parents stole from me."

The phenomenon of passing is among the least-discussed issues within the community of the black elite. It is a source of great shame and annoyance, even among those people who recognize its necessity and its benefits from an economic perspective. Although I have known a few people who pass in order to gain short-term advantages such as better professional connections in business networking organizations or in order to purchase a co-op in a particular apartment building, most accomplish this by simply omitting information about themselves rather than actively lying about their family identity. The issue has less relevance in today's black community, because while most affluent blacks in America had historically also been light-skinned, today there are large numbers of blacks who are able to gain admission to top academic institutions, as well as to top employers, regardless of the shade of their Negroid complexion. As more and more opportunities open for blacks in housing and employment, the primary remaining reasons that certain blacks will choose to pass are those that determine their ability to maintain close intimate relationships with people who would otherwise not embrace them.

AFTERWORD

The issue of class has always been a difficult and polarizing subject among blacks in America. Some have argued that the mere discussion of class differences within the black community serves to undermine the possibility of unifying a race that has long faced other serious challenges. Many others would insist that to disregard class differences among black Americans would be to treat the population as a monolith—a group with only one experience and only one perspective.

Just as we have acknowledged and studied the class distinctions among white immigrants of European backgrounds—the Irish, the German Jews, the Italians—I believe it is necessary to examine all of the socioeconomic classes among African Americans. A complete review of American history demands that we understand the contributions that have been made by the black elite because they are as authentically "American" as the Irish elite and the Jewish elite, and because they are also as authentically "black" as the urban working class blacks that historians and sociologists so consistently research.

A great proportion of the individuals that I interviewed—whether they were former debutantes, leading physicians, or powerful *Fortune* 500 executives—initially expressed a desire to remain anonymous. While ultimately most of them finally agreed to speak to me on the record, many of them insisted that whites and blacks of a different class would attack them for acknowledging the accomplishments of their families or of themselves. Many had been raised to apologize for their success and for their ambition, even though equally accomplished relatives had preceded them by several generations. While they willingly spoke about success among their black colleagues who participated in their ninety-year-old fraternities, cotillions, social clubs, or summer resort activities, they remarked that such discussion was unwelcome among outsiders. The older of my subjects—particularly those who grew up in southern segregated towns—feared that such discussion would cause whites to take away their property—as was done to the Church family in Memphis and other individuals in various cities. The younger of my subjects worried that the urban black poor would label them "sellouts" or "Uncle Toms" simply because they had gained academic and financial success. Such is the ambivalence of growing up with a foot in two worlds.

From the time of Madam C. J. Walker, the nation's first self-made female millionaire, through the one hundred years that followed, there have been hundreds of "black elite stories" filled with triumph and tragedy. Each story serves to inform us that generalizations about black people and black elite people cannot be made.

While some members of the black elite have occasionally engaged in such divisive behavior as placing too much emphasis on family lineage, membership in certain clubs, and complexion—even to the extent of passing for white—the group has, since its beginnings in the 1870s, celebrated one important thing: contributing to American culture. It is a group that values intellect, success, and tradition. And while they may have arrived in this country as slaves or free men and women from Africa, the West Indies, or Europe, their accomplishments and contributions were achieved on American soil. Making the climb from slavery and blatant discrimination to wealth and achievement is what the promise of America is supposed to be about. The families of the black elite embody the best of the American dream. For this reason, the story of the black upper class is a story of America.

ACKNOWLEDGMENTS

One of the greatest challenges of writing a book about a group as insular as the black upper class is gaining access to the key individuals who can offer insight into the families, social clubs, and institutions that have quietly, yet decidedly, shaped the group's long history. Even though I grew up belonging to many of the black organizations that are described in the book, and although I sought the assistance of many archives and library collections, the core of my research was gathered from the individuals who know, or are related to, the many families that I've profiled.

To gain this access over the last several years, I needed to rely on family members, childhood friends, professional colleagues, and others who opened their address books and combed through personal records, clipping files, past guest lists, and current social organization membership rosters. I am immensely grateful to all of these people, and I am sorry that I can name only a few of them here.

In Atlanta, I was particularly aided by Dr. Asa Yancey, Marge Dunbar Yancey, Eloise Murphy Milton, Eleanor Milton Johnson, Ella Gaines Yates, Portia Scott, Keith Chaplin, Teresa Artis, Lisa Cooper, Stephen Gura, Jeannie Goldie Gura, Teresa Wiltz, Yvonne Jackson Wiltz, Joy San

Walker Brown, Janice White Sikes of the Auburn Avenue Library on African American Culture, and Dr. Carole Merritt of the Herndon Home.

In Detroit, much help was offered by Judge Damon Keith, Mary Agnes Miller Davis, Deborah Fitzgerald Copeland, Robin Hamilton Sowell, Joseph Brown, Michael Goodin, Leslie Hosey, Nancy Tappes Glover, Albert Thomas, Marian Thomas, Vincent Thomas, and Marguerite Gritenas.

In Chicago, many personal accounts were provided by Truman K. Gibson Jr., Dempsey Travis, Maudelle Bousfield Evans, Theresa Fambro Hooks, Dr. James Jones, Jetta Norris Jones, Ronnie Rone Hartfield, Eleanor Chatman, Charles Montgomery, Dedra Davis, Vivian Durham, Fern Jarrett, and Vernon Jarrett.

In Washington, D.C., a great deal of information was provided by Marjorie Holloman Parker, Benaree Pratt Wiley, ViCurtis Hinton, Bebe Drew Price, Alice Randall, Carolyn Syphax-Young, Charlotte Schuster Price, Alberta Campbell Colbert, Cynthia Mitchell, Judge Henry H. Kennedy, Dr. Joan Payne, Dr. Alethia L. Spraggins, Savanna Clark, Paul Thornell, Paquita Attaway, Ersa Poston, and Adele Logan Alexander.

In my parents' hometown of Memphis, I was able to spend many hours with Ronald and Marianne Walter, who serve as a bridge between old and new society with their ties to Jack and Jill, the Links, the Junior League, and so many other organizations in that city. Others who gave of their time during my many visits and phone calls were Frances Hayes, Alma Roulhac Booth, Dr. James Ford, Barbara Ford Branch, Shelly Branch, Erma Clanton, Alex Coleman, Erma Lee Laws, H. Arthur Gilliam, Earl Douglas, Dr. Anita L. Jackson, Rita Stotts, Blanche Edwards, Dr. William H. Sweet, Margaret Mayfield Rivers, Calvin Walk, Keith McGhee, Ralph White, my cousins Addie Griffin Owen, Anna Griffin Morton, and Dr. Angela Owen Terry, and my dear Uncle Leotha and Aunt Earlene, who knew practically everyone who ever graduated from Booker T. Washington High School.

In Los Angeles, I was given special assistance by Ivan Houston, Jewel Cobb, Thomas Shropshire, Teresa Clarke, Phyllis Campbell Alexander, James Bond, Heather Bond Bryant, and Bernard Kinsey.

For the history and families of Philadelphia, I was helped immensely by Gwynne Wright, Boyd Carney Johnson, Dr. Nellie Gordon Roulhac, Ellie Dejoie Jenkins, Norman Jenkins, Dr. Melvin Jackson Chisum, and Valaida Walker.

To collect data on Nashville, New Orleans, and Tuskegee, guidance came from Patsy Campbell Petway, George Busby, Rose Busby, Gail Busby, Corinne Schuster, Leatrice McKissack, Viola King, Alice Randall, Dr. James Norris, and Marian Taylor.

In the New York City area, the place I call home, I received an immense amount of help from longtime society columnist from the *Amsterdam News* Cathy Lightbourne Connors, Dr. James "Rump" Jones, Ada Fisher Jones, Dr. George Branche, Dr. Matthew Branche, Dr. Wyatt Tee Walker, Dr. Moran Weston, Alvin Hudgins, John and Ernesta Forster Procope, Harold and Helena Doley, and my mentor, Percy Sutton.

When it came to my discussions of the famous black resort areas in Martha's Vineyard, Sag Harbor, and elsewhere, I was assisted greatly by Chester and Gladys Redhead, George and Mary Gardner Lopez, Alelia Nelson, William Pickens, Adrienne Lopez Dudley, restaurateur Barbara Smith, Dr. Thomas Day, Barbara Brannen, Judy Henrique, Ernie Hill, Doris Guinier, Phyllis Murphy Stevenson, Earl Arrington, Robert W. Jones, Doris Pope Jackson, Dr. Beny Primm, Jacquelyn and William Brown, Jack Robinson, Doris Stewart Clark, and Judge Albert Murray.

In my research on Jack and Jill and on the experiences of elite children who participated in cotillions, exclusive camps, and attended the top private boarding schools or colleges, I relied on Barbara Collier Delany, Nellie Thornton, Shirley Barber James, Ilyasah Shabazz, Gwynne Wright, Eric Chatman, Judge Henry M. Kennedy Jr., Elsie Ashley, Yvonne Ashley Galiber, Orial Banks Redd, Paul Redd, Alberta Campbell Colbert, Beatrice Moore Smith-Talley, Berecia Canton Boyce, Sam Watkins, Marsha Simms, Dolly Calhoun Williams, Herman Robinson, Kimberly Webb, Rikki Stubbs Hill, Dr. John Evans, Jacqueline Forster, Eileen Williams Johnson, and my good friend Loida Lewis.

In my research and discussions of women's organizations, I was aided by Patricia Russell-McLoud, Dr. Marcella Maxwell, Audrey Thorne, Portia Scott, Anita Lyons Bond, Toni Fay, Phyllis Murphy Stevenson, Anna Small Murphy, Dr. Mirian Calhoun Hinds, Nellie Arzelia Thornton, Bebe Drew Price, Ruth White, Jamil French, Evelyn Reid Syphax, Paquita Attaway, Ersa Poston, Betty Shabazz, and Barbara Anderson Edwards.

To assist me in my understanding of the role of elite men's groups like the Boulé and the Guardsmen, I relied on Harvey C. Russell, E. Thomas Williams, Anthony Hall Jr., Rev. Harold T. Lewis, Winder

Fisher, Ernest Prince, Boyd Carney Johnson, Dr. Melvin Jackson Chisum, Dr. James "Rump" Jones and Ada Fisher Jones, Theodore Payne, Dr. George Lopez, Dr. Alfonso Orr and Dorothy Orr, George Busby, and my cousin Dr. Robert Morton.

I would not have gathered my information and turned it into an actual manuscript without the resources and calm atmosphere provided by the Schomburg Center for Research in Black Culture, Howard University Moorland-Spingarn Research Center, Joint Center for Political and Economic Studies, Harvard University Library, Princeton University Library, Fordham University Library, New York Public Library, Greenburgh Public Library, Mount Pleasant Library, Chappaqua Library, Scarsdale Public Library, and White Plains Public Library.

Special thanks is owed to Dr. Mirian Calhoun Hinds, Margaret Morton, Dauna Williams, Robin Schlaff, Beth Radow, Steve Emanuel, Andrew Siegel, Dolores Harris, Arthur Eulinberg, Boyd and Eileen Johnson, Ronald Walter, Lawrence Hamdan, Jordan Horvath, Brad Roth, James Grasfield, Jay Ward, Searcy O. Grahame, Anne and Al Gottlieb, Anna Small Murphy, Gail Campbell Busby, and Ann and Andrew Tisch, who went beyond the call of duty as friends and advisers in supporting me during the writing of this book.

My editor, Gladys Justin Carr, welcomed me into a home where everyone believed in this book—even when I needed all those extra weeks for revisions. Deirdre O'Brien, Erin Cartwright, Cynthia Barrett, Stephanie Lehrer, Steve Sorrentino, and Cathy Hemming gave me the encouragement that a writer always needs. My colleagues at *U.S. News & World Report*, Lee Rainie, Stephen Smith, and Mort Zuckerman, recognized the need for a book like this.

My agent, Esther Newberg, and her ICM colleagues Jack Horner, Nicole Clemmons, and Jeff Jackson have done a wonderful job at allowing me to keep my attention on the research and the writing. And during the process, I could not have had two better friends than my assistants, Andrea J. Heyward and Deborah Wheelock Taylor, who spent many months keeping track of historic photographs and maintaining notes and messages from my sources, and Adrienne Lewis, who kept things running smoothly through all the craziness.

My parents, Richard and Betty, offered me the kind of information and advice that is necessary when pursuing a project as "inside" as this. Their great devotion to the Links, Jack and Jill, and Martha's Vineyard, along with their memory of historic moments in black culture reminded me of the value of such a book. And my best cheerleaders, Sheri Betts and my brother Richard, gave generously of their time and advice.

My deepest gratitude is owed to that woman I first met in the halls of Harkness Commons: my brilliant wife, Pamela Thomas-Graham. Even in the midst of negotiating her way through a complex business career, a new mystery series, and our complicated pregnancy, she continues to have the strength to demand intellectual honesty from all that we do. Gordon and I both know that she is the best friend, mother, and mentor that a man could ever have.

L.O.G.
Chappaqua, New York
November 1998

Lawrence Otis Graham is a nationally known attorney and commentator on race, politics, and class in America. A contributing editor at *U.S. News & World Report*, he is the author of twelve other books, including *Member of the Club* and *Proversity: Getting Past Face Value*, as well as articles in the *New York Times, Essence*, and *Glamour*.

A graduate of Princeton and Harvard Law School, Graham is best known for appearing on the cover of *New York* magazine after leaving his Manhattan law firm and going undercover as a busboy at a discriminatory country club in Greenwich, Connecticut. He also appeared on the cover for his story "Harlem on My Mind."

In addition to teaching African American studies at Fordham University and American government at Dutchess Community College, Graham has worked at the White House and the Ford Foundation. His affiliations include the Council on Foreign Relations, Council on Economic Priorities, Westchester Holocaust Commission, American Red Cross, NAACP, Boy Scouts, 100 Black Men, Urban League, and Rotary International. He is a popular commentator on diversity and politics and has appeared on more than two hundred TV programs.

An alumnus of Jack and Jill, the son of a Link, and a member of the Boulé, Graham is uniquely qualified to write about the black upper class. His wife, Pamela Thomas-Graham, is a management consultant and the author of the Ivy League Mystery Series. She is a graduate of Harvard College, Harvard Business School, and Harvard Law School. They have a son and live in Westchester County, New York.

PHOTOGRAPHY CREDITS

Pages 1–2	The Walker Collection of A'Lelia Bundles
Page 3	Top: The Walker Collection of A'Lelia Bundles; bottom: Courtesy of the Herndon Home
Page 4	Courtesy of the Herndon Home
Page 5	Top: Courtesy of Dempsey J. Travis; middle, bottom: Courtesy of Nellie Gordon Roulhac
Page 6	Top: Courtesy of Anna Small Murphy; bottom: Courtesy of Phyllis Murphy Stevenson
Page 7	Top: Courtesy of Anna Small Murphy; middle: Martha's Vineyard Historical Society; bottom: UPI/Corbis-Bettmann
Page 8	Top: UPI/Corbis-Bettmann; middle: Courtesy of Ersa Poston; bottom: Courtesy of Mary-Agnes Miller Davis
Page 9	Top: Courtesy of Mary-Agnes Miller Davis; bottom: Angel Franco/*NY Times* Pictures
Page 10	Courtesy of Camp Atwater
Page 11	Top: Courtesy of Mary-Agnes Miller Davis; bottom: Courtesy of E. Thomas Williams
Page 12	Top, middle: Courtesy of E. Thomas Williams; bottom: Courtesy of Harry and Barbara Delany
Page 13	Top: Courtesy of Dr. Asa Yancey; bottom: *Chicago Tribune* Photo
Page 14	Top: UPI/Corbis-Bettmann; middle, bottom: From the author's personal collection
Page 15	Top: UPI/Corbis-Bettmann; middle, bottom: Courtesy of Loida Nicolas Lewis
Page 16	Top: From the author's personal collection; bottom: Courtesy of Harold Doley

INDEX